THE JOY OF PHYSICS

THE JOY OF

PHYSICS

ARTHUR W. WIGGINS

CARTOONS BY SIDNEY HARRIS

530
WIG
4475968

Published 2011 by Prometheus Books

The Joy of Physics. Copyright © 2011 by Arthur W. Wiggins Cartoons copyright © 2011 Sidney Harris. All rights reserved. No part of this publication may be reproduced, stored in a retrieval system, or transmitted in any form or by any means, digital, electronic, mechanical, photocopying, recording, or otherwise, or conveyed via the Internet or a Web site without prior written permission of the publisher, except in the case of brief quotations embodied in critical articles and reviews.

Cover illustration by Sidney Harris
Cover design by Grace M. Conti-Zilsberger

Inquiries should be addressed to
Prometheus Books
59 John Glenn Drive
Amherst, New York 14228–2119
VOICE: 716–691–0133
FAX: 716–691–0137
WWW.PROMETHEUSBOOKS.COM

15 14 13 12 11 5 4 3 2 1

Library of Congress Cataloging-in-Publication Data

Wiggins, Arthur W., 1935–
The joy of physics / By Arthur W. Wiggins; Cartoons by Sidney Harris.—Revised.
pages cm.
Includes bibliographical references and index.
ISBN 978–1–61614–453–1 (pbk. : alk. paper)
1. Physics—Popular works. 2. Physics—Experiments—Popular works. I. Harris, Sidney, 1933– illustrator. II. Title. III. Title : Joy of physics.

QC24.5.W55 2011
530—dc22

2010048462

Printed in the United States on acid-free paper

Contents

ACKNOWLEDGMENTS

In writing this book, the encouragement of colleagues and friends was extremely beneficial. Joy Friedler, Paul Winston, Gerald Faye, and Dave Stoddard all gave helpful advice in the early going. Huge thanks are also in order for Al Varone, who started this ball rolling years ago. My cartoonist friend Sidney Harris inspired me with fabulous caricatures, and Eugene Mann contributed dynamite graphics to please readers' eyes.

The challenge from my agent, Louise Ketz, was to write something she could enjoy. Does this do it? Thanks are also appropriate for Prometheus's sharp-eyed editor, Linda Regan. It's tricky business to tighten a manuscript and encourage the author at the same time, but she carried it off. Tremendous credit is in order to my wife, Barbara, who showed amazing patience and understanding.

Last, but certainly not least, I wish to thank the thousands of students who have put up with my goofy brand of humor over the years. Hang in there. Joy happens.

Casey at the Bat

—by Ernest Lawrence Thayer

. . .

And somewhere men are laughing,
and somewhere children shout,
But there is no joy in Mudville—
mighty Casey has struck out.

FOREWORD

Many people see little joy in physics, because, like mighty Casey, they have struck out.

In high school, brave students struggle to learn physics, but the subject often remains mysterious, and some students even fail the course. Others believe physics is only for a select group and won't even give it a try. The result is that students are more likely to feel fear or frustration than joy. My own experience was significantly different. In my high school, the physics teacher was close to retirement and he approached the subject mildly by focusing on lab experiments, which, I thought, were fun.

In college, it was a different story. I ran pell-mell into the classic stumbling block: the language in which physics is framed, mathematics. Although the two subjects were often taught without much connection, I had excellent math professors and I was able to make the necessary integration without taking too big a hit in my GPA. For others, whose math wasn't up to snuff or who had trouble getting math and physics to work together, physics might as well have been a class in Sanskrit.

It doesn't have to be that way. Math is certainly a primary tool, but physics concepts are also accessible through examples, analogies, and experiments, as this book will demonstrate. A similar approach works for chemistry, as ably shown by Cathy Cobb and Monty L. Fetterolf in their book *The Joy of Chemistry*. Just as there are many overlaps between physics and chemistry, the two *Joys* have substantial similarities.

You, dear reader, are assumed to be responsible enough to handle experiments safely and to be in possession of a wealth of information about modern civilization. I will address you as an equal and pitch discussions as if you were a friend or a college student curious about physics. The general approach here will be to start with the simple and familiar and move to the complex and unfamiliar. Examples from ordinary life, mini-biographies, hands-on activities, clear illustrations, and occasional cartoons by America's premier science cartoonist Sidney Harris are all designed to make this book accessible and fun. To paraphrase Benjamin Disraeli, people write the book they want to read, and that's what I've done here. The treatment is not exhaustive (or exhausting, hopefully) like a textbook or an encyclopedia but rather illustrative. If you need more detailed knowledge, I've listed a few excellent textbooks in the "Additional Resources" section at the end of the book.

What makes physics so worthwhile? For me, it's a combination of things:

- Physics' enormously ambitious goal: to understand the workings of the entire universe from the smallest to the largest. This appeals directly to people's strong innate curiosity, which we all have.
- Hands-on experiments. Connecting reality with ideas is very satisfying.
- Physicists. They are people who have figured out all these concepts from scratch, often while living interesting and complicated lives. I'm fascinated by the creativity and energy of human beings.

For me, all of the above constitutes the joy of physics. So let us begin.

INTRODUCTION—THE SCIENTIFIC METHOD

> "The whole of science is nothing more than a refinement of everyday thinking."
>
> —Albert Einstein

> "You can observe a lot by just watching."
>
> "It's tough to make predictions, especially about the future."
>
> —Yogi Berra

The goal of physics is to achieve a general understanding of the physical workings of the entire universe. The whole universe, that is. If that seems enormously ambitious, that's because it is. As the typical backseat questioner would ask: Are we there yet? No, but substantial insights have been achieved. How will we get farther? The motivating power behind physics and all of science, for that matter, is *the scientific method.* That's what this chapter is about.

At the most basic level, the scientific method involves a dynamic interaction between reality and ideas. Neither one by itself is sufficient to produce a durable approximation of truth, but together each becomes very powerful. By analogy, suppose you see an ad for the latest jeans. They look great, and the description sounds just like what you wanted. But is that enough for you to place an order? Wouldn't it be better to go into a store and look at the color, touch the fabric, and try them on to see how they look and fit? So it is with the scientific method. Ideas, however appealing, need a reality check.

To put this into perspective, let's examine the workings of science's method. One way of viewing it is to organize the activities of science into a sequence of steps.

Observation requires that a natural occurrence be perceived by human senses or some suitably designed sensor. This step is based in reality and requires accurate measurements of the event and documentation of what was observed so others can repeat it.

Observation: Some specific happening in physical reality is sensed.

Hypothesis (H): A general idea of the cause of all such events is created.

Prediction: If we presume H to be true, then some other happening is forecast.

Experiment: A specific happening in physical reality is sensed.

Compare Experiment with Prediction
- **match** => H is supported, but not proved
- **doesn't match** => H must be modified

The *hypothesis* is created in the world of ideas. To communicate the hypothesis, it must be expressed in language that carefully defines its terms. The language may be a conventional one but often it is mathematics—a distinctly foreign tongue to some (more about that later). Note that a hypothesis is designed to be general so that it may be applied anywhere in the universe and at any time.

To make a *prediction*, the hypothesis must be related to another situation besides the original observation. Based on that hypothesis, a forecast of some other physical happening must be made through deductive logic.

The *experiment* step is set up to match the conditions of the prediction and requires accurate measurements of whatever occurs.

Comparing experimental results with the prediction has two possible outcomes:

1. Experimental results match the prediction. This supports the hypothesis but cannot prove it, since the experiment is specific and the hypothesis is general.
2. If the experimental result doesn't match the prediction, this shows the hypothesis to be false, and it must be modified to accommodate this new result or else discarded.

The logic behind this process probably seems quite familiar. It is so ingrained into our culture that many people refer to it as common sense. For example, suppose you wake up to a loud noise in the middle of the night (Observation). Perhaps the first idea that flashes through your sleep-numbed mind is that cats are easily bored in the middle of the night (Hypothesis) and your cat has knocked something over in its quest to find entertainment (Prediction). If you can muster the energy, you may get up to look for something on the floor and notice a guilty-looking cat slinking away (Experiment). If you find what you predict, your hypothesis is supported, but if you don't find any such thing, you'll need a new hypothesis.

Here's a more scientific example:

Observation: An apple falls and hits Isaac Newton on the head.

Hypothesis: Conceptual. All masses in the universe attract all other masses, and the force of attraction depends directly on the masses and inversely on the distance between them.

Mathematical: $Fg=\frac{Gm_1m_2}{r^2}$ (G = universal gravitational constant, m_1 and m_2 = masses,

r = distance between masses)

Prediction: Conceptual. Earth attracts the Moon and keeps it orbiting in an approximately circular path.

Mathematical. Working through the algebra implies that the time required for the Moon to complete one orbit = 27.3 days.

Experiment: The Moon's orbital period is measured to be 27.3 days.

Compare: Experimental evidence matches the prediction, so Newton's law of universal gravitation is supported but not proven.

These examples make the scientific method appear nice and neat, but this is just a rational reconstruction. In reality, the steps may be performed by different people in different places at various times that are not always in chronological sequence, and sometimes by experimenters attempting to accomplish something vastly different. As you know, human endeavors don't always produce results that have been anticipated through reason. That's one of the things that make human beings interesting. Of course, there is the inevitable fine print.

SYSTEMS OF UNITS

Both observations and experiments require accurate, repeatable, and easily communicated measurements. In order to carry them out, a standardized system of units, called the metric system, has been devised:

length, measured in meters (m);
mass, kilograms (kg);
time, seconds (s); and
charge, Coulombs (C).

Many countries use the metric system, although there are still pockets of resistance, such as the United States.

One of the conveniences of the metric system is the use of prefixes that stand for powers of ten to express subdivisions and multiples of the basic units. Here are a few examples:

Nanometer = nm = 10^{-9} m
Micrometer = μm = 10^{-6} m
Millimeter = mm = 10^{-3} m
Centimeter = cm = 10^{-2} m
Kilometers = km = 10^{3} m
Megameters = Mm = 10^{6} m
Gigameters = Gm = 10^{9} m

This system makes conversions from one metric unit to another extremely simple. For example, if you wanted to know the number of centimeters in 1 kilometer, it is simply a power of ten, which is easily obtained directly from the prefixes. So, the answer would be 100,000. Contrast this with the English system: How many inches in 1 mile? You'd have to know the conversions from miles to feet and feet to inches. Then, you'd need a calculator to obtain the result: $5{,}280 \times 12 = 63{,}360$. Another advantage of the metric system is that its power of ten basis fits in well with *standard scientific notation*, the way numbers are usually expressed in science:

$$5{,}280 = 5.28 \times 10^3$$

Conversions become particularly easy. For example, if you wanted to convert 5,280 km to cm, you could write this as 5.28×10^3 km = ? cm. Substituting the numerical values for the prefixes and using the rules for combining exponents yields 5.28×10^3 km = 5.28×10^8 cm. (When multiplying, simply add exponents.) The only part of the number affected by the conversion is the exponent; the number out front (the coefficient) remains intact.

Mathematics

Language must express the ideas contained in the hypothesis in an understandable fashion. But multiple languages pose a problem because translating from one language to another introduces inaccuracies. A single universal language would be ideal, but the closest thing we've got to a truly universal language (Esperanto is not widely used) is mathematics. Like so many things, math has good and bad points. One bad aspect of mathematics is that most people commonly use only arithmetic and a little algebra whereas many subdivisions of math (arithmetic, algebra, trigonometry, and calculus) are used in physics. Thus, when physics' hypotheses are stated in mathematical terms, the reaction of many people is: That's Greek to me. To avoid this problem, hypotheses within this book will be stated in both English and mathematics.

On the plus side, mathematical statements are usually written in general symbolic terms, which match the hypothesis. Specific predictions are then made by substituting numbers into the equations. The deductive logic (true premises infer true conclusions) built into this process preserves the truth inherent in the hypothesis, so comparison between experimental results and prediction generates a valid test for the hypothesis.

Mathematics is such a convenient and concise language that many physicists' papers and articles rely heavily on it, further distancing their work from people whose mathematics education is minimal. Personally, I have noted my own tendency to lapse into "math-speak" when trying to explain a concept, so this has become a familiar problem.

Physics' Relationship with Other Fields

Physics and chemistry overlap in many ways, since they both study atoms, molecules, and their interactions. Chemistry studies these topics in more depth, while physics ranges wider, treating still-smaller particles as well as larger aspects of the universe. College courses in physical chemistry and chemical physics are often quite similar. The active fields of biophysics, geophysics, and astrophysics attest to the strong interrelationships between physics and the other sciences.

Technology constantly borrows and uses physics' ideas to invent the vast array of gadgets and instruments that dominate our modern culture. This particular interaction has brought physics into close encounters with ethics, producing results that even have huge consequences for the future of civilization (more later).

Physics is so thoroughly embedded in our civilization that it often seems to be part of the fabric of ordinary life. As we proceed, some of these connections will be made explicit, but I'm sure you will find others as well. As a first example, the hit movie *Jerry Maguire* featured discussions about possibilities, but who will ever forget the famous line that demanded experimental evidence: "Show me the money." That's just what we're about to do.

EXPERIMENTS

The whole point of experiments is to check how well reality matches physics' ideas. In this book, experiments begin each chapter, so reality will be fresh in your mind when reading how physics' ideas explain each topic under discussion.

At an operational level, you, the experimenter, will follow this format:

- Perform a physical action—the experiment.
- Measure and record information obtained during the experiment.
- Repeat the measurement several times to minimize erroneous results.
- Make calculations using measured information.
- Compare calculated results to what was anticipated based on physics' principles.

While most of the experiments can be accomplished by one experimenter, some of the experiments would benefit from a team effort, so your powers of persuasion may be tested along with the physics.

All the experiments (and the smaller experiments that we will call *experimentinos*) have been tested, some over many years of teaching. Since they all involve inexpensive apparatus, patience is often required for them to function properly. All of them work, but some require more repetition than others. It's like riding a bicycle. Once you get the hang of it, it's easy.

I urge you to perform the experiments rather than just reading them. It gives your mind concrete images to build ideas around and makes understanding much easier. To me, reading physics without doing the experiments is like reading Shakespeare without attending the plays. You can do it, but you lose much of the effect Shakespeare intended.

Safety is a constant concern, so heed the precautions listed in the instructions for several of the experiments. Eyes are particularly vulnerable, so safety glasses are often recommended. I urge you to use them, because it is better to be safe now than sorry later. Since I presume you are a responsible adult, you should treat these experiments with appropriate caution. After all, the items involved are not toys. Actually, some of them are toys, but you will not be playing with them.

A WORD ABOUT CALCULATORS

Although some of the experiments in this book will require calculations, before you spring for the latest model calculator with graphing capabilities, giant displays, 232 arithmetic functions, and assorted bells and whistles, here's my word of advice: *DON'T*.

All you'll need is addition, subtraction, multiplication, division, and square root (√), which are readily available in inexpensive calculators. Far be it from me to discourage economic activity, but I have formulated Wiggins's rule: Never pay more for a calculator than you would for a pizza. People often purchase way more calculating power than they actually need and they then may become confused by all the excessive options and capacity. Buy whatever calculator you wish, but remember that only simple functions will be used for the experiments in this book.

APPARATUS NEEDED FOR ALL EXPERIMENTS*

safety glasses for all motion experiments	stopwatch
flying disc (Frisbee)	high-bouncing ball (SuperBall)
metric tape measure	half-ball popper (see Experiment 2)
2 small springs (not very tight)	spring scale (calibrated in newtons and grams)
yo-yo	magic spring (Slinky)
2 substantial straws	balloon
multimeter to measure resistance, current, and voltage, can be digital or analog	breadboard (see Experiment 20 for details)
4 small resistors (700–1,000 Ω)	9 V battery
4 wires with alligator clips at ends	LED (see Experiment 22)
small but strong magnet	coil of magnet wire (more than 100 turns)
small compass	small lens (like a pocket magnifier)
diffraction grating (more than 100 lines/inch)	small laser pointer
2 polarized slides	6 CDs (not reusable)
CD player with window	playground teeter-totter
shake flashlight	50 coins
2-liter plastic soda pop bottle	garden hose
golf club or broom	flexible foam ball or foam brush or sponge

Plus normally available household items such as an object hanging from a rearview mirror; a pencil; a book; some tape; a glass; water; pieces of aluminum foil and paper; a large, empty room; a piece of sandpaper; sunshine; a shiny soupspoon; a small, flat mirror; a water sprayer; a wire coat hanger; a microwave oven; a piece of cardboard; a piece of waxed paper; a cup of chocolate chips; a marker; a small piece of cellophane; and a small amount of glycerin.

*Although all the items above may be obtained from various sources—local or online—I will assemble a few Physics Lab-in-a-box for your convenience.[1] Check www.eBay.com for details.

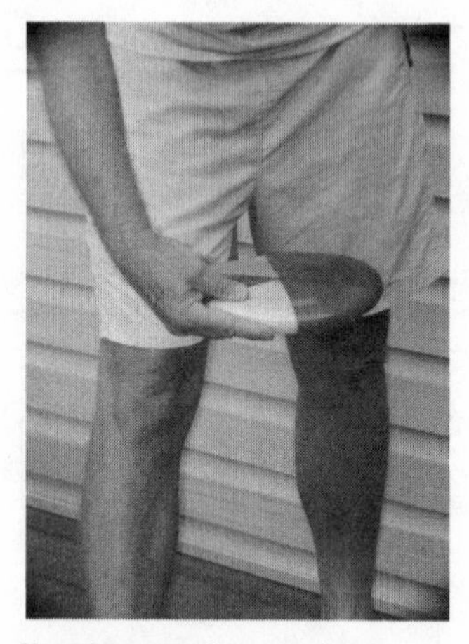

Experiment 1: Identified Flying Objects

Equipment needed: disc, stopwatch, measuring tape (metric)

Human-powered flying discs have been around for more than fifty years, supposedly originating when New England college students flung inverted pie tins from the Frisbie Baking Company. Commonly called Frisbees (a registered trademark like Kleenex), they have been appropriated by all kinds of people, such as dog trainers, sports nuts (disc golf), and physics professors trying to turn a perfectly respectable work-avoidance technique into an educational experience.

The fundamental idea is to throw the disc so that it rotates as it flies forward. The rotation (discussed under the appropriate physics heading of *angular momentum*) helps keep the disc stable so it flies in a fairly straight line, although curved paths are sometimes desirable and fun. Many different release techniques are used, but the simplest one involves holding the disc parallel to the ground and flinging it with a flip of the wrist from a backhand position (see the illustration above). The disc flies straightest if thrown low and flat; its path then remains in a horizontal plane.

The point of this experiment is to measure distance and time in order to calculate velocity. You'll need a flinger, a catcher, a timer, and a distance measurer. While some of these functions may be combined, at least two people would make the experiment easier, more fun, and possible.

Procedure

- One person flings the disc.
- Start the stopwatch as the disc is released.
- Another catches the disc.
- Stop the stopwatch simultaneously with the catch, measuring the time in seconds (s).
- Measure the distance from release to catch in meters (m).
- Repeat ten times.
- For each throw, calculate the velocity: distance traveled (m) / time (s).
- Calculate the mean velocity: the sum of all the velocities / the number of throws.
- Compare your mean velocity to the typical range of values: 3 m/s to 10 m/s.

Throw numbers	Distance, meters (m)	Time, seconds (s)	Velocity = distance/time (m/s)
Example	10	2	5
1			
2			
3			
4			
5			
6			
7			
8			
9			
10			
Mean velocity =			

The mean velocity is found by adding the velocities of all ten trials and dividing by ten.

Experiment 2: Updated Cannonballs

Equipment needed:
half-ball popper,
measuring tape (metric)

Half-ball popper, bottom half

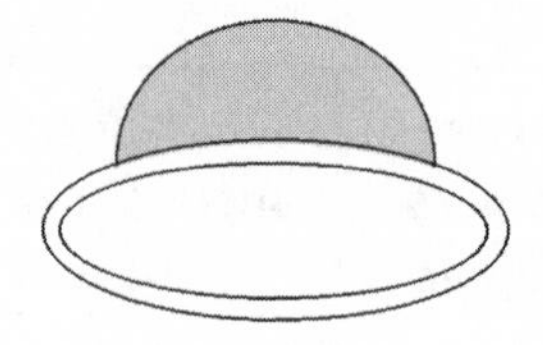

Turn inside out, hold by edges

Cannonballs are awkward to handle and their launching mechanism is especially difficult to operate unless you happen to have ready access to a pirate ship or a fortress. Of course, there is an element of danger involved in launching anything. So this experiment uses a projectile that is made of rubber and is self-launching. This reduces the danger, but you must still *make sure it doesn't hit you or someone else in the eye. Safety glasses are recommended and caution must be exercised.*

To make this projectile, you will need a racquetball. Cut it in half, then sand the edges a bit. If you sand too little, each half won't turn inside out, and if you sand too much, the half will turn inside out and stay that way. If you sand just enough, this half-ball will remain inside out for a short time, and then launch itself in whatever direction it's aimed. It makes a popping sound, so it is called a half-ball popper.

Procedure

- Prepare the half-ball popper.
- Set up a *rigid* supporting plane at a 45° angle to launch the half-ball popper.
- Turn the half-ball popper inside out (see illustration) and place it on the rigid supporting plane.
- Launch the half-ball popper and record the horizontal distance traveled.
- Repeat ten times.
- Find the mean launch velocity and save it for comparison with Experiment 6.

Launch number	**x = Horizontal distance, (m)**	**gx = (9.80)(x), $\sqrt{(m^2/s^2)}$**	**Launch velocity = $\sqrt{gx}$, (m/s)**
Example	3.5	34.3	5.86
1			
2			
3			
4			
5			
6			
7			
8			
9			
10			
			Mean launch velocity =

Chapter 1.
HOW THINGS MOVE—POSITION, VELOCITY, AND ACCELERATION

"To understand motion is to understand nature."[1]
—Leonardo da Vinci

One of physics' first problems was extremely practical: at what angle should you launch a cannonball to hit a particular target? Certainly this doesn't sound as grand or sophisticated as analyzing the whole universe, but it does fit nicely with chemistry's roots: attempting to turn base metals into gold. Cannons came into the picture when people went after someone else's gold.

Although physics' early history is rich in characters and their stories, we will focus here on motion from a more modern perspective. Not only does motion form the basis of much of physics' later developments, but it also has a direct connection to human experience. We are surrounded by moving things and experience a great deal of personal motion—even it it's not always in the right direction.

Let's start with the simplest motion first.

MOTION ALONG A LINE (ONE-DIMENSIONAL MOTION)

Suppose you live on a road that runs straight east and west. A dog sitting on your road could be located by how far it is from your house. Since physics uses the metric system, the dog might be found, say, 39 meters (almost 130 feet) east of your house. This is called the dog's position. Graphically, the street is represented by a single line, called the x-axis, and your house is located at the origin. East is conventionally regarded as positive and west as negative.

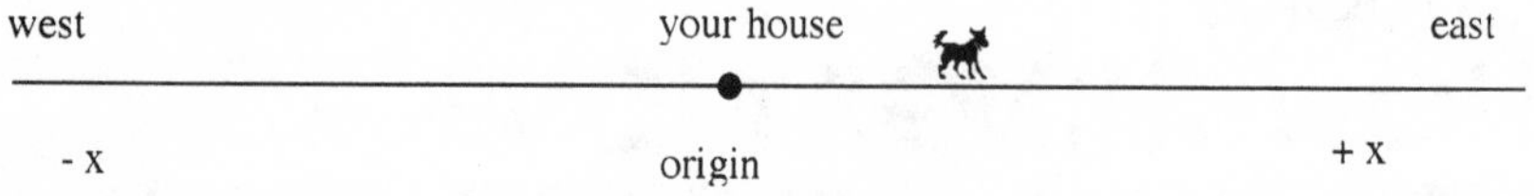

west your house east

- x origin + x

dog position

If a body changes its position, that change is called a displacement. Representing this mathematically, the earlier position is called x_1, read as x sub 1, while the later position is x_2, and the displacement is $x_2 – x_1$, represented by Δx, read delta x.

To summarize conceptually and mathematically, and including units:

> Displacement is the change in an object's position.
>
> $$\Delta x = x_2 - x_1$$

The meter is a convenient unit for length measurement since the height of most humans is between 1.5 m (5 ft) and 2 m (6 ft 7 in), the length of a normal stride is almost 1 m (3 ft), and a stack of forty books like this one would be 1 m (3 ft) tall.

For example, if the dog moved from its 39 m spot to a position 51 m (170 ft) east of your house, the dog's displacement would be 51 m − 39 m = 12 m.

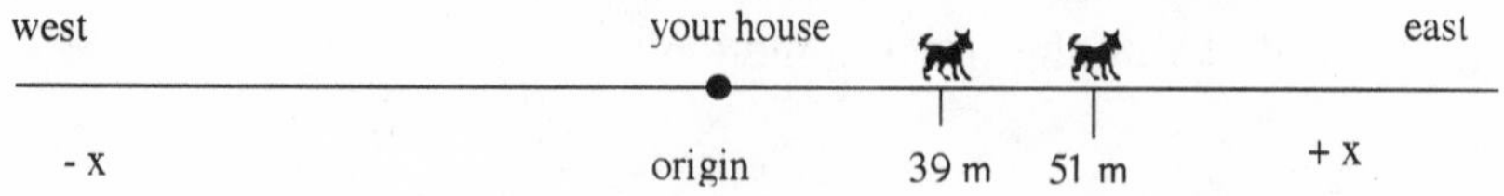

west your house east

- x origin 39 m 51 m + x

dog displacement

Next, the elapsed time must be taken into account. The *velocity* of a body tells how fast a body's position is changing.

> Velocity is the distance traveled by a body divided by the time required to travel it.
>
> $$v = \frac{\Delta x}{\Delta t} \quad \text{units = meters/seconds = m/s}$$

A velocity of 1 m/s (3 ft/s) corresponds to a leisurely stroll (Frankenstein set this standard), while 10 m/s (30 ft/s) is achieved by 100 m (100 yd) dash champions. Your flying disc tosses are usually somewhere between 3 and 10 m/s, as you found in Experiment 1. Speed and velocity are used interchangeably in linear motion, but in two- or three-dimensional motion, there is an important difference—and that is direction.

Velocity is such an integral feature of modern life that automobiles have a device to measure their speed directly, the speedometer. The speedometer's units (miles/hr = mph or km/hr) differ from the scientific standard m/s, but the conversion is quite simple: 60 mph is about 100 km/hr or almost 30 m/s (44 ft/s). Riding in a car at speeds of 0 to 30 m/s seems pretty ordinary, but some bodies travel a lot slower. For example, human hair grows at a few centimeters per month and currents in the Earth's mantle cause earthquakes by dragging crustal plates at a speed of a few centimeters per year. Much faster speeds are also possible: passenger jets fly more than 200 m/s (600 ft/s); the fastest airplane travels at 3,000 m/s (9,000 ft/s); Earth's motion around the sun requires 30,000 m/s (90,000 ft/s); and the ultimate speed is that of light, 300 million m/s (900 million ft/s).

Returning to the example of our friendly dog, if it takes the pooch 4 seconds to travel from 39 m to 51 m, his velocity is 12 m / 4 s = 3 m/s (9 ft/s). This is only an average velocity, since the dog might have run the first 5 meters in 1 second, taken 1 second for a bark break, then run the last 7 meters in 2 seconds.

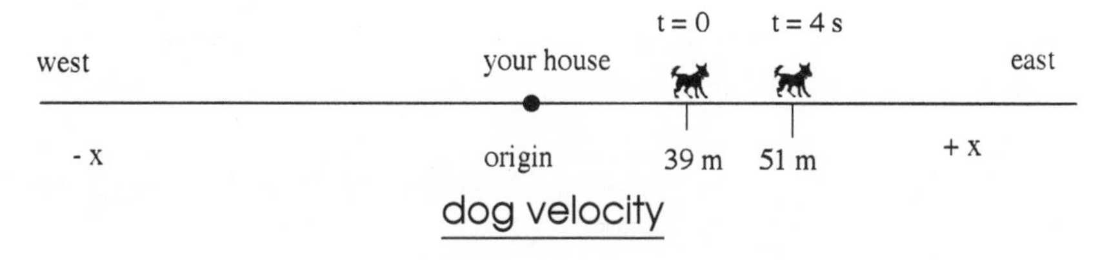

dog velocity

Next, what about the case where the velocity changes? The rate of change of a body's velocity is called *acceleration.*

> Acceleration is the change in a body's velocity divided by the time required to change it.
>
> $$a = \frac{\Delta v}{\Delta t} \quad \text{units = meters/sec/sec} = m/s^2$$

Speeding up yields a positive acceleration, while slowing down makes the acceleration negative. Perhaps that's why the gas pedal on your car is called the accelerator. But why isn't the brake called a decelerator? (a mystery we won't solve here).

A value of 1 m/s^2 is a mild acceleration corresponding to having your car accelerate from 0 to 60 mph in 28 seconds. (That's not much of a jack-rabbit start, but it's easy on gas.) A world's record 0 to 60 mph acceleration is approximately 10 m/s^2, about the same acceleration as a dropped object near the Earth's surface falling because of gravity. That special acceleration of freely falling bodies is called the acceleration due to gravity and given the symbol g. For calculations, $g = 9.8\ m/s^2$ is used. Humans have built-in limitations on the amount of acceleration we can stand. Blackouts and possible internal damage can occur with accelerations much beyond 60 m/s^2 (also called 6 g's because it is 6 times the acceleration due to gravity).

If our friendly dog travels at 3 m/s, then is observed 5 seconds later traveling 6 m/s, it accelerated at a rate of (6 m/s – 3 m/s)/5 s = 0.6 m/s^2.

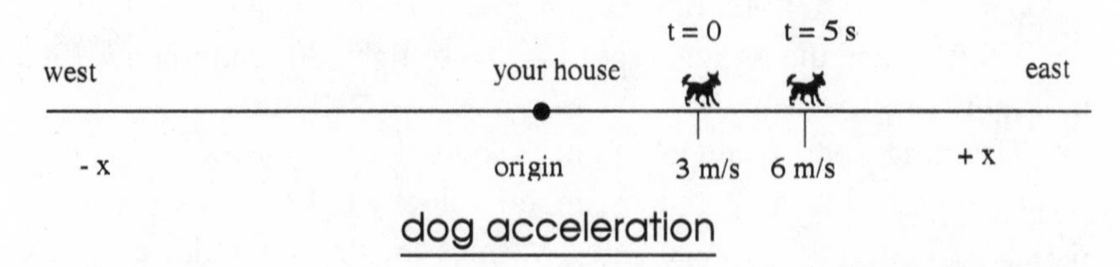

dog acceleration

Combining the position, velocity, and acceleration in a slightly different way.

> An accelerating body's velocity is changed by an amount equal to the acceleration times the elapsed time.
>
> $v = v_0 + at$, where v_0 = velocity at time = 0

The distance traveled is also affected by acceleration:

> The position of an accelerated body is increased by its velocity times the elapsed time plus half the acceleration times the elapsed time squared.
>
> $$x = x_0 + v_0 t + \frac{1}{2} at^2$$
>
> where x_0 and v_0 are the body's position and velocity at time = 0

> Another equation may be obtained by eliminating time from the x and v equations:
>
> $$v^2 = v_0^2 + 2a(x - x_0)$$
>
> where x_0 and v_0 are the body's position and velocity at time = 0

This equation is particularly useful in solving problems in which time is not involved.

Understanding motion along a line is a good start, but it's not enough. Even a dog escapes its straight line occasionally.

Motion in Two or Three Dimensions

Two-dimensional motion in a *horizontal* plane includes chess pieces moving on a board, a hockey puck slid on the ice, or even a dog running in a flat backyard. The motion's direction has now become important and is sometimes specified in compass cardinal points. Motion in a *vertical* plane is perhaps more interesting because Earth's gravity pulls everything vertically downward. Examples of this kind of motion encompass not only the classic cannonballs but also movements made in many sports, such as baseball, basketball, football, golf, soccer, and tennis. In principle, analysis of the motion of any of these examples is simple. Dividing the motion into two parts, horizontal (x) and vertical (y), the analytical tools for one-dimensional motion are applied twice, once for each dimension. The resulting path of the ball, called its *trajectory*, is clearly curved, undoubtedly corresponding to your ball-playing experience.

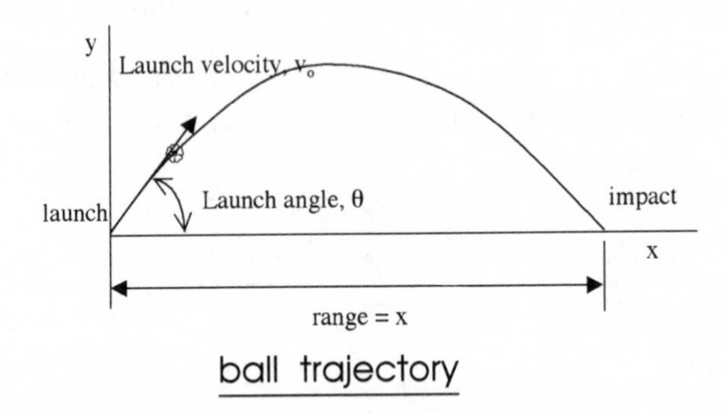

ball trajectory

Using the equations developed earlier, the range, x, can be found if you know the launch velocity, v_o, the launch angle, θ, and the acceleration due to gravity, g. It is:

> $(x = v_o^2\sin, 2\theta)/g$, where sin is the shorthand notation for sine = opposite/hypotenuse

This equation can be solved for the launch angle as a function of the range and launch velocity. Or, knowing the range and angle, you can find the launch velocity, as you saw in Experiment 2. So, could you use this knowledge to become a technical adviser, telling pirates where to aim their cannons, or a caddy for Tiger Woods, advising him on the angle his club should have? Not quite. Other factors are at work besides gravity. In reality, interactions with air make all shots fall short of their expected range or even curve right or left, turning their motion into three-dimensional rather than two-dimensional movement. So, what causes bodies to move? The short answer is: forces. The longer answer is contained in the next chapter.

Experiment 3: Weight 'Til the Sun Shines, Nellie

Equipment needed: spring scale calibrated in newtons and grams and several objects (see table)

The purpose of this experiment is to measure the weight and mass of several household objects so you can become familiar with standard physics units.

Procedure

Hang an object from a spring scale calibrated in newtons (weight) and grams (mass) and record your values in the table below:

Object	Weight (newtons)	Mass (grams)	Mass = grams/1,000 (kilograms)
shake flashlight	1	100	0.1
set of keys			
half-ball popper from Exp. 2			
yo-yo from Exp. 5			
metric tape measure from Exp. 1			
CD			
other			
other			

Chapter 2. WHY THINGS MOVE—FORCES AND THEIR EFFECTS

Now that we've seen some of the effects of motion, it's time to investigate their causes.

As we all know from experience, a body sitting still doesn't move all by itself. It takes effort to get it going. Officially, that's called a *force*. But is it enough to start a body moving? Will a body keep moving even if you stop pushing it or must the force be exerted continuously? These simple-sounding questions are actually extremely difficult and have perplexed very intelligent people for quite a long time.

Let's start with someone whose many thoughts were recorded: Aristotle (384–322 BCE). Though Aristotle was a keen observer, systematic experiments never appealed to him. This talented philosopher had ideas about an incredibly wide range of subjects, including politics, logic, history, ethics, the arts, biology, and nature. Many of his ideas were first-rate. He differed substantially from his teacher Plato on some ideas, but when it came to natural philosophy, he agreed with him. They both regarded the heavens as perfect, as opposed to the imperfections all around them. They saw this world as made of four elements (fire, air, earth, and water) and theorized that each element had natural tendencies that governed its motion. Based on his observations, Aristotle taught that heavy bodies fall faster than lighter ones, and keeping a body moving required continual action by an agent. "A body in motion can maintain this motion only if it remains in contact with a mover," he said. Aristotle's writings were lost for a long time but were discovered by Arabs around 1200 CE, hidden in a cave. They were then translated into Latin and fascinated scholars immediately. Plato's teachings had dominated Western thought for a long time, but Aristotle's newly discovered writings were highly regarded, perhaps overly so. During the Renaissance, the works of Aristotle were taught almost on a par with the Bible. But that was when the questioning began in earnest.

The first effective challenge to Aristotle's natural philosophy (the earlier name for the study of the universe) was made by Galileo Galilei (1564–1642). Galileo was one of the chief architects of the Scientific Revolution, in which experimental evidence became the ultimate test of an idea's truth. He was well aware that Aristotle's ideas lacked experimental support. In 1609 a former student wrote Galileo a letter, telling him that Dutch scientists had ground glass into the shape of lenses and used them as eyeglasses, microscopes, and spyglasses. Galileo was fascinated. He had his instrument maker create a spyglass to his specifications, which was then sold to Venice's Senate. (The Senate members' interest undoubtedly centered on more worldly considerations, such as who was entering their harbor.) Galileo then commissioned a better spyglass and immediately turned it upward to study the sky. He discovered things never before seen: moons of Jupiter, craters on the Moon, and spots on the Sun. These discoveries dealt serious blows to Aristotle's notions about the perfection of heavenly bodies. As if that wasn't enough, Galileo used two different techniques to demolish Aristotle's ideas about heavy bodies falling faster than lighter ones. The first way was by means of a thought experiment. If a brick falls and suddenly splits in two, will each piece suddenly begin falling half as fast or will they continue to fall at the same rate? It didn't make sense that their rate would change, so Aristotle's ideas weren't logical. Galileo's other trick, which has no direct historical evidence, was having his assistants drop cannonballs of different mass from the Leaning Tower of Pisa. Onlookers could see for themselves that the bodies fell at the same rate and hit the ground simultaneously. As a measure of Galileo's attitude toward authority and truth, he may have invited university colleagues and church dignitaries to witness his experiments and compare the evidence of their own senses with the teachings of Aristotle. Continuing his assault on Aristotle's physics, Galileo used another thought experiment to address the question of whether a continuous force is needed to keep a moving body in motion. If a ball rolls down an incline, its speed increases as it goes down the slope, but if the ball encounters a horizontal segment, it rolls with a constant speed until friction finally stops it. Galileo reasoned that bodies move at a constant speed as long as no force (including friction) acts on them. This was called the principle of *inertia.* Not only did Galileo's work propel the scientific revolution, but it also set the stage for the next development.

Isaac Newton Mini-Biography[1]

The year 1642 was a phenomenal one. At its beginning, Galileo died, and at its end, Isaac Newton came into existence—but just barely. According to his mother, Hannah Ayscough (pronounced Askew) Newton, premature baby Isaac would have fit into "a quart pot." But Isaac's health was only one of several worries for his mother. Isaac's father, also named Isaac, had died three months earlier. So here was a yeoman farmer's widow, nursing a premature infant and trying to run a sizable farm, at the beginning of the violent English Civil War (1642–1651). But things got better for her quickly. Within two years after Isaac's birth, she received a marriage proposal from a sixty-three-year-old childless widower clergyman, Barnabas Smith. It was an offer she couldn't refuse. The wealthy pastor Smith promised to take care of young Isaac's future by providing him with income from a property, which he would get when he reached maturity. In the meantime, Isaac was to be cared for at the family home, Woolsthorpe, by his grandparents Ayscough. They had come to run the farm while Isaac's mother moved in with Reverend Mr. Smith, just two miles away. Newton's mother bore Smith three children in six years. Mr. Smith then died in 1653, the same year Newton's grandfather, James Ayscough, died. By now a wealthy widow, Hannah Ayscough Newton Smith moved back to Woolsthorpe to live with her younger children (Mary, Benjamin, and

Hannah), her mother, and Isaac, now ten years old. In less than a year, Isaac was sent off to school in Grantham, five miles away. But he didn't live at home; he boarded with the Clark family. There are mixed reports about his success at school, but he clearly enjoyed his leisure time in which he built toys, windmills, clocks, and furniture. Eventually, his mother decided Isaac needed to learn how to run the family farm, so she brought him home and assigned a trusted servant the task of making a farmer out of him. This enterprise failed, because Isaac preferred to read books or build clocks rather than run a farm. His uncle William Ayscough finally persuaded Hannah to let Isaac attend his alma mater, so Isaac went to Trinity College, Cambridge, in 1661.

At Cambridge, Isaac met another Isaac, Isaac Barrow. Barrow was twelve years Newton's senior and an extremely well-rounded and highly regarded scholar. Prior to 1663, Barrow had a faculty appointment in Greek. However, Reverend Henry Lucas left a number of charitable donations at his death, including four thousand books and a substantial endowment for a faculty position in mathematics at Cambridge. Barrow became the first Lucasian Professor of Mathematics at Cambridge. His advice to students was to work independently and keep natural philosophy and mathematics at the forefront of their studies. Barrow's private research interests included the mathematics of derivatives (the rate of change of a mathematical function), which Newton absorbed eagerly.

Just as Newton was awarded his BA, England's public health problem grew worse, so public gatherings were banned and universities were shut down (though church services continued) to minimize the spread of the plague. Newton returned to Woolsthorpe in late 1665. Perhaps his shiny new degree impressed his mother, because she gave up her quest to teach him farming. Newton was free to work on projects of his choice, but this time it wasn't clocks and windmills. Armed with the latest knowledge he could glean from Cambridge's books, Newton set himself larger goals and concentrated on them mightily. "I keep the subject constantly before me," he said, "and wait 'till the first dawnings open slowly, by little and little, into a full and clear light." No one knew how long the university shutdown would last, but it turned out to be almost two years. So Newton was able to accomplish a lot (and more will be discussed shortly):

- He invented a mathematical technique called fluxions, which became calculus.

- He formulated three general laws explaining the motion of bodies.
- He conceived the law of universal gravitation.
- He systematized the general procedure by which science operates.

Upon returning to Cambridge, Newton applied to continue his studies by becoming a Fellow of the University, and Isaac Barrow was appointed to examine him to determine his worthiness. It's difficult to imagine Barrow's response when he saw the fruit of Newton's two years of effort. Newton was admitted to the Fellowship handily and awarded an MA the following year.

In 1669 Isaac Barrow resigned his faculty appointment to become Charles II's chaplain. He recommended the post be filled by his former prize student, so Newton became the second Lucasian Professor of Mathematics. Newton was working primarily on light and he invented and built a telescope using a mirror rather than a lens. This reflecting telescope was superior to anything else in existence at the time. Although Newton was naturally reserved, Isaac Barrow suggested he demonstrate the telescope at a meeting of the Royal Society. Until then, Newton had written little and stayed out of the public eye, but Barrow was persuasive and Newton agreed. The telescope caused a sensation at the meeting. Partially in response to the Royal Society's enthusiasm, Newton published his theory about light and color the following year. Things looked rosy for Newton, but things didn't quite turn out as well as it seemed they might.

Almost ten years earlier, the Royal Society (which had originally been the Society for the Promoting of Physico-Mathematical Experimental Learning) had been formed. One of the officers was the curator of experiments, whose job was to demonstrate three or four significant experiments for each (weekly) meeting. At first, the society had no funds to pay a salary for this function or even to buy materials for the experiments. This seemingly impossible task was undertaken by Robert Hooke, however. He was a very talented experimenter who worked primarily with microscopes, though he enjoyed the challenge of preparing a wide variety of experiments. Hooke had built a reflecting telescope (which was far less effective than Newton's) and had many ideas of his own about light and color. When Newton's theories were published, Hooke was extremely critical and claimed that the parts of Newton's theory that were correct were stolen from him, and that the other parts were wrong. Newton didn't take kindly to the criticism and carried on a semipublic feud with Hooke that lasted thirty years and ended only when Hooke died in 1703.

Newton's natural reticence to publish his works was reinforced by the Hooke controversy, but he responded to questions and requests from friends, sometimes with the stipulation that his name not be used in publication. In 1686 Edmond Halley (later of comet fame) visited Newton to talk about the Moon, Kepler's laws, and gravity. Finding that Newton had already solved the problem of gravity but hadn't published his findings, Halley talked him into preparing his work for publication at Halley's expense. *Principia Mathematica Philosophiae Naturalis* was published in 1687. The *Principia*, as it became known, included differential and integral calculus (called fluxions by Newton), three general laws of motion, universal gravitation, and more. It established Newton as a preeminent scholar and generated both controversy and enormous respect. Within a few years, Newton seemed to lose interest in motion and began working on alchemy and biblical chronology, about which he published many more papers than on scientific topics. Briefly, he became a member of Parliament and was active in the political life of the university. In 1696 he was appointed to be Warden of the Mint (later Master), moved to London, and worked in the mint offices located in the Tower. Much to the chagrin of established bureaucrats, Newton took the ceremonial position seriously and helped reform the monetary system of England by cleaning out corruption and prosecuting counterfeiters. His personal scientific research had ended, but he was elected head of the Royal Society in 1703 and reelected each year until his death in 1727. In his last years, an enormous amount of time was spent in a controversy with Gottfried Leibniz about who invented calculus first.

Newton was short of stature and became quite stout in his later years. Fellows of the university were not allowed to marry, and Newton never did. He was always kind to his half brother and sisters and their children. Newton's niece, Catherine Barton, became his London secretary, which made for a lively and well-run household. In his lifetime, Newton had two serious illnesses, with symptoms of insomnia and anxiety. Some speculate these were bouts of depression or even nervous breakdowns. His absentmindedness was legendary. He was so focused that if he left a room to get a drink for a friend, he might forget to return and the friend would find him working on some problem. One famous story concerned dismounting his horse at the foot of a steep hill and then arriving at the top of the hill holding the bridle but no horse, which had slipped away unnoticed. Newton became very wealthy and distributed most of his money to family members before his death. He died from a bladder stone, and a subsequent autopsy showed the presence of large amounts of mercury in his body, perhaps a result of his alchemy experiments.

Though he was notoriously cantankerous, his assessment of himself is remarkably humble: "If I have seen further it is by standing on the shoulders of giants."[2]

In the *Principia*, Newton's laws concerning the general motion of bodies were stated in Latin. Here we will translate and express them in modern English:

Newton's First Law: *Every body continues in its state of rest, or of uniform motion in a straight line, unless it is compelled to change that state by forces impressed upon it.*

$$\text{If } F = 0, \text{ then } a = 0$$

This is just a restatement of Galileo's principle of inertia. If no net force acts on a body, it will not accelerate; it will move with a constant velocity.

Newton's Second Law: *The acceleration produced by a net external force acting on a body is directly proportional to the magnitude of the force and inversely proportional to the mass of the body, and the acceleration is in the same direction as the net external force.*

$$\vec{F}_{net\text{-}external} = m\vec{a} \quad \text{units} = \text{kg m/s}^2 = \text{newtons} = \text{N}$$

The second law provides a major tool by which the motion of any body may be determined as long as the forces that act on the body are known. To apply this principle, a graphical technique called a *free body diagram* is used (see the next chapter for examples). The body to be analyzed is sketched, then the forces acting on the body are drawn and added. Their directions are taken into account by using mathematical quantities called *vectors*. The net force is determined and set equal to the mass times the acceleration. To solve for the acceleration, equations from the prior chapter are used to predict the subsequent velocity and position of the body. The unit of force, obtained by multiplying the units of mass and acceleration, has been named the newton, in honor of Isaac Newton. One newton is the equivalent of a 1 kg mass accelerating at 1 m/s^2. This is not a very large force, as demonstrated in Experiment 3.

Newton's Third Law: *To every action there is always opposed an equal reaction; or, the mutual actions of two bodies upon each other are always equal, and directed to contrary parts.*

$$\vec{F}_{12} = -\vec{F}_{21}$$

Two bodies involved in a single interaction exert forces on each other that are equal in magnitude and opposite in direction. If you push on a wall, the wall pushes back just as hard but in the opposite direction. This law is sometimes applied to societal situations where its appropriateness is questionable, as in saying one politician's action prompts an equal and opposite reaction from another.

Newton's genius was evident in the way he used a period of isolation, away from the academic world, to summarize and generalize the work of prior natural philosophers. He then integrated those ideas with his own original ones into a general framework for analyzing the motion of bodies. Although they are stated as laws, don't be deceived into thinking these are unbreakable laws that apply in every situation. Newton himself knew that experimental support is a necessity for scientific hypotheses, and there were possibly areas beyond his experience in which his ideas might not apply. We now know that extremely small bodies and bodies that travel at speeds approaching that of light are beyond the limits of Newton's laws.

Newton integrated the description of motion (kinematics) with the causes of motion (dynamics) by linking them with his second law, F = ma. So acceleration is caused by forces. The next question is, where do forces come from? Forces, including Newton's analysis of gravity, will be the subject of the following chapter.

Experiment 4: Dropsy

Equipment needed: ball, measuring tape calibrated in meters, stopwatch

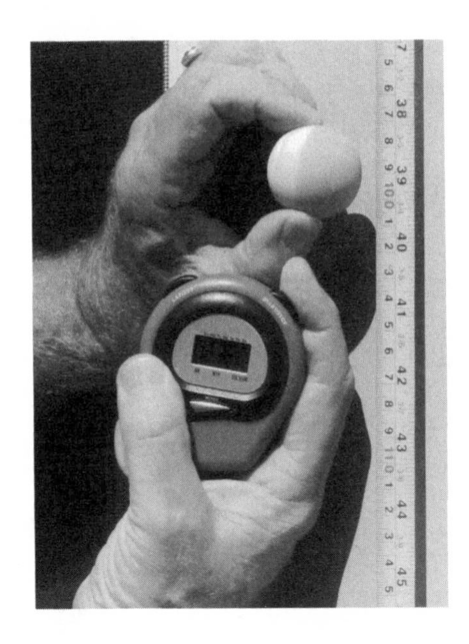

The purpose of this experiment is to measure the distance and time of a freely falling object and calculate the object's acceleration. As long as air friction is small enough to be ignored, your result should match the acceleration due to gravity, 9.80 m/s^2.

Procedure

- Set up a metric tape measure vertically and hold a ball at 2 meters (6 feet) above the ground.
- Record the height of the ball above the ground.
- Drop the ball and start the stopwatch simultaneously.
- Stop the stopwatch at the instant the ball hits the ground.
- Record the time elapsed.
- Calculate the acceleration of the ball, according to $a = 2y/t^2$.
- Repeat for a total of ten trials.
- Find the mean acceleration and compare it to the accepted value of 9.80 m/s^2.

Trial	Vertical distance = y (m)	Time elapsed = t (s)	Acceleration = $2y/t^2$ (m/s^2)
Example	2	0.65	9.47
1			
2			
3			
4			
5			
6			
7			
8			
9			
10			
Mean			

Chapter 3. SOURCES OF FORCES— THE BIGGIES IN NATURE

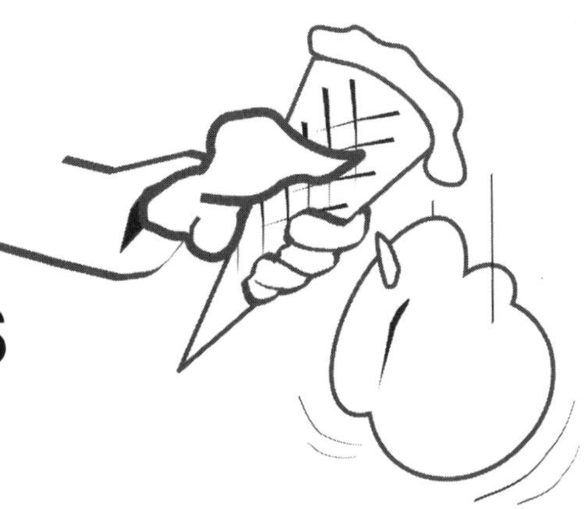

So, how about forces? Where do forces come from? Several fundamental forces have been identified in nature, and a number of other forces are derived from the basic four. Newton not only set up the rules for force causing acceleration; he analyzed the most familiar force: gravity.

Gravitational Force

Back on the farm in Woolsthorpe, while Newton was avoiding the plague, he wrote in his diary about noticing apples fall from trees. It seemed to him that gravity was the cause of an apple's fall, but he began wondering about gravity's reach. Might the Moon continue in its orbit because it is gravitationally attracted to the Earth? Using the mathematics he had just derived, Newton substituted values for the quantities involved, but the resulting prediction for the distance between the Moon and Earth differed from the then-accepted value by more than 10 percent. Although Newton remained confident of his idea, he knew something wasn't quite right, so he set it aside. Twenty years later, Edmond Halley revived Newton's interest in gravity. He also told Newton about a more accurate value for the Earth-Moon distance that had been recently measured. It turned out that the new distance matched Newton's prediction very closely.

Universal Gravitation

Every body in the universe with mass attracts every other body with mass, and the force of attraction depends directly on the product of the masses and inversely on the distance squared between the masses.

$$F_{gravitational} = \frac{Gm_1m_2}{r^2}\text{, where } G = \text{gravitational constant} = 6.67 \times 10^{-11}\,\text{Nm/kg}^2$$

An aspect of Newton's genius was his recognition that gravity extended to large distances as well as ordinary ones. (The application of gravity to very small distances is currently a topic of research.)

Since we live on the surface of the Earth, the gravitational attraction between Earth and anything on its surface is given a special name: *weight.* Newton's law of universal gravitation takes on a special form for weight:

$$F_{gravitational} = \frac{Gm_1m_2}{r^2} \Rightarrow \text{weight} = w = \left[\frac{Gm_{Earth}}{r^2_{Earth}}\right] m_{body} = m_{body}g$$

where g = acceleration due to gravity = $9.80\ \text{m/s}^2$

Some authors, especially those of chemistry texts, treat mass and weight as if they were synonymous. The authors of *The Joy of Chemistry* make no mistake about this point, and I wish all authors would be as meticulous. Certainly mass and weight are strongly related, but, as you know, even close relatives are seldom interchangeable. Here are three differences between weight and mass: *units*—weight is measured in newtons and mass in kilograms; *direction*—weight is a force that points toward the center of the Earth, mass has no direction; *universality*—weight is specific to the surface of planet Earth, mass is the same wherever a body may be located.

Contact Forces

Another familiar force occurs when two bodies are touching. If you were to place this book on a table, the table would exert a force, supporting the book so it would not be pulled to the floor by gravity. As long as the table is flat, the force exerted by the table would be vertical, perpendicular to the table's surface. This is referred to as a *normal* force, because *normal* means "perpendicular" in mathematics. To help apply Newton's second law, *free body diagrams* are drawn, in which all forces acting on a body are shown in their respective directions. In free body diagrams, a normal force is given the symbol N. To complete a free body diagram for a book on a table, the book's weight (w) would also be shown, acting vertically downward:

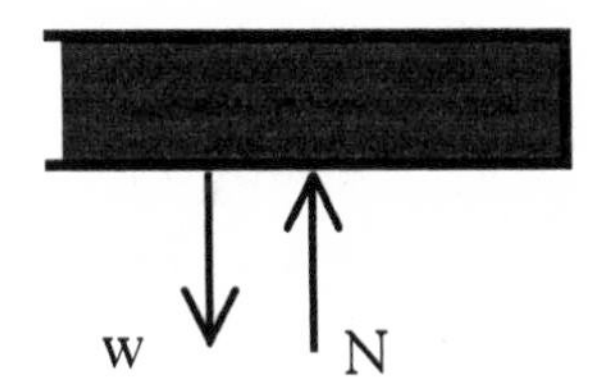

book free body diagram1

If these are the only two forces acting, the diagram is complete. It shows that the book is an example of Newton's first law in action: No net force, no acceleration.

Another force that may be present when bodies are in contact is *friction*. Friction acts along the contact surface between two bodies and may occur if there is relative motion or even if the motion is possibly going to happen. The latter is referred to as impending motion. Friction opposes motion, whether it is real or impending. In free body diagrams, a friction force is represented by f.

Returning to the example of the book on the table, if you push the book to the right with a force F, a frictional force is generated that opposes the applied force, and the diagram becomes:

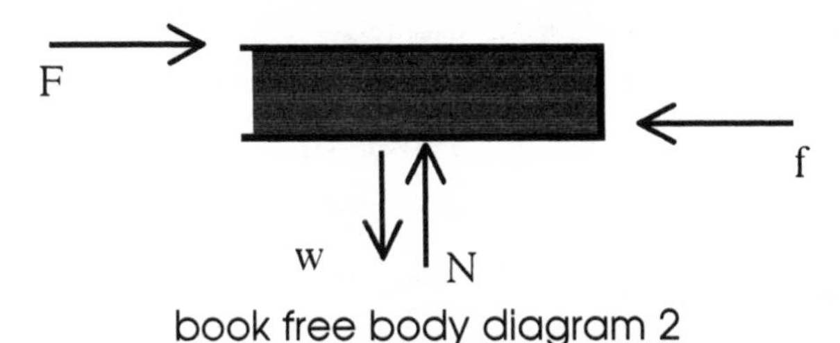

book free body diagram 2

Friction is motion's worst enemy. Not only does it oppose motion, but it adjusts so that it matches exactly whatever force is trying to cause motion, from zero all the way up to the maximum amount of force friction is capable of exerting. You push with 2 newtons, friction pushes back with 2 newtons. You push with 3 newtons, friction pushes back with 3 newtons, and so on. As if that's not bad enough, if you change the direction of the applied force, friction adjusts by changing its direction. When the applied force finally exceeds friction's ability to push back, the book will start to accelerate. But the frictional force still has one trick left. It doesn't always remain at the maximum value that opposed the initial motion. Once motion actually starts, the frictional force often reduces, making it easier to maintain motion. You know this from pushing a heavy object like a desk across a floor. It's hard to get the motion started, but once the desk moves, it becomes much easier to push.

Because the frictional force is so tricky, you might be tempted to think friciton would be particularly complicated and difficult to formulate mathematically. Luckily, there is a simple relationship that is approximate, but reasonably accurate.

Maximum frictional force opposes the real or impending motion and is directly proportional to the formal force between the two bodies in contact and a coefficient, depending on the surfaces of the bodies involved, and whether the motion is impending or real.

$f \leq \mu N$ opposing real or impending motion,

where μ is a friction coefficient, dependent on the materials of the bodies in contact and whether the motion is impending (static) or real (kinetic) and the normal force between the bodies, N

The following table lists representative values of friction coefficients for different bodies in contact, averaged over many different sources.

Materials	**Static friction coefficient**	**Kinetic friction coefficient**
Wood on wood	0.4	0.3
Waxed ski on snow	0.05	0.05
Rubber on concrete (dry)	1.0	0.8
Rubber on concrete (wet)	0.6	0.5
Glass on glass	0.95	0.4
Steel on steel (dry)	0.7	0.6

Steel on steel (lubricated)	0.12	0.07
Teflon on teflon	0.04	0.04
human joints	0.01	0.004

Friction can actually be helpful. The act of walking involves trying to force your foot backward. Friction opposes this and pushes you forward. Similarly, your car's engine tries to spin the tires backward. Think about what happens when friction is absent.

Experimentino 1

In your car, if you have something hanging from your rearview mirror (fuzzy dice, baby shoes, a crystal, etc.), have a passenger estimate the angle the string makes with the vertical during acceleration or deceleration. If the angle is just under 30° (26.6°), the acceleration or deceleration is half of g.

These ideas about contact and friction forces were known to varying degrees by Leonardo da Vinci, Galileo, and Newton. But there is a lot more to contact than meets the eye. Newton needed to get off the shoulders of giants and go to a level so tiny that even Hooke couldn't see it in his microscope. From a modern perspective, we know that matter consists of molecules that are in turn made of atoms, and they have electron clouds as their outermost layer (see chapter 26). What we see from our human-sized perspective as contact between bodies is actually electron clouds of one body's molecules overlapping electron clouds of the other body's molecules. Since electrons are negative, we don't need a lot of electrical knowledge to realize that there is a repulsive force generated. What we thought was a mechanical force is actually an electrical force, so contact is not one of the fundamental forces.

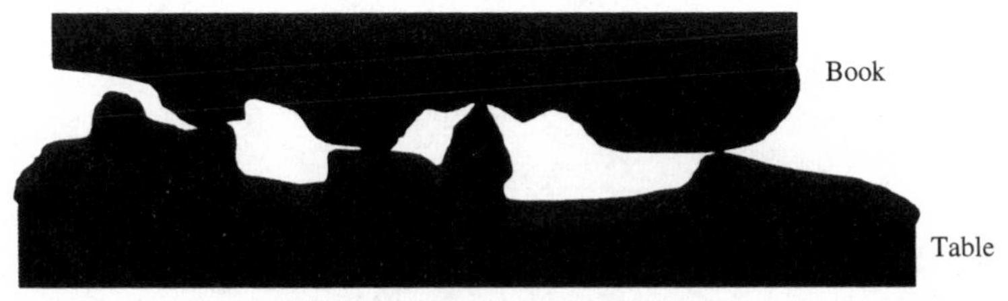

molecular view of contact

Similarly, friction becomes understandable in terms of dragging one set of molecular hills and valleys across another. At the molecular level, even extremely smooth surfaces have built-in roughness. Electrical bonds are made and broken by the molecules of the bodies. Might molecules of a body be sheared off by electric forces caused by relative motion? That's what sandpaper is all about. The role of lubricants also becomes clearer in molecular terms. A liquid would tend to fill the valleys, making the sliding smoother. Also, if the lubricant molecules were quite different from the molecules of the bodies, fewer bonds would be formed and broken.

ADDITIONAL MECHANICAL FORCES

In analyzing the motion of ordinary bodies, another class of forces is provided by such things as ropes, strings, cables, and chains. Pulling on a string applies force to the body to which the string is attached and puts the string in tension. In a free body diagram, tension forces are given the symbol T, and always act along the string. For example, if you tie a string around this book, place it on a tilted table, and pull upward along the slope, the free body diagram would look like this:

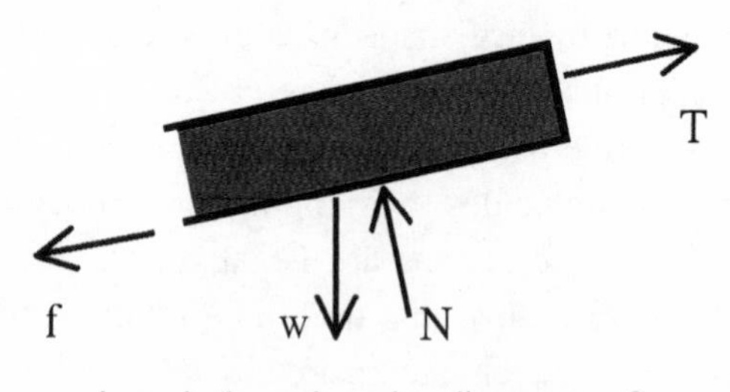

book free body diagram 3

At the molecular level, these forces are again electric in nature, since the rope molecules attract each other electrically. Thus, these forces are not fundamental either.

Elastic Force: Many materials stretch a little in tension and shrink a little in compression, then return to normal when the force is released. The ideal example of an elastic material is a spring, in which the amount of stretch is directly proportional to the force applied. Electric bonds between molecules are the source of elastic forces, so they are also not fundamental.

Electric Force: Charged bodies exert electric forces on each other. Since most bodies in our everyday experience are not charged, electric forces seldom show up in ordinary free body diagrams. Electric forces are very important in electric circuits and electronic gadgets, however.

Magnetic Force: Very few materials are naturally magnetic, so magnetic forces are seldom seen in ordinary experience outside of refrigerator magnets. Magnetism is so closely linked to electricity that the two are treated as one force—electromagnetism—and constitute the second of the four fundamental forces.

Strong Nuclear Force: This force affects only extremely small particles that are packed close together in a small volume, such as the nucleus of an atom. Nevertheless, the strong nuclear force is extremely important in the overall scheme of physics (and society).

Weak Nuclear Force: As its name implies, this fourth fundamental force is weaker than the strong nuclear force and has an even shorter range.

Physics in the Culture

Forces and their effects on motion are so deeply understood by our culture that the incongruity is good for a laugh when a cartoon character steps off a cliff and doesn't fall until he realizes there is nothing supporting him.

Still, the dictates of gravity are often ignored when it suits us. In the movie *Total Recall*, Arnold Schwarzenegger's character spends some time on Mars, but the gravitational pull seems to be the same as on Earth. Since the acceleration due to gravity on Mars is only 38 percent of what it is on Earth, wouldn't it have been funny (and difficult for the special effects people) to see Arnold and all the other characters lift heavy weights even more easily than usual and leap around when they tried to walk?

UFO sightings also imply the existence of some very interesting forces. Some sightings report that these unknown objects hover a while, then move away extremely rapidly, disappearing from view within seconds. The acceleration required for such a maneuver would be huge, and if a reasonable mass were involved, the force would be enormous. Further, the internal structure of the UFO's crew would have to be substantially different from ours. Humans black out beyond accelerations of 6 g's; at more than 10 g's, internal organs would begin tearing loose from their moorings. At accelerations much beyond 10 g's, soft internal tissues would be squashed against bones and become liquid, and the bones would break. These points pose a problem for alien abductions. Humans whisked away in such a fashion would have their innards turned to soup.

Another pseudoscience/physics connection is provided by astrology. The mechanism by which the Sun, Moon, and planets in the solar system influence humans at birth has never been identified. If we consider only the forces known to physics, the strong and weak nuclear forces must be ruled out because they are too short-range to influence anything over astronomical distances. The electromagnetic force is stronger but requires charges to operate, which leaves out both planets and humans. Only gravitational force remains. Using the masses and distances for nearby astronomical bodies, the Sun and Moon exert the largest gravitational forces, but the doctor that delivered the baby and the nearby medical device that goes "Ping" exert stronger gravitational forces on a newborn than any planet.

Then, there's the classic *Star Wars* story that features the omnipresent Force. The Force is rather vague but extremely fundamental in that galaxy far, far away. By now, you may be wondering if physics represents the Dark Side or the Light Side of The Force. Stay tuned: more information is on the way.

Experiment 5:
Horizontal Globalization

Equipment needed: yo-yo, straw, measuring tape calibrated in meters, stopwatch

The goal of this experiment is to whirl a yo-yo in a horizontal circle and measure enough quantities to check the standard physics relationship for circular motion.

Procedure

- Unroll the string from the yo-yo and make a mark 50 cm (20 in) along the string, measured from the center of the yo-yo.
- Cut the straw so that you have a piece about 10 cm (4 in) long.
- Feed the yo-yo string through the straw.
- Attach the metric tape measure (weight from Experiment 3) to the end of the yo-yo string (this may require a paper clip).
- Holding the straw vertically, twirl the yo-yo just fast enough so the mark on the string is located at the top edge of the straw.
- Using the stopwatch, measure and record the time for one revolution of the yo-yo. (You may wish to time ten revolutions and divide by 10.)
- Record the measurements and repeat ten times.
- Determine the yo-yo's mean tangential velocity, $\frac{2\pi r}{t}$.

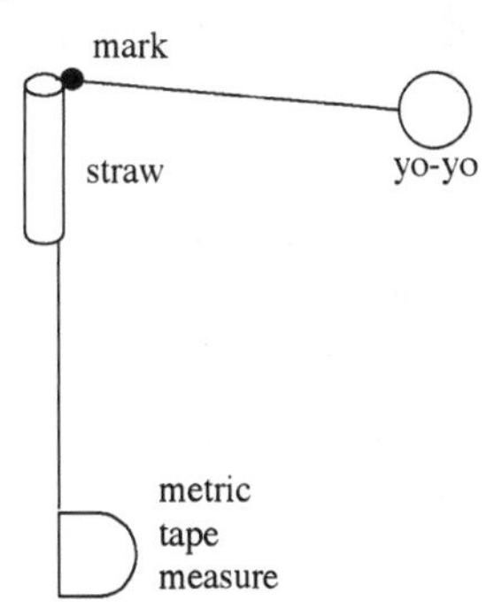

- Compare the weight of the metric tape measure [newtons] to the (yo-yo mass[kg])(mean tangential velocity[m/s])^2/(distance from yo-yo center to mark[m]).

If these quantities match to within 10 percent (friction, timing uncertainties), the concept is supported.

Trial	**Time for one revolution, s**	**Tangential velocity, 2pr/t, m/s (r = 0.5 m)**
Sample	0.38	1.37
1		
2		
3		
4		
5		
6		
7		
8		
9		
10		
Mean		
Compare mv^2/r to the weight of the metric tape measure (Sample: 1.37 newtons compared to 1.3 newtons)		

Chapter 4. ROUND AND ROUND IT GOES, AND WHERE IT STOPS . . . CIRCULAR MOTION

"That inconstant moon that monthly changes in her circled orb . . ."

—William Shakespeare, *Romeo and Juliet*

Armed with some understanding of motion and its causes, let's examine an extremely familiar motion, a body moving in a circle. As an example, we'll take a traffic circle, also known as a *roundabout*. Usually, drivers don't travel the whole way around, but sometimes they do make a complete circle, possibly more than once (that would be me). You might almost think of this as being similar to motion along a line, except the line has been bent into a circle.

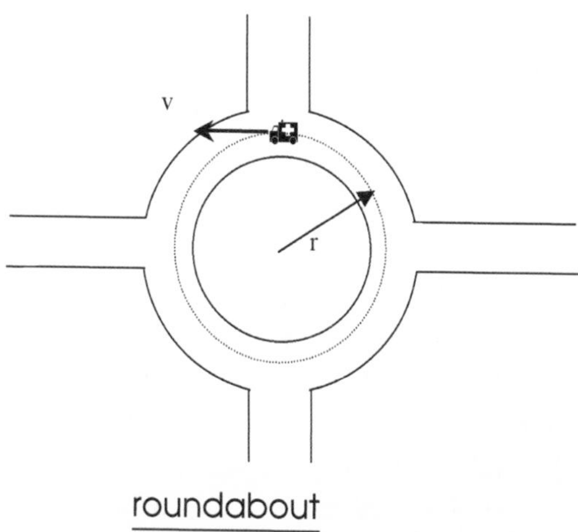

roundabout

Many of the ideas developed to deal with linear motion still apply. If you proceeded with caution around the roundabout and watched your speedometer carefully, you could maintain a constant speed. You can perform this experiment as a purely intellectual exercise.

> The speed of a body moving in a circle is the total distance traveled divided by the time it takes to complete one circuit.
>
> $$v = \frac{2\pi r}{t}, \text{ where } r = \text{circle radius and}$$
> $$t = \text{time to travel around the circle}$$

Although the car's speed is constant, its velocity is not, since the direction of the car is constantly changing. At any instant, if the car quit moving in a circle, it would go straight (Newton's first law), so the velocity's direction is tangent to the circle and is referred to as *tangential* velocity.

Since the velocity changes, this is the condition for accelerated motion. The magnitude of the acceleration is given by:

> The acceleration of a body in circular motion has a magnitude that is directly proportional to the square of the tangential velocity and inversely proportional to the circle's radius.
>
> $$a_{\text{circular}} = \frac{v^2}{r}, \text{ where } v = \text{tangential velocity}$$
> $$\text{and } r = \text{circle radius}$$

The direction of the acceleration is toward the center of the circle and is referred to as *centripetal* acceleration. Centripetal means center-seeking, so you can think of a body in circular motion as being constantly accelerated toward the center to keep it from flying off in a straight line. Returning to our roundabout example, as you proceed in a circle, you must be accelerating. But Newton's second law says a force is needed to accelerate a body. This is called a *centripetal* force.

A centripetal force is necessary to maintain circular motion and depends directly on the mass of the body and the square of the body's tangential velocity, and inversely on the radius of the circle. The centripetal force's direction is toward the center of the circle.

$$F_{centripetal} = \frac{mv^2}{r}, \text{ where } m = \text{body's mass, } v = \text{tangential velocity, and } r = \text{circle's radius}$$

direction = toward circle's center

So how do you accomplish this acceleration? Not by stepping on the accelerator. That would make your car go faster and possibly lead to legal trouble or worse. No, it's much simpler. You change the direction of the velocity by turning the steering wheel. But turning the steering wheel doesn't qualify as a force that causes acceleration, since it isn't external. Ever have a situation where turning the steering wheel didn't produce a change in the direction of the velocity? Snowy, icy, wet conditions reduce the friction coefficient to such a small value that the frictional force is insufficient to turn the car. It's friction that supplies the force to turn the car. In fact, circular motion cannot occur unless a centripetal force is provided by one of the forces we've described.

Sometimes you'll read or hear about a *centrifugal force*. This term is easily confused with *centripetal* force, but it's more than just word befuddlement that irritates me. Think about sitting in a car that's traveling in a roundabout to the left at a pretty good pace. If you're in a righthand seat, you'll find yourself near the right door in short order. The erroneous way of looking at this is to say that there is a force that pushes you to the right, and this force is called a centrifugal force. This is flat-out *wrong*. There is no force pushing you away from the center of the circle. In fact, there is a force pushing you toward the center of the circle. It is supplied by the door. The reason you get pushed on by the door is that your body is a good law-abiding citizen and obeys Newton's first law, which says you will move in a straight line until some external force causes that situation to change. That external force is caused by the door, and it pushes you toward the center of the roundabout, maintaining your circular motion. If the door didn't push on you, you would proceed straight and fly off tangentially. (Notice there is no experimentino set up for you to open your car door to test this. That wouldn't be a good idea.)

Any one of the forces we saw in the last chapter is capable of providing the force to maintain circular motion. Let's look at examples of each force causing a body to move in a circle.

GRAVITY

Ever see (or do) the demonstration where a bucket of water is swung in a vertical circle without spilling a drop? If you whirl the bucket fast enough, there is little chance of a spill, but how slow can you go? The critical point is the top of the arc. At the slowest possible tangential speed, the water just begins to come loose, and the bottom of the bucket exerts no contact force. Here's a free body diagram of the water in the bucket at the top of the loop:

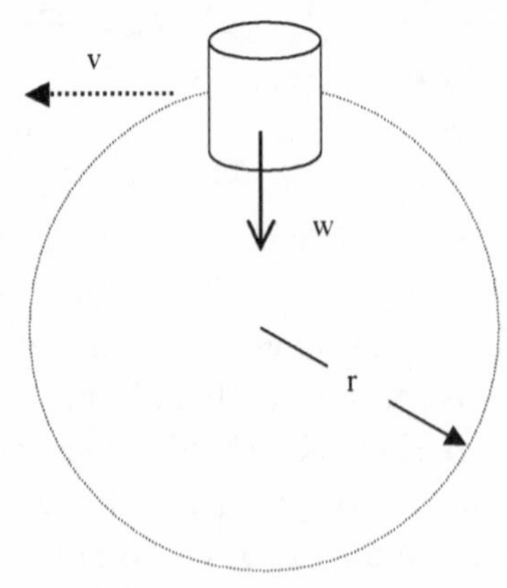

whirling water bucket

The weight, $w = mg$, supplies the force that maintains circular motion, so $mg = mv^2/r$. Solving for v, $v = \sqrt{gr}$. Of course, predictions are not the last word. If you want to carry the scientific method to the next step, you'll need a bucket and some water. . . . Never mind, this could get messy, and no one wants to get hit by a flying bucket.

A much larger example involving gravity is provided by the Moon's orbiting the Earth in a nearly circular orbit, and the Earth's orbiting the Sun similarly. (Although the orbits are elliptical, they are very close to being circular.) In fact, only Mercury and the dwarf planet Pluto have orbits that differ by a noticeable amount from circles. Newton was the first to recognize the gravitational force as the force that maintained circular motion in the solar system. Incidentally, the dominance of circular orbits is understandable in terms of the solar system formation. Materials orbiting the Sun in noncircular orbits tended to collide with each other, which helped form the planets.

Starting in 1957, thousands of artificial satellites have been launched into Earth orbit. Although satellites generally orbit elliptically, some extremely useful ones are in circular orbits. Wouldn't it be interesting to have a satellite that moves around the Earth just as fast as the Earth spins on its axis? Such a satellite would seem to hover above one spot on the Earth as if it were mounted on a gigantic tower. Connect this with the satellite dish antennas that you often see pointed at a single spot, and you have a picture of what is called a *geosynchronous* satellite. Once the satellite is launched into orbit, gravitational attraction between the satellite and the Earth keeps the circular motion going. Since the center of the geosynchronous satellite's orbit must be at the Earth's center, such satellites orbit above a point on the Earth's equator. This maintains a fixed position above a point on the Earth. It turns out that such satellites orbit at an altitude of about six Earth radii, so that's where the antennas must point.

FRICTION

An example related to the traffic roundabout is provided by a car rounding a flat curve. Without some centripetal force, the car would continue to travel straight. So turning the wheel generates a friction force perpendicular to the wheels, as shown in the free body diagram:

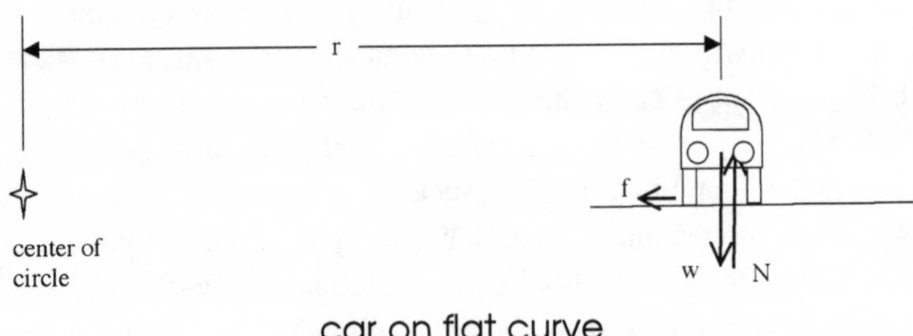

car on flat curve

Vertically, the car doesn't accelerate, so w = N. But horizontally, the car has a net frictional force that provides the centripetal force to keep the car accelerating its way around the curve. Using the previous mathematical relationships yields the requirement that $v^2 = \mu gr$. Thus, a traffic engineer, knowing the radius of the curve, and making an assumption about the friction coefficient between the tires and the road surface, can calculate the speed at which cars can negotiate a curve safely. If you are ticketed for speeding on a curve, you could argue about your high friction coefficient tires, but, as they say, "Tell it to the judge."

Normal Force

The situation changes a bit if the curve is banked. As the free body diagram show:

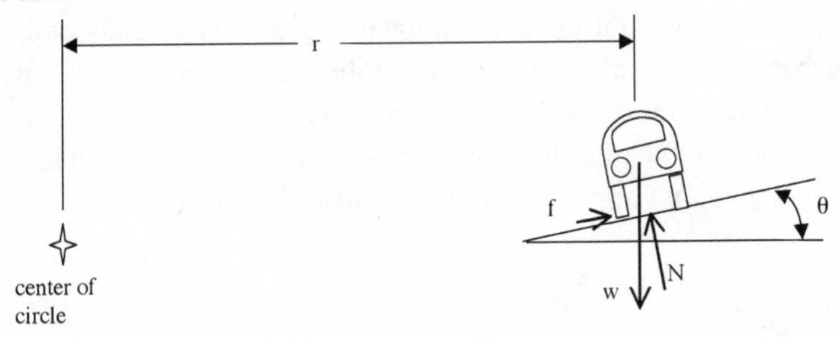

car on banked curve

The only force that has a horizontal component, and thus could maintain circular motion, is the normal force generated by the ground. For the case shown in the diagram, if the car traveled very slowly and there was no friction, it would slide down the slope. Friction puts an end to that possibility by pointing up the slope. Since friction points away from the center of the curve, it doesn't help maintain circular motion. On the other hand, if the car traveled too fast, its tendency would be to slide up the curve. So what does the friction do in that case? It changes direction and opposes motion up the plane. There is one speed at which the car can negotiate the curve with zero friction required. Setting $f = 0$ and solving $F = ma$ yields $v^2 = gr \tan\theta$. Surely you've noticed that some banked curves can be negotiated effortlessly, or you've seen photos of steeply banked ovals where cars are tested at high speed without any need for friction.

Another situation where the normal force provides the force necessary to maintain circular motion is on a carnival ride, sometimes called the Rotor, Graviton, or a similar name. It is a large cylinder that you enter and stand against a wall. The cylinder starts to spin, and you feel pushed against the wall. Then the floor drops away, but you don't fall. Most people feel as if there is a centrifugal force that pushes them into the wall. Here's the free body diagram:

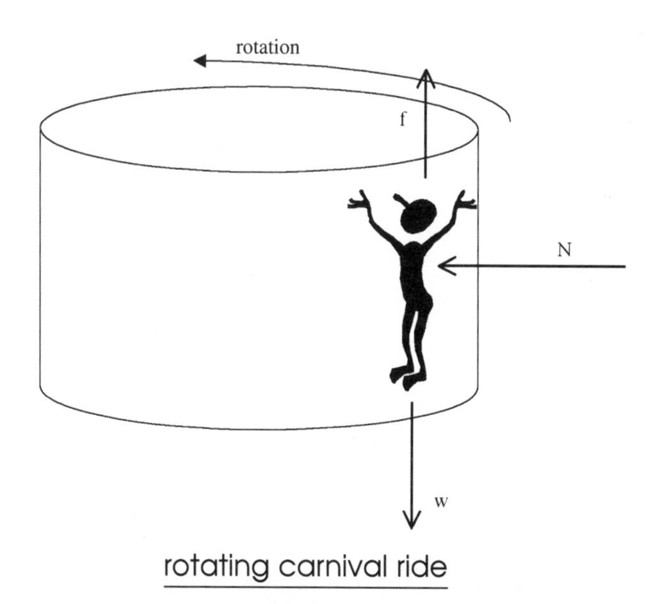

rotating carnival ride

In the horizontal direction, the only force acting on you is the normal force exerted by the wall. This is the centripetal force that maintains circular motion. What about a centrifugal force? No such force acts on you.

Some of the confusion derives from the fact that you exert a force on the wall. That force does point away from the center, but since it acts on the wall, not you, it doesn't affect your motion. The force exerted by the wall on you and the force exerted by you on the wall are an action-reaction force pair, covered nicely by Newton's third law. You don't fall vertically because of friction, since it balances your weight. If you use the equations developed previously, it turns out that the friction coefficient needed to keep you from sliding down is given by $\mu = gr/v^2$. So if you try to get on one of these carnival rides wearing a slippery Teflon jacket, the carnival workers might not let you enter.

TENSION

As you saw in Experiment 5, whirling a yo-yo in a horizontal circle is an example of tension providing centripetal force.

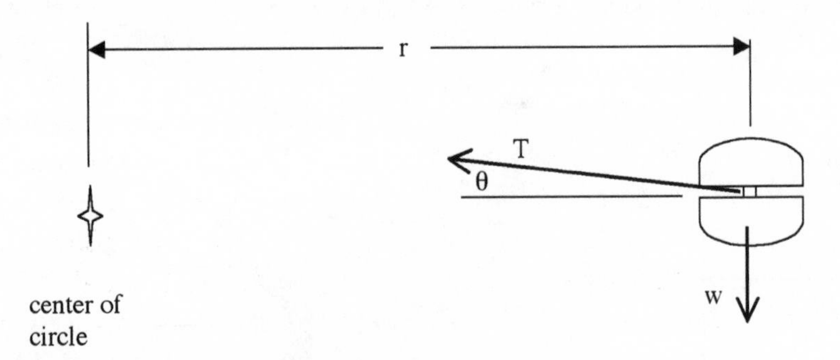

yo-yo free body diagram

As long as the angle is small, $T \approx mv^2/r$. The string tension depends on the yo-yo mass, speed, and distance.

Since $v = \dfrac{2\pi r}{t}$, all quantities can be measured and the relationship checked.

An extremely useful example of circular motion is provided by the *centrifuge*. This device contains several tubes mounted radially—in an almost horizontal plane, so they can be rotated rapidly around a vertical axis located at the center of the device.

If a particular tube containing liquid and heavier particles is spun at high speed, the heavier particles attempt to travel in a straight line, which will carry them to the bottom of the tube. Friction with the liquid slows their travel a bit, but fast rotation speeds cause heavy particles to settle fairly quickly. This process speeds sedimentation and is useful in separating more dense materials from less dense ones. One example of this separation, which has been in the news recently, is the separation of isotopes of uranium. This relates to nuclear reactors and/or bombs (see chapter 26).

Physics in the Culture

Amusement park rides can be very entertaining, and, interestingly, some parks have Physics Days, when high school students make measurements, perform calculations, and write reports about the physics of their experiences. One of the most popular rides is a roller coaster that includes a loop-the-loop, where you go around and upside down. Rapid circular motion is thrilling and even seems like astronaut training.

As another entertaining experience, I'm sure we've all seen a cartoon in which one character whirls something (a carrot, perhaps) around his head, then gets distracted, and walks away, while the object continues spinning. This invariably gets a good laugh, even from small children. Do you suppose it's because they know there must be a centripetal force in order to maintain circular motion? I'd love to think so, but probably not.

Experiment 6: Outgoing = Incoming

Equipment needed: half-ball popper, measuring tape calibrated in meters

In the absence of frictional force, the principle of conservation of energy says that a body's kinetic energy can be turned completely into gravitational potential energy. The purpose of this experiment is to measure a half-ball popper's potential energy and use energy conservation to find the launch velocity of the popper and compare it to the launch velocity found in Experiment 2. If the velocities match, the energy conservation principle is supported.

Safety glasses are recommended. Make sure you handle the popper carefully so it doesn't hit you in the eye.

PROCEDURE

- Set the metric tape measure vertically, extending it to at least a distance of 2 m (about 6 ft).
- Use the half-ball popper from Experiment 2 and turn it inside out. Handling it by the edges, place it on a rigid horizontal surface.
- When the half-ball popper pops straight up, note the height to which it rises.
- Record the height in the following table.

- Calculate the velocity based on energy conservation: $v = \sqrt{2gh}$, where $g = 9.8\ m/s^2$.
- Repeat for a total of ten trials.
- Find the mean velocity.
- Compare the mean velocity to the launch velocity found in Experiment 2.

Trial	**Height (m)**	**$v = \sqrt{2gh}$ (m/s), where $g = 9.8\ m/s^2$**
Sample	1.8	5.94
1		
2		
3		
4		
5		
6		
7		
8		
9		
10		
Mean		

Chapter 5.
THE ULTIMATE FOUR-LETTER WORD—WORK

"This is not fortune's work neither, but Nature's . . ."
William Shakespeare, *As You Like It*

Those natural scientists (the terms *physics* or *physicists* weren't used until 1715) who followed Newton applied, extended, and simplified his efforts. The first extension of Newton's ideas that we will examine is an extremely powerful tool, and though it is a familiar word, it has a different meaning than usual: *work*.

If a body has a force applied to it and moves through some distance, the product of the force and distance constitutes the definition of the physics quantity *work*.

> The work done on a body is proportional to the force acting on a body and the distance moved by the body in the direction of the force.
>
> work = $W = F\Delta x$ units = Nm = joules = J
>
> where F = net force acting
> and Δx = distance moved by the body in the direction of the force

To get a feeling for the size of the unit of work, you could accomplish 1 joule of work by pushing a hockey puck along frictionless ice with a 1 newton force for a distance of 1 meter (3 feet). Since a 1 newton force is about the equivalent of the weight of a set of keys, the joule is not a very large quantity.

Because of work's restrictive definition, there are several instances that might seem as though work has been done, but work in the physics sense is not done.

No F: Continuing with the hockey puck example, look at what happens after you stop pushing. If you presume the ice is frictionless, the puck would be moving along smartly in the direction you pushed it. But no work is being done, since it isn't being pushed.

No Δx: In another sports setting, suppose you have lifted a heavy weight high over your head and are holding it there, enjoying the admiration of the audience. Are you doing any work?

No. While you're holding it over your head, the weight isn't moving any distance, so no physics work is done, regardless of how you may sweat while keeping that heavy weight in the air.

F and Δx perpendicular: In all the examples of circular motion we noted in the last chapter, the centripetal force is directed toward the center of the circle, while the motion of the body is tangent to the circle. Since these two directions are perpendicular to each other, the body doesn't move at all in the direction of the force, so no physics work is done by any of the forces that supported circular motion.

You may also notice that work can be either positive or negative. Negative work is done when the body moves one way even though the force acting on the body points the opposite way. This happens every time you apply the brakes in your car. The friction force points to the rear, even though your car continues to move forward.

ENERGY

By now, you may be wondering what might be the point of this quantity work (some people spend a lifetime pondering this question). Physics has an answer. Let's return to the hockey puck. What happens to the puck after you perform work on it? It moves, rather than remaining at rest. The work has been transformed into motion of the body. Using the equations developed earlier yields the quantity:

$$W = F\Delta x = \frac{1}{2}mv^2$$

The quantity $\frac{1}{2}mv^2$ is called *kinetic energy*. This relationship between work and energy is called the *work-energy theorem* and is a powerful tool in physics. (Technically, this is called translational kinetic energy, since the body is considered to be a point moving without rotation. We will see a related form, called rotational kinetic energy, in chapter 8.)

Kinetic energy is energy of motion and is directly proportional to a body's mass and the square of its velocity.

$$KE = \frac{1}{2}mv^2 \qquad \text{unit = joules}$$

Returning to the weight lifter example, we've seen that no work is done while the weight is stationary. What about the work done to lift the heavy weight over your head? If you lift slowly and without acceleration, the force you apply exactly balances the weight, so the work you do equals the negative of the work done by the gravitational force. The work doesn't turn into kinetic energy, but you could make that happen easily enough: all you've got to do is to let go of the weight. The weight will fall and it will gain kinetic energy. So what happened to the work? You could think of it as being stored energy that could be converted to kinetic energy. The work became a different kind of energy. Lifting a body against gravity builds *gravitational potential energy.*

Gravitational potential energy is directly proportional to a body's mass, the acceleration due to gravity, and the vertical height above a reference level.

GPE = mgh, where m is the body mass, g is the acceleration due to gravity, and h is the vertical height above a reference level — units = joules = J

It's not always possible to generate potential energy. Forces that perform work that can be converted to kinetic energy are called *conservative forces.* Let's look at each of the forces from chapter 3: The *normal force* generates no potential energy because motion occurs perpendicular to the force's direction, so zero work is done. *Friction* generates work, but it is negative and produces no potential energy because you can't get frictional work to turn into kinetic energy. In fact, friction decreases a body's energy—more about this shortly. The *elastic force* does generate potential energy because the work you must do to compress a spring is stored in the spring and can be reclaimed as kinetic energy by simply letting the spring loose. This is called *elastic potential energy. Electromagnetic forces* generate potential energy, as does the *nuclear force.*

Here's a general definition of potential energy, represented symbolically by PE: Potential energy is energy of position. Conservative forces are defined to be forces that allow potential energy to be produced, whereas nonconservative or dissipative forces produce negative work and zero PE.

This brings us to a very powerful principle, the *conservation of total energy*.

If no dissipative forces are present, the total energy of a closed system of bodies is constant at any point in the motion of the bodies.

> For a closed system of bodies with no dissipative forces,
> $$(KE + PE)_1 = (KE + PE)_2$$

If there *are* dissipative forces (friction is usually the major dissipative force) in the system, the dissipative work subtracts from the total energy.

If a system includes dissipative forces, the difference in the total energy between any two points is equal to the dissipative work done between the points.

> For a closed system of bodies, if dissipative forces are present,
> $$\Delta(KE + PE) = W_{dissipative}$$

This is an extremely powerful principle and allows difficult problems to be analyzed more easily. For example, consider a pendulum swinging back and forth:

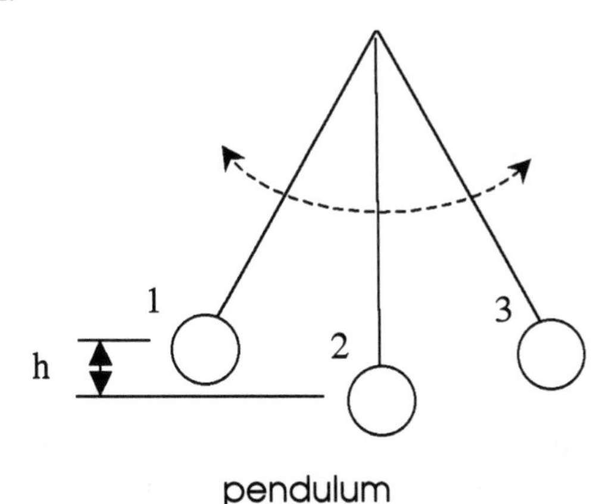

pendulum

If we ignore any friction at the top support, there are no dissipative forces present, so the total energy at 1 is the same as the total energy at 2 or 3. Solving the energy conservation equation directly yields $v_2^2 = 2gh$. Using the relationships for circular motion, Newton's second law, and trigonometry to analyze a pendulum's motion will yield the same result, but energy conservation is much quicker and more direct.

On a much larger scale, consider a rocket ship leaving a planet. What launch velocity would be required so that the rocket could escape the planet's gravitational pull? This is called the *escape velocity*. Using the general form for gravitational potential energy, and solving the energy conservation equation, yields the result $v^2 = 2GM/R$, where G is the gravitational constant, M is the planet's mass, and R its radius. For Earth, the escape velocity is 11 km/s, for Mars 5 km/s, and 620 km/s for the Sun. For that strange body known as a black hole, this analysis produces the impossible result that the rocket needs to exceed the speed of light in order to escape. Since that violates the universal speed limit—nothing can surpass the speed of light—this begins to reveal some of the strange properties of black holes.

POWER

In society, lots of people want power, but when they get it, it seems temporary and elusive. In physics, it's not elusive at all. *Power* is simply the rate of doing work. The faster you work, the more power you've got. Or, since work and energy are interchangeable, power is also how fast energy is transformed from one kind to another.

> Power is the rate at which work is done, or the rate at which energy is transformed from one kind to another.
>
> $$P = \frac{W}{\Delta t} = \frac{\Delta E}{\Delta t} \qquad \text{unit} = \text{J/s} = \text{watts} = \text{W}$$

Power's unit, the watt, can be demonstrated by applying the 1 newton force to push the hockey puck a distance of 1 meter and accomplishing this task in 1 second. As you can see, the watt is not a very large unit. If you're an auto enthusiast, you may know the British unit for power, the *horsepower*, which is 746 watts. The watt may be a more familiar unit in an electrical context, since it is stamped on lightbulbs and hair dryers. Aren't watts a better unit than horsepower? What does your 1,500-watt hairdryer have to do with horses?

Experiment 7:
Follow the Bouncing Ball

Equipment needed: high-bouncing ball, metric measuring tape

Collisions vary widely, from perfectly elastic ones where no kinetic energy is lost to the perfectly inelastic variety, where colliding objects stick together. The purpose of this experiment is to test a high-bouncing ball to see how close it comes to being perfectly elastic.

PROCEDURE

- Set up the metric measuring tape in a vertical direction.
- Hold the ball at a height of 1 m (3 ft) above a rigid horizontal surface.
- Drop the ball and note the height to which it rebounds.
- Record the rebound height in the table below.
- Find the fraction of the original height the ball regained in its bounce.
- Repeat for a total of ten trials.
- Find the mean of all ten trials.

Trial	Original height, m	Rebound height, m	Original height/ rebound height
Sample	1	.85	.85
1			
2			
3			
4			
5			
6			
7			
8			
9			
10			
Mean			

Chapter 6.
OOPS! COLLISIONS—IMPULSE AND LINEAR MOMENTUM

"I would not waste my life in friction when it could be turned into momentum."

—Frances Willard
(Women's Christian Temperance Union president)

Originally, Newton worked with momentum in formulating his laws of motion. *Momentum* is the product of mass and velocity. As originally framed, Newton's second law says:

> The time rate of change of the momentum of a body is directly proportional to the net force acting on it, and the direction of the momentum change is in the same direction as the net force.
>
> $$\vec{F}_{net} = \frac{\Delta \vec{p}}{\Delta t}$$

For a single body with constant mass, the change in momentum is simply mass times acceleration, and Newton's second law takes its familiar form. But if this relationship is generalized to a system of two or more bodies that interact only with each other, all forces are internal to the system, so there is no net external force. This implies that the momentum of such a system doesn't change over time. The constant nature of momentum gives physics a powerful new principle, the *conservation of linear momentum*:

> For a system of bodies that interact only with each other, the total momentum of the system remains constant over time.
>
> $\vec{p}_{\text{total@time 1}} = \vec{p}_{\text{total@time 2}}$ for a system of bodies that interact only with each other

This is a very useful relationship that leads to valuable insights and simplifies the analysis of difficult problems.

For example, consider a rifle about to fire a bullet. As you set up your rifle at a shooting range and aim it at the target, the momentum of the rifle-bullet system is clearly zero. Yet, when you pull the trigger, the bullet that emerges from the rifle's barrel carries forward momentum. For the total system's momentum to remain zero, the rifle must acquire backward momentum, enough to balance the bullet's forward momentum. Does this happen? Anybody who has ever shot rifles knows about *recoil*. That's why beginning shooters are trained to hold the rifle stock firmly against their shoulder. The momentum of the rifle drives it backward into the holder and can result in a nasty whack if it is allowed to build up some speed before it hits you.

As physicists love to do, let's start with a simple case before we reveal the complications. Two bodies engage in a pure contact interaction, in short, a *collision*. As long as other forces may be ignored, the total momentum of such a system before a collision is the same as the total momentum of the same bodies after the collision. Illustrating this situation with a diagram, the bodies' velocities before collision are represented by v, while the velocities after the collision become v′:

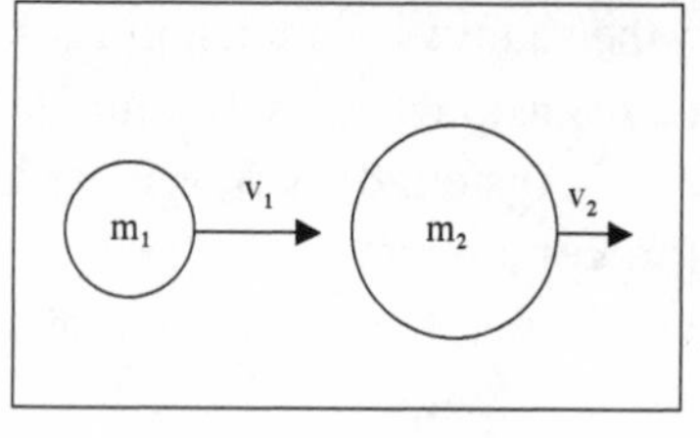

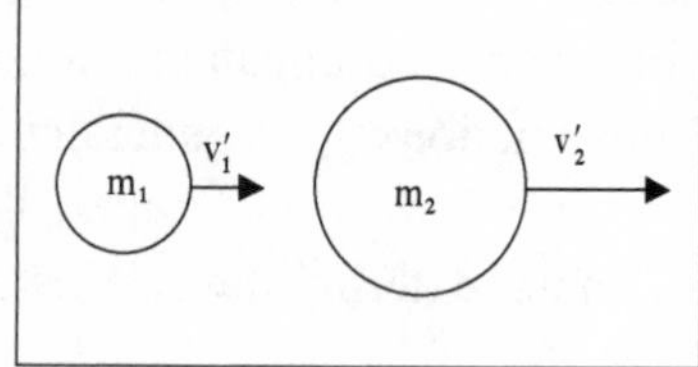

collision

Applying momentum conservation yields the equation $m_1v_1 + m_2v_2 = m_1v_1' + m_2v_2'$. In a typical collision, the masses and velocities before the collision are known, and the velocities after the collision are desired. Since momentum conservation is only one equation, two unknown velocities cannot be obtained. Some additional information is required, which must be provided by details of the collision. Although many different collisions are possible in reality, two particular cases illustrate the limits and provide the additional relationship necessary.

Perfectly Elastic Collisions

This idealized collision is one in which no kinetic energy is lost; the sum of the kinetic energies of the colliding bodies remains constant:

$$\frac{1}{2}m_1v_1^2 + \frac{1}{2}m_2v_2^2 = \frac{1}{2}m_1v_1'^2 + \frac{1}{2}m_2v_2'^2$$

This condition provides a second equation that allows the velocities of the bodies after the collision to be determined. After a lot of algebraic manipulation, this equation becomes $v_1 - v_2 = -(v'_1 - v'_2)$. The implication of this relationship is that the relative velocities of the two bodies has the same magnitude before and after the collision, but the sign change says that body 2 is going faster than body 1 after the collision, as you would expect. As long as there are no potential energy changes during the collision, a perfectly elastic collision conserves both energy and momentum. Examples of (almost) perfectly elastic collisions include a metal ball rebounding from a metal plate or an extremely elastic ball (SuperBall) colliding with another. Experiment 7 gives an example of a collision that is close to being perfectly elastic and that would have been perfect had the ball bounced back to its original height. Other examples include those executive toys with metal balls on strings that swing back and forth and collide repeatedly or pool or billiard ball collisions.

Perfectly Inelastic Collisions

The worst collision (from an energy loss standpoint) occurs when two bodies stick together and move away from the collision with a common velocity. Momentum conservation describes this collision completely, since the common velocity after the collision is the only unknown. The momentum conservation equation becomes: $m_1v_1 + m_2v_2 = (m_2 + m_2)v'$. Examples of perfectly inelastic collisions include a railroad car colliding and coupling with another or an auto being hit with a juicy snowball. Clearly, energy is lost in perfectly inelastic collisions. For example, if one railroad car couples with another car (originally stationary) of equal mass, half the original car's kinetic energy shows up as kinetic energy of the joined bodies.

So far, the discussion has focused on bodies that are moving along a single line, but real collisions can be much more complex. Bodies may have offset velocities, producing a glancing collision, or there may be a collision from the side. These are all analyzed by momentum conservation, but they become more complicated mathematically.

For a simple case of some interest to motorists, consider the case of a rear-end collision between a moving car called an *impactor* and a *target* vehicle that was originally stationary.

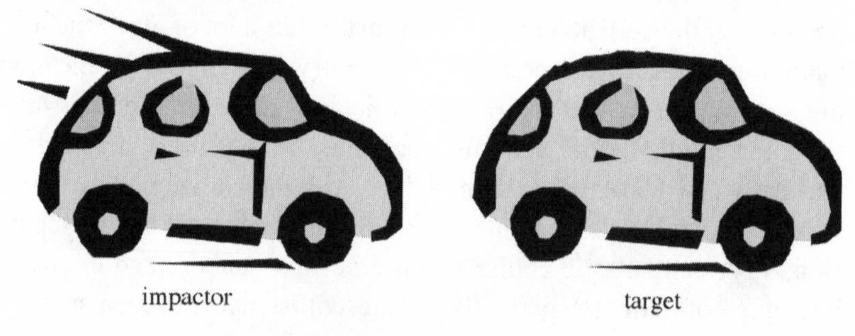

colliding cars

Rear-end collisions between vehicles of equal mass:

- Perfectly elastic collision: $v'_{target} = v_{impactor}$ => Like the executive toy, all the impactor's momentum and energy is transferred to the target vehicle.

- Perfectly inelastic collision: $v'_{target} = 1/2\ v_{impactor}$ => Some of the impactor's kinetic energy is lost in the collision.

Rear-end collision with impactor having twice as much mass as target vehicle:

- Perfectly elastic collision: $v'_{target} = 4/3\ v_{impactor}$ => The target vehicle travels 33 percent faster than the original speed of the impactor.
- Perfectly inelastic collision: $v'_{target} = 2/3\ v_{impactor}$ => The target vehicle still doesn't go as fast as the impactor's original speed.

Rear-end collision with impactor having half as much mass as target vehicle:

- Perfectly elastic collision: $v'_{target} = 2/3\ v_{impactor}$
- Perfectly inelastic collision: $v'_{target} = 1/3\ v_{impactor}$

These examples seem to favor more massive vehicles, but additional characteristics like rollover possibilities and fuel economy must also be considered (see www.nhtsa.dot.gov and www.epa.gov).

Automobile crash analysis sometimes involves forensic experts, crime scene investigators, and evidence technicians. They determine the degree of elasticity of a car crash, estimate the friction coefficient, measure the length of skid marks, and eventually find an approximate value for the speed of a vehicle prior to the crash. There are often serious legal implications of all this physics.

Let us return to Newton's original formulation,

$$\vec{F}_{net} = \frac{\Delta\vec{p}}{\Delta t}$$

This may be rewritten as

$$\vec{F}\Delta t = m\Delta\vec{v}$$

Viewed in this context, $m\Delta v$ is the familiar change in momentum, but $F\Delta t$ is a new quantity, called *impulse*. This is called the *impulse-momentum relation*:

> the impulse acting on a body changes the body's momentum, according to the equation $F\Delta t = m\Delta v$

Since an auto collision has a fixed momentum change (constant mass, change in velocity from original speed before impact to zero after), anything that lengthens the impact time will reduce the force of impact, since $F\Delta t = F\Delta t$. The role of automobile air bags is to increase the impact time during an accident, thus reducing the force of impact. The crush zones built into autos at the front and rear also increase the impact time and absorb energy at the same time. Much of modern vehicle safety engineering depends on the physics of impulse and energy.

PHYSICS IN THE CULTURE—SPORTS

TV sports commentators often discuss momentum and momentum changes in games. I wonder if they really mean the product of a team's mass and speed or if this is just a subtle plug for physics. Actually, impulse and momentum play a large role in any sport in which some object is struck: baseball, golf, hockey, racquetball, and tennis. Coaches and trainers in each of these sports emphasize that the best results are obtained when a player follows through after hitting the ball/puck. Skeptical beginners often wonder how any action on their part can influence the motion of the ball or puck when its flight has already begun and there is no longer any contact with the club/stick/racquet. Sometimes they cite the ridiculous example of a putter trying to wiggle the club to apply "English"—that is supposed to change the direction of an errant putt several meters away. While the example is very graphic, it's not really the major point from a physics perspective. Here's another example from golf that gets closer to the physics of follow-through. Why doesn't a 100 kg (220 lbs) golfer drive the ball twice as far off the tee as a 50 kg (110 lbs) golfer? You might think the doubly massive golfer could apply twice as much force to the ball, but there is clearly more to this story. Let's analyze the club head and ball as two bodies colliding:

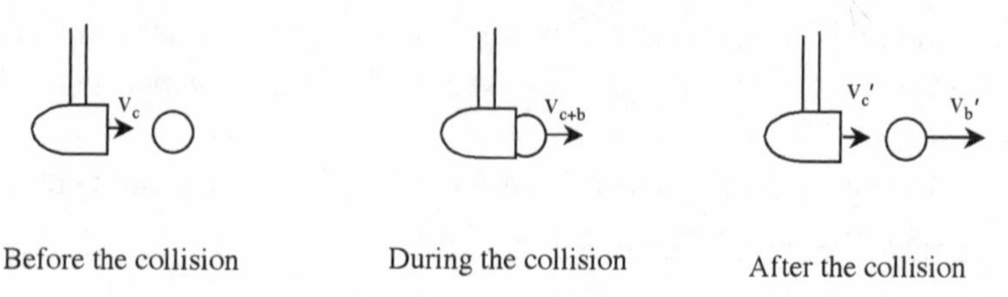

golf ball and club colliding

The collision begins as an inelastic one, since the ball sticks to the club head for a brief period while the ball compresses like a spring. Then the ball jumps forward and loses contact with the club head after the collision, carrying momentum and energy with it. As long as the club and ball are in contact, $m\Delta v = F\Delta t$, so either a *large* force or a long contact time makes the velocity change *large*. In this context, follow-through means keeping the club in contact with the ball as long as possible. The smaller golfer may develop a swing that maximizes contact time. The bigger golfer may exert more force, but if his follow-through allows the club head to slow down prematurely, the contact time will be smaller. Thus, the smaller golfer may apply almost as much impulse, giving the ball nearly the same velocity increase.

Discussing golf might make you pause and think about how else physics might apply to that humbling game. There are three keys to a successful golf swing: rotation, rotation, and rotation. But that's the subject of the next chapter.

Experiment 8:
You Say You Want a Revolution?

Equipment needed: CD player with window, CD, marker

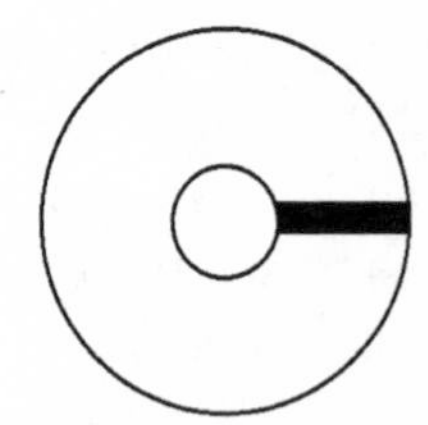

CDs rotate at different speeds, depending on whether the song (information) is located near the center of the disk or near the edge. These speeds differ by a factor of two and should be observable if the CD player has a window for observation.

PROCEDURE

- Start playing a song at the beginning of the disk—this is located near the center.
- Estimate the speed of rotation of the disk—this should be faster rotation.
- Skip to the song at the end of the disk—this is located near the outer edge.
- Estimate the speed of rotation of the disk—this should be slower rotation.

The speeds are probably too fast to time with a stopwatch, but their differences should be apparent.

Chapter 7.
SPINNING WHEELS, GOT TO GO ROUND

"As long as the world is turning and spinning, we're gonna be dizzy and we're gonna make mistakes."
—Mel Brooks

So far, we've idealized bodies as points and studied how they move: in a straight line, along a trajectory curved due to gravity, or around in a circle. But real bodies aren't points. They come in all sizes and shapes and move in seemingly complex paths. As an example, how about a circus knife thrower? Rather than a knife, if the thrower tossed a ball that hit the wall next to his fearless assistant, we'd be able to analyze the ball's motion using the techniques discussed in the last several chapters. But that's not very entertaining. The knife can't act like a point and just thud into the wall facing any old direction. It has to arrive point first and stick there, quivering slightly (or is that the assistant's job?).

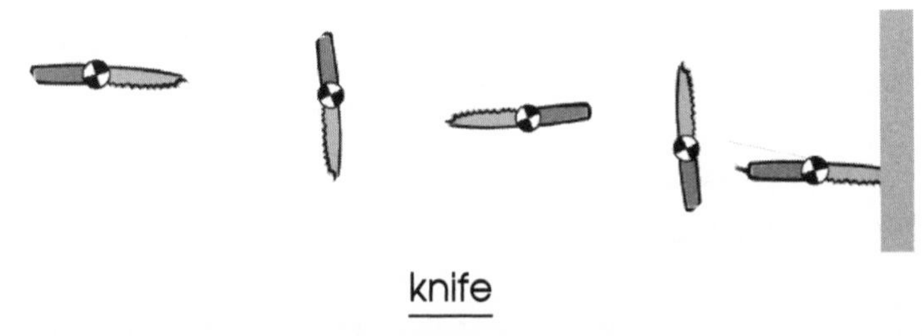

knife

In the illustration, one special spot on the knife has been marked with the symbol:

Although the knife rotates as it travels, that particular spot follows the same path that the knife would have followed if it had been a point mass. This special spot is called the *center of mass*, abbreviated CM. This concept is especially useful, since the general movement of a real body can be treated as two separate motions: the motion of the center of mass, which moves as if it were a point mass, and a rotation of the body about an axis through its center of mass.

Our plan here will be to find out how to locate the center of mass of a body and then to undertake a description of the rotation of a body about a fixed axis, which is called rotational kinematics. (In the next chapter, we'll investigate the cause of rotational motion called rotational dynamics and, finally, we'll allow the center of mass to move and complete the analysis of general motion.)

Locating the Center of Mass

For real bodies that are symmetric, the center of mass is easy to find: it is located at the center of the body. For bodies that are not symmetric, it gets a little harder. There is a practical way to find the center of mass for a real physical body.

Experimentino 2

Take the half-ball popper from Experiments 2 and 6 and hold onto one edge with a pointed object like a pencil. The location of the half-ball popper's center of mass is at the intersection of a vertical line through the support point and a line through the popper's axis of symmetry. It would look like this:

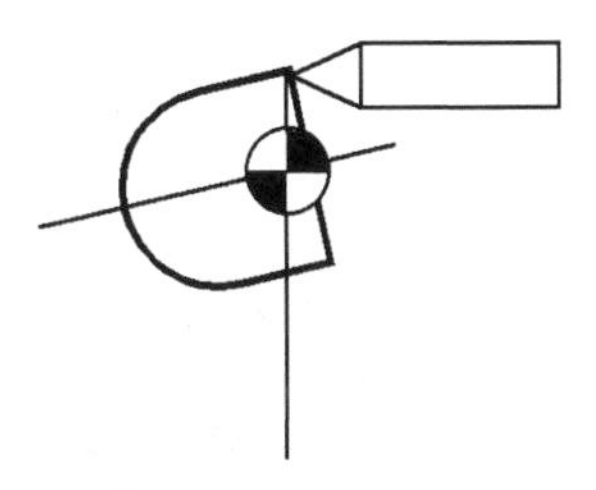

half-ball popper CM

For a more complicated body with less symmetry, the CM is found by first hanging it from one point, drawing a vertical line, then hanging it from another point and constructing another vertical line. The center of mass is located at the spot where the lines intersect.

To find the center of mass analytically, one technique is to divide the body into smaller pieces whose centers of mass are known, then find the CM of the whole body as the weighted average of the pieces. For complicated bodies that can be described mathematically, finding the center of mass involves integral calculus, which is familiar to many college students (thanks to Newton and Leibniz).

Another term that describes a similar quantity is the *center of gravity*, or CG. The only time the CM and CG differ is the case where the body is so large that the acceleration due to gravity is significantly different from the top of the body to the bottom.

One familiar example of the utility of the center of mass is in automotive crash tests. Production model cars are crashed in several different tests and each crash is recorded by about a dozen high-speed movie cameras (1,000 frames/second), some of which are mounted on the car itself. Inside the car are crash test dummies,[1] designed to simulate human beings in size, weight, and physiology. Inside the dummies are sensors, which measure acceleration, force, and deflection. The outside of the dummies has the center of mass symbol for various parts of the body, so the motion of each part can be tracked by a frame-by-frame analysis of the film.

ROTATION OF AN ENTIRE SOLID BODY AROUND AN AXIS THROUGH THE CM

The biggest difference between a solid body rotating and a point mass whirling around in a circle is that the point mass has only one tangential velocity; but for the rotating solid body, the tangential velocity of any point depends on how far it is located from the axis. You know this from merry-go-round rides. The closer you stand to the center, the slower you go. Instead of looking at a point, consider a line that extends radially outward from the axis. As the body rotates, the entire line moves. Rotation in a direction opposite to the rotation of a clock is called counterclockwise (CCW) and is considered positive. The opposite rotation, clockwise (CW), is considered negative.

The position of the line at any time during the rotation could be described by the angle the moving line makes with an arbitrary fixed line.

The angle is defined as the ratio of the length of the arc divided by the radius of the body.

$$\theta = \frac{\text{arc length}}{\text{radius}} \quad \text{units} = \text{m/m} = \text{radians}$$

Note that this angle's units actually cancel, leaving the angle technically with no unit at all. This lack of unit problem was remedied by Lord Kelvin's brother, James Thomson, a mathematics professor. He first used the term *radian* in print in 1873. It is a handy way to measure because the arc length is found simply by multiplying the radian measure of an angle by the radius. However, there are several other measures of an angle. Perhaps the most familiar is the *degree*, °, which is much older than the radian; it probably originated with the Babylonians hundreds of years BCE. Why did they choose the degree, 360 of which make a complete circle? Six equilateral triangles make up one complete hexagon, which has the same angle as a whole circle. If each equilateral triangle has 60° angles, this makes 360°. Besides, 360 is nicely divisible by almost all the single-digit integers (poor 7 is excluded) without a remainder. Converting from degrees to radians and back isn't too difficult, since there are 2π radians or 360° in a complete circle. This means that 1 radian is $360/2\pi \approx 57.3°$. Some calculators have three different modes for measuring angles; these are listed as DRG. D stands for degrees, as noted above, and R means radians, but G often stumps people. It stands for *grads*. This measure divides the right angle (90°) into one hundred parts. Thus, 1 grad is the same as 0.9°. Grads originated at the adoption of the power-of-ten-based metric system, when French metric pioneers got carried away and tried to decimalize everything, including the right angle. They even tried to divide the day into decimal units, but it wasn't accepted.

Since a rotating body's angle changes, time must be taken into account. This is accomplished by defining the rate of change of angle with time as the *angular velocity*, ω.

The angular velocity, ω, is the rotating body's change in angle divided by the time it takes to change the angle.

$$\omega = \frac{\Delta\theta}{\Delta t} \quad \text{units} = \text{rad/s}$$

The angular velocity tells how fast the angle changes as a body rotates. The concept is similar to linear velocity, but the official unit, rad/s, is not familiar to most people, just as linear velocity's unit, m/s, wasn't well known. What we need is a familiar object that spins at a rate of 1 rad/s. There are plenty of spinning bodies in ordinary life, but the unit for measuring their rotational speed isn't rad/s, it is rpm. Revolutions per minute measures the angular velocity of your car's engine and is shown on the tachometer on the dashboard of many cars. It turns out that 1 rad/s is the same angular speed as about 9.5 rpm. Since most cars idle at 500 to 1,000 rpm, if you open the hood of your car and watch the engine turn over, you'd be observing an angular velocity of 50 to 100 rad/s—too fast. Planet Earth spins on its axis, but that speed is less than 0.0001 rad/s—too slow. We need something between these speeds to demonstrate 1 rad/s. How about vintage turntables that play vinyl records, as shown by the graphic at the beginning of the chapter? The speeds at which these turntables rotated were 78 rpm, 45 rpm, 33 1/3 rpm, and (rarely) 16 2/3 rpm. Before this technology was supplanted by the CD, perhaps the most popular speed was that of long-playing (LPs—up to thirty minutes) records, 33 1/3 rpm—just over 3 rad/s. That's closer, but not close enough. Perhaps the best way to observe a rotational speed of 1 rad/s is with a rotating food serving tray called a lazy Susan. (You may wonder who Susan was. One possibility: the rotating tray might have been invented by either Thomas Jefferson or Thomas Edison; each had a daughter named Susan. Another possibility: the name has a "catchy" sound and appealed to marketers.) At a rotational speed of 1 rad/s, the tray would rotate slowly enough that even Susan would be able to select food samples as it slowly spun.

A very simple relationship exists between the tangential velocity of any point on a rotating body and the angular velocity of the body: $v = \omega r$. In terms of a lazy Susan, food near the outer edge would have a larger tangential velocity than food near the center. You may also have noticed this relationship while riding merry-go-rounds.

This tangential velocity relationship has a direct connection to the spinning of a CD, as you saw in Experiment 8. The pits and lands on the underside of the CD are read at a constant rate, which means that the CD must rotate faster for a track near the center of the CD and slower near the outer edge in order to keep the tangential velocity constant. This brings up the next idea. Just as linear velocity wasn't necessarily constant, neither is angular velocity.

The time rate of change of angular velocity is called the *angular acceleration* and is given the symbol α.

> The angular acceleration of a rotating body is the change in angular velocity divided by the amount of time required to make the change.
>
> $$\alpha = \frac{\Delta\omega}{\Delta t} \quad \text{unit} = \text{rad/s}^2$$

Let us return to the technology museum for the turntable that rotated at 3 rad/s. If it took three seconds to reach its operating speed, its angular acceleration would be 1 m/s^2. Not very quick. On the other hand, a computer hard drive that may operate at 5,400 rpm (570 rad/s) and gets up to speed in less than two seconds has an angular acceleration of almost 300 rad/s^2.

Analyzing rotational motion has introduced several new symbols: θ, ω, and α, thus illustrating one of the difficulties people have in learning physics—the plethora of symbols. Chemistry, for some, is bad enough, with ninety-two naturally occurring elements, each with a unique symbol. But at least chemistry's symbols all use the Roman alphabet. The newest quantities we've seen here are all Greek (hence, some people's complaint that physics is Greek to them). The root of the difficulty is based on the fact that physics uses single-letter symbols as abbreviations for quantities, and the twenty-six letters in the Roman alphabet don't go nearly far enough to cover all physics' quantities. As physics grew, new symbols were needed for new quantities, which led to a giant option: (A) stay with the Roman alphabet exclusively and use two letters to represent quantities, the way chemistry does; or (B) keep one symbol per quantity but add Greek alphabetic symbols. The way it turned out, physics uses only a few two-letter symbols, and Greek letters are much more common. Also, physics began to use subscripts: x_1, v_o, F_{net}, and many more. If it had been up to me, I would have opted for two letters, but no one asked me.

You may recall that linear motion had definitions for x, v, and a that are closely paralleled by the ones for θ, ω, and α. Using similar logic yields relationships for the angle and angular velocity as functions of time:

An angularly accelerating body's angular velocity is changed by an amount equal to the acceleration times the elapsed time.

$$\omega = \omega_0 + \alpha t, \text{ where } \omega_0 = \text{angular velocity at time} = 0$$

The angle rotated is also affected by acceleration:

The angular position of an angularly accelerated body is increased by its initial angular velocity times the elapsed time plus half the angular acceleration times the elapsed time squared.

$$\theta = \theta_0 + \omega_0 t + \frac{1}{2}\alpha t^2, \text{ where } \theta_0 = \text{angle at } t = 0 \text{ and } \omega_0 = \text{angular velocity at } t = 0$$

If these relationships are combined algebraically and the time eliminated, another equation is obtained that doesn't contain time explicitly:

$$\omega^2 = \omega_0^2 + 2\alpha(\theta - \theta_0), \text{ where } \theta_0 = \text{angle at } t = 0 \text{ and } \omega_0 = \text{angular velocity at } t = 0$$

This set of equations allows angular motion to be described completely.

Experimentino 3

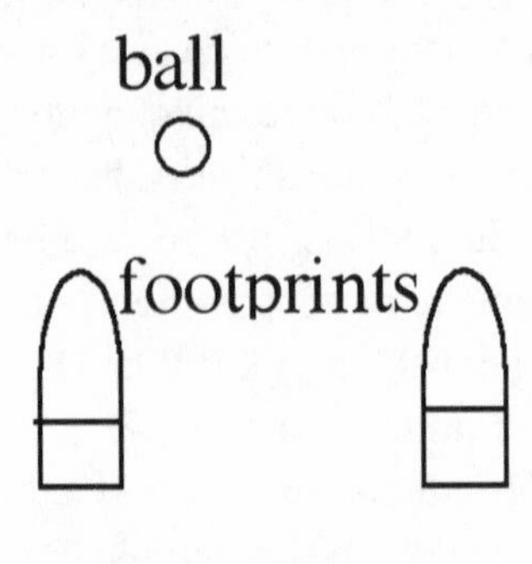

It's time to test the three rotations that make up a golf swing. If you have access to a golf club (or a substitute, like a broom), grip it in the normal way and address an imaginary ball as the diagram shows.

Hold the club as if you were going to strike the (imaginary) ball, then:

- Rotate your shoulders and hips clockwise around a vertical axis through the center of your body. Move the club head in a horizontal plane. If you then rotate counterclockwise, the club head will move into a position to strike the ball.
- Rotate your arms clockwise around a horizontal axis through the center of your chest, pointing out of the front of your body. If you are right-handed, your left arm can stay straight, but the right arm must bend at the elbow. This motion raises the club head. If you then rotate counterclockwise, the club head moves downward, in a position to strike the ball.
- Rotate your hands clockwise, around an axis slightly above horizontal, through the midpoint of your wrists, pointing outward from the front of your body. Your left wrist can pivot easily, but the right one is somewhat hampered by its position. This motion raises the club head vertically, and when you rotate counterclockwise, it returns the club head so it can strike the ball.

The big trick to a golf swing is to coordinate these three rotations so they occur in the right order; this gives the club head maximum velocity at ball impact and sets up a follow-through that maximizes the collision time. Quite a tall order, which explains why so few of us can excel at such a demanding game.

"When I want a long ball, I spin my hips faster," said golfing great Jack Nicklaus.[2]

So what is the cause of rotation? The short answer is torque. For the longer answer, see chapter 8.

Experiment 9: Mo' Yo-Yo

Equipment needed: yo-yo, metric tape measure, stopwatch

This experiment again involves the use of that ancient toy, the yo-yo. Although probably invented by the Chinese, the first pictorial record of the yo-yo turned up on Greek pottery around 500 BCE. Sporadically popular all around the world, the yo-yo figured in the French play *The Marriage of Figaro* in 1792. Figaro shows his nervousness in a scene in which he plays with a yo-yo. When another character asks what it's good for, Figaro replies, "It is a noble toy, which dispels the fatigue of thinking." In this experiment, rather than dispel thinking, the yo-yo will help our physics thought by demonstrating a body that undergoes translational and rotational motion at the same time.

PROCEDURE

- Completely unroll the yo-yo and measure the length of the string.
- Record the string length (m) in the data table below.
- Fully wind the string and hold the yo-yo at least 1 m (3 ft) above the ground.
- Release the yo-yo and start the stopwatch simultaneously.
- At the instant the yo-yo reaches the bottom of its travel, stop the stopwatch.
- Record the elapsed time (s) in the data table below.
- Repeat for a total of ten trials.
- Find the mean time of travel (s).

Applying energy conservation principles to the yo-yo's motion yields the equation:

$$g = \frac{h}{t^2}(GF)$$

where g is the acceleration due to gravity, 9.80 m/s^2, h is the string length, t is the elapsed time, and GF is a geometric factor that is a function of the yo-yo's overall diameter, the diameter of the inner axle that holds the string, and the distance between the two halves of the yo-yo. This factor can range from 10 to 20. For a small yo-yo, use a value of 13 for the geometric factor. Calculate the value for g based on your experiment results and compare it to the accepted value of 9.80 m/s^2.

Trial	**Time to unroll completely (s)**
Sample	1.13
1	
2	
3	
4	
5	
6	
7	
8	
9	
10	
Mean	

Record string length = h ____________ (m)
Calculate 13h/t^2 and compare your result to g = 9.8 m/s^2
Sample: h = 0.95 m, g = 9.7 m/s^2, within 1 percent.

Chapter 8. FORCE WITH A TWIST—TORQUE

"If you can argue about torque and horsepower for more than an hour, you might be a gearhead."
—Jeff Foxworthy

Now that a body's rotation has been described (rotational kinematics), we'll next explore the cause of angular acceleration (rotational dynamics), then we'll put it all together: a body's center of mass moves like it's a point, and the body may rotate around its center of mass.

The Cause of Rotation

For translational motion, Newton found the key relationship: a body accelerates because of force, according to the relation F = ma. Now that we've seen that rotational motion has an angular acceleration, α, it would seem likely that there would be a similar relationship for rotation: (forcelike quantity) = (masslike quantity) α. First, let's find out about the *masslike quantity*.

Experimentino 4

The goal of this experiment is to take two bodies of the same mass and see which one is easier to rotate. You'll need 2 CDs (or DVDs) that will be unusable after we finish (sacrificed for physics), plus sixteen coins, which will not lose their value, and a method of sticking the coins to the CDs (any kind of tape will do). Fasten the coins to both sides of the disks, producing a pattern like this:

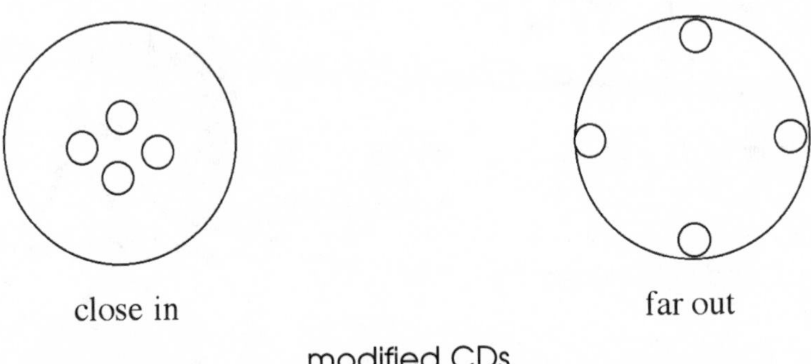

modified CDs

Once the disks are prepared, you'll need a long, gentle slope for them to roll down. A board or shelf would do nicely, preferably about 1 m (3 ft) long. Since both disks have the same mass, if you simply laid them on their sides, they would accelerate equally down the slope. But if you turn them on edge and let them roll down the slope, would their angular acceleration be the same? The best way to find out is to set them next to each other on the slope and let them roll. (If you're a gambling person, you might want to bet on the outcome.) The winner of the race will be the one that is easier to roll. It may take a few tries to get them to roll straight, but persevere.

The masslike quantity discussed above clearly involves more than just the amount of mass. The other factor is the location of the mass. This masslike quantity is called the *moment of inertia*. It is actually the sum of each individual mass that makes up the body times its distance (squared) from the axis of rotation. The disk with the coins pasted close to the center is clearly easier to roll, since it has a smaller moment of inertia. On the other hand, the disk with the coins pasted far from the center has a larger moment of inertia, and its corresponding angular acceleration is less. Calculating moments of inertia can be an interesting (and lengthy) mathematical exercise that may involve integral calculus. Just to give a few examples, the following table gives the moments of inertia of several bodies, if we presume they have uniform mass distributions and their rotation is about their axis of symmetry.

Body	Moment of inertia
Hoop	mr^2
Solid disk	$1/2\ mr^2$
Sphere	$2/5\ mr^2$

Next, we need to look at the *forcelike quantity*.

A simple way to approach this is to check the units of the right-hand side of the equation. $I\alpha$ has units of $kgm^2\ rad/s^2$. Since the radian is a nonunit, the unit of the entire quantity is the Nm. That differs from the unit of force—the newton—because of the additional length unit. This is called the *torque*. It is the force applied multiplied by the distance between the force and the rotation axis, measured perpendicular to the force's direction. Your mental picture should be of a wrench.

A wrench is used to cause rotational motion of a bolt or nut. A force is applied at some distance from the axis of rotation, and the bolt turns. This product of force and distance from the axis (often called the moment arm or lever arm) is called *torque* and given the symbol τ. Special wrenches, called torque wrenches, have a dial that reads the torque directly, allowing bolts to be tightened to specified levels. Summarizing,

> The torque acting on a body is directly proportional to the body's moment of inertia and angular acceleration.
>
> $\tau = I\alpha$ unit = Nm (not joules)

The unit of torque is the Nm. Technically, this is the same as joules, but Nm are used to avoid confusion with energy. In the English system, the unit for torque is the foot pound, and the conversion is 1 Nm = 0.86 foot pound. The newton meter is not a very large unit and is often exceeded by mild wrenching practiced by weekend handymen (and handywomen). Plumbers are often called upon to exert far larger torques to free corroded fasteners. They achieve large torques by the simple expedient of attaching a pipe to the wrench, increasing the lever arm and, therefore, the torque, substantially. Other tools that take advantage of torque include Allen wrenches, often used in bicycle repairs.

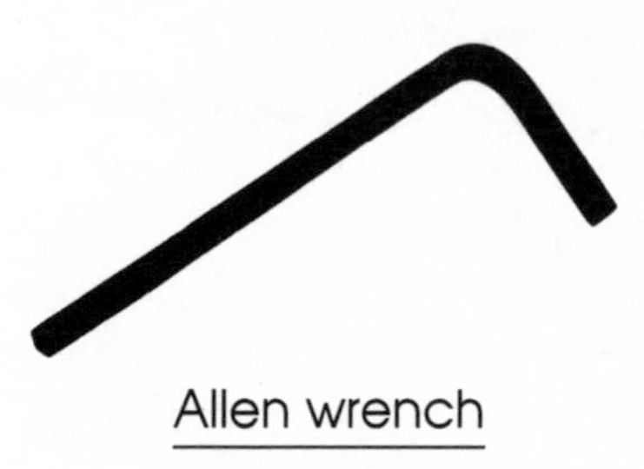

Allen wrench

The Allen wrench has a hexagonal shape and can be used in either orientation to give more or less torque, yet still fit within whatever clearance is available. Other tools that utilize torque in their function include screwdrivers and pliers. Special ones built with fatter and longer handles require less force to generate the same torque. Some kitchen tools are also billed as easy to grip. They feature oversized handles and high-friction surfaces to get the required torque with less applied force.

A bicycle chain applies force to a gear attached to a bicycle wheel, causing it to turn. You may notice the different gear diameters. The larger the gear size, the more torque is generated even though the same force is applied by the chain through the action of the pedal (which also exerts the torque). Generally, the higher the gear number, the smaller the gear diameter and hence the smaller the torque applied. Conversely, when you shift to a lower gear number, the gear has a larger diameter and you exert more torque on the wheel, allowing you to climb hills more easily.

If you follow auto performance statistics, you know that torque is part of an ongoing argument among car buffs. What's better, horsepower or torque? It's a variation on the old beverage commercial where people can't decide if more taste beats less filling. Torque spins the wheels, and if your concern is to pull heavy loads or climb steep hills, you need torque at low engine speeds (rpm). If you wish to make the fastest time between red lights (without violating the speed limit, of course) or you drive at high rpm, horsepower is probably more important.

Even the simple act of opening a door involves torque. Viewed from the top, an opening door is a rotation, and requires a force to accomplish it. Normally, you exert a force on the part of the door farthest from the hinge. If you pushed anywhere else, it would require more force to achieve the same torque.

I have noticed an unusual rotating system a few times over the years: the pen rotator. Someone sits with pen in hand, seemingly ready to write, when all of a sudden the pen makes a complete rotation, then returns to its original position, ready to write. To accomplish this trick, the person doing the rotating must have applied a torque with his thumb, but it must have been exactly the right amount and in the right place to overcome the frictional torque provided by his fingers.

An example of a very large torque in action is provided by the Moon as it orbits the Earth. (See the endnotes for the Giant Impact hypothesis of the Moon's formation.)[1] As you know, one side of the Moon always faces the Earth. The Moon rotates synchronously with its orbital path. The reason for this is a torque that is related to gravity and tides. The Moon is flexed by the Earth because the side closer to Earth feels more gravitational force than the far side. Early in the Moon's existence, it was more fluid, and this flexing produced a heavy area that distorted the Moon from a perfect sphere. To put it another way, the Moon is sufficiently large that its center of mass is different than its center of gravity.

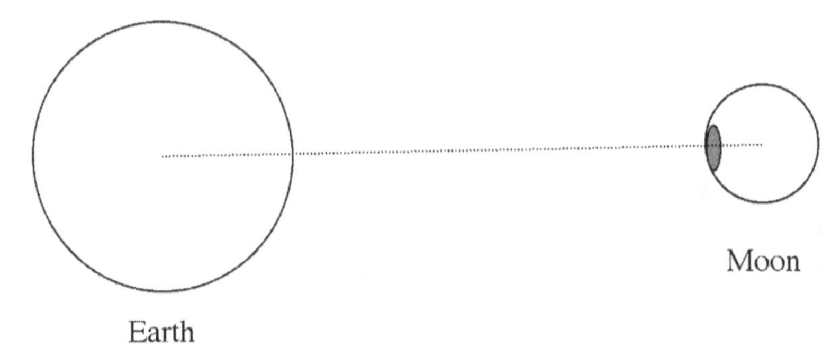

Earth-Moon system

If the heavy spot were not facing the Earth, it would create a torque that would rotate the Moon until the heavy spot was along the line that joins the centers of the Earth and the Moon. This situation is common enough that most major moons of the solar system have synchronous rotation.

An example of a very small torque is provided by a contact lens that corrects for a condition known as astigmatism. Contact lenses that correct vision abnormalities were designed to float on the surface of the outermost part of the eye, the cornea. Early contact lenses could rotate on the cornea without making any difference in the corrective power of the lens. But people with astigmatism have different radii of curvature of their corneas in the horizontal and vertical planes. So, astigmatics needed different corrections depending on which plane of vision was being corrected. Thus, lenses that rotated didn't correct their vision properly. To fix this problem, ballast was added to the bottom of the lens. If the lens rotated from the preferred position, a torque would be generated that would rotate the lens back into the proper orientation.

Just as Newton's work on translational motion was extended to momentum and energy, rotational motion is extended to angular momentum and angular kinetic energy.

Angular Momentum

Linear momentum is defined as mass times velocity, so you might suspect angular momentum would be similar in its definition, and it is: moment of inertia times angular velocity = mvr for a particle in circular motion. While linear momentum is a vector, angular momentum is just a bit simpler—it has two possible directions, clockwise or counterclockwise. The units of angular momentum are kgm^2/s, which are given no special name. The power of this concept lies in the principle of conservation of angular momentum:

> For a system of bodies that interact only with each other, the total angular momentum of the system is constant.
>
> $$L = I\omega = \text{constant}$$

Although there are many examples from sports, perhaps the most famous one features the ice-skater. If an ice-skater performs a scratch spin, she starts out with her arms outstretched and rotates by making a torque with her skates scratching the ice. When she is spinning as fast as she can, she stops scratching and pulls in her arms. The result? She spins faster. From an angular momentum standpoint,

$$I_1\omega_1 = I_2\omega_2$$

A similar situation holds for sports like gymnastics, diving, or boarding. If you want to turn somersaults, make your body small to minimize its moment of inertia and thus maximize the rotational velocity (this is often called the tuck position). Divers slow down their rotation in order to enter the water as vertically as possible by straightening out their bodies. This maximizes the moment of inertia (the pike position). In practice, this is done so quickly that there is no time for conscious thought, only reaction.

An example of a perfectly inelastic collision of rotating bodies is provided by a common playground apparatus, the merry-go-round. To get it going, you apply a torque by pushing the edge until you've got it going fast enough. Then you stop pushing. The merry-go-round coasts a long time before stopping due to the frictional torque in the bearings. Before friction works very long, suppose you hop on. Your mass adds to the moment of inertia of the merry-go-round, so the increase of moment of inertia must be balanced by a decrease in angular velocity, according to $I_1\omega_1 = I_2\omega_2$, so the immediate effect of your jumping on would be to slow the merry-go-round's rotational velocity. What would happen if you moved from the edge of the merry-go-round toward the axis of rotation? The merry-go-round would speed up, just like the ice-skater.

A spinning top or gyroscope undergoes an interesting motion, based on angular momentum. If the top wasn't spinning, any tilt of its axis would create a gravitational torque that would cause it to tip over. But if it is spinning, it has a large angular momentum around its axis. Any tilt of the top produces a gravitational torque, and that causes the angular momentum vector (and hence the spin axis) to rotate around its original orientation. This is called *precession*, which becomes more pronounced as the angular velocity decreases due to friction at the support point.

A large example of angular momentum conservation is provided by the solar system. Prior to the formation of the Sun and the planets, the material that eventually formed the solar system existed in the form of a large cloud of gas and dust called the protosolar nebula. This cloud was probably rotating slowly, since its random velocities didn't exactly average out to zero. Gravity worked on these particles, and the cloud shrank, reducing its moment of inertia. Comparing it to the case of the ice-skater, we know what happened next. The spin rate went up, and the cloud collapsed into a disk. This situation continues today in that the planets lie mostly in a disk (called the plane of the ecliptic) and revolve around the Sun counterclockwise as viewed from above.

Rotational Work and Rotational Kinetic Energy

Just as translational work was the product of force applied and distance moved in the direction of the force, the rotational world has an analogous situation.

> Rotational work is directly proportional to the torque applied to a body and the angle moved by the body in response to the torque.
>
> $$W_{rotational} = \tau\theta \qquad \text{units} = \text{joules} = \text{J}$$

Further, a rotating body has energy of motion that corresponds to translational kinetic energy, only this is now called *rotational kinetic energy*, and depends on the body's moment of inertia and rotational velocity.

> Rotational kinetic energy is directly proportional to a body's moment of inertia and angular velocity squared.
>
> $$KE_{rotational} = \frac{1}{2} I\omega^2 \qquad \text{units} = \text{joules} = \text{J}$$

The principle of conservation of total energy must be modified by the addition of this term:

> If no dissipative forces are present, the total energy of a closed system of bodies is constant at any point in the motion of the bodies.
>
> For a closed system of bodies with no dissipative forces,
> $$(KE_{translational} + KE_{rotational} + PE)_1 = (KE_{translational} + KE_{rotational} + PE)_2$$

We are now in a position to analyze the complete motion of a body—translational and rotational. As an example, let's consider the yo-yo from Experiment 9:

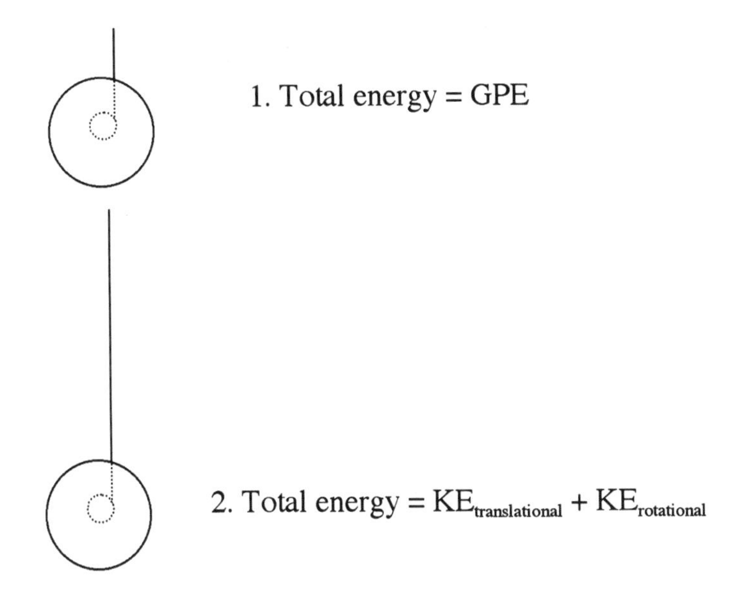

Ignoring any frictional effects, the total energy must be constant, so mgh = $1/2\ mv^2 + 1/2\ I\omega^2$.

Bowling provides an example from sports. A child just beginning to bowl might start the ball rotating as it proceeds down the alley by applying a torque to the ball by pushing on its top with a force directed toward the pins. As the ball rolls along, the center of mass would proceed down the alley and the ball would rotate around its center of mass. So the rotational work done by the child would be turned into both rotational and translational kinetic energy. As long as the friction coefficient was sufficient, the contact point between the ball and the alley would have no relative motion, so no frictional work would subtract from the energy. This condition is referred to as rolling without slipping. The translational speed of the ball just matches the rotational velocity times the radius of the ball.

A stronger and more experienced bowler might throw the ball with side spin or no spin at all. Since slipping definitely occurs, the ball skids for a while as it proceeds down the alley. But the skid doesn't last. After a while, the ball begins rolling without slipping. The reason for this change is the frictional torque that develops because of the slipping. This torque rotates the ball until the tangential speed of the ball's edge matches the speed of the center of mass.

Examples of rotational motion abound in sports. There's the person who tosses his tennis racket into the air with a little spin, so that it makes one complete revolution and lands so the grip is in his hand (that would be me). Or the street magicians who keep sticks or a kind of yo-yo moving and rotating with a pair of control sticks called juggling sticks, devil sticks, or diabolos. Or the spectacular baton twirling performances at sporting events that involve many different moves and performers as well as high tosses and multiple batons. If all this movement and twirling is making you dizzy, take heart. The next chapter is about bodies that don't move at all.

Experiment 10: Teeter-Totter

Equipment needed: teeter-totter, metric tape measure, human volunteers

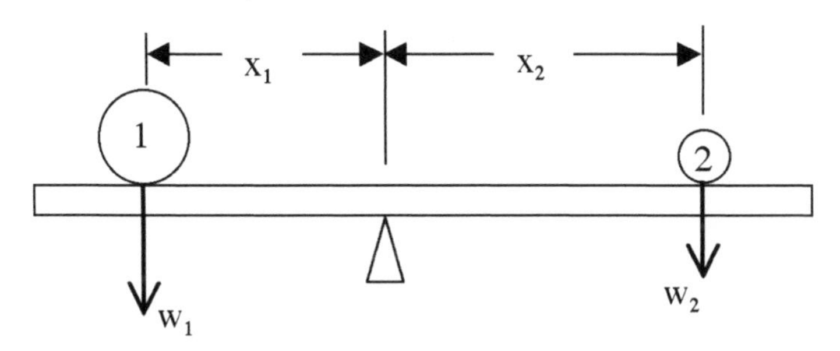

A teeter-totter with a horizontal board is an example of a body in equilibrium. The purpose of this experiment is to balance two people on a teeter-totter and make measurements to support the physics principle that they exert equal torques around the pivot point.

Procedure

- Obtain the weights of bodies 1 and 2, preferably from the same scale.
- Balance bodies 1 and 2 on the teeter-totter so that the board is horizontal and no one's feet are touching the ground.
- Measure the distance from the center of the contact area of each person to the pivot point.
- Calculate the torques $(w_1)(x_1)$ and $(w_2)(x_2)$.
- Compare the two torques to see if they are equal.

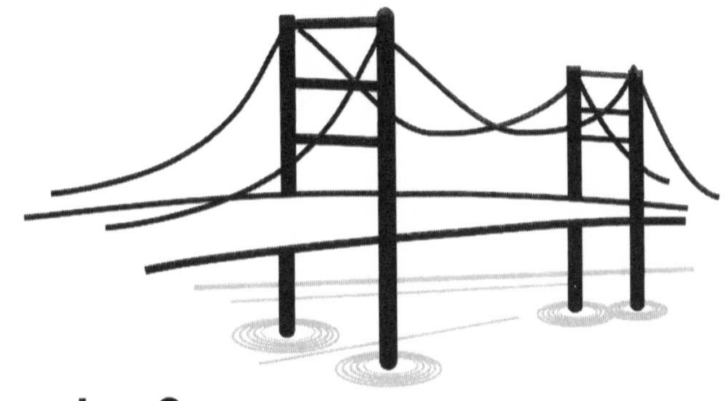

Chapter 9. THE STRANGE CASE OF THE BODY THAT DOESN'T MOVE—STATIC EQUILIBRIUM

"Professor wants to make sure we remember to use free body diagrams in our statics problems, so he talks into one of those electronic repeating parrots and lets it sit there chanting 'Free Body Diagrams!' . . . over and over again. Got some good laughs out of the class. Guess you had to be there."

—student of Professor Epstein

After all that motion in the last few chapters, bodies that don't move might seem too tame. On the other hand, maybe the hustle and bustle are too much, and a little quiet time might be in order. Take bridges and buildings and comfy chairs, for example. What is the physics that describes a body that doesn't move? Since forces cause motion, it would seem, at first glance, that all we need to do is to apply no net force to a body at rest and it won't move. Technically, this is called *static equilibrium.*

First condition for static equilibrium:
The net force acting on a body must be zero.

This is a good start, but it's not the whole story. An experimentino will show why.

Experimentino 5

Put a book (preferably not this one, since you're reading it) on a table in front of you. Exert two equal forces on the book at the same time: (1) Push the right side of the book away from you. (2) Pull the left side of the book toward you. The net force is zero, but does the book move? The whole book doesn't move translationally, but it does rotate. This tells us that there must be an additional requirement for static equilibrium.

Second condition for static equilibrium:
The net torque acting on a body must be zero.

A *dimensioned free body diagram* helps to apply these conditions. Experiment 10 shows a partial drawing of this type. Clearly, the counterclockwise torque caused by mass 1 must be balanced by a clockwise torque on the part of mass 2. In a more extreme version of this situation, replace the teeter-totter with a *lever*. Mass 1 could be a huge boulder, and at mass 2's location there could be someone exerting a force to move the boulder. As long as the moment arm x_2 is long enough, the boulder's weight could be lifted because the torques would be equal. Classically, a lever pivots about a pivot point, called a *fulcrum*, and the ratio of the lever arms is called the *mechanical advantage*.

A similar example is provided by painters on a scaffold that is supported at the ends:

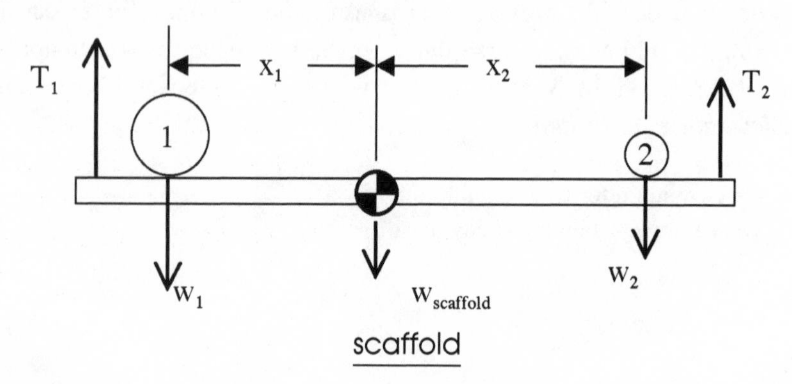

scaffold

Setting the net force to zero implies that the supports exert as much force upward as the weights of the painters and the scaffold, but only the sum of the support forces may be found from the first condition. In working with torques, the choice of a pivot point is completely arbitrary, since it rotates about no point at all. So if torques are taken around one of the support points, then one force has no moment arm, and the other force can be found immediately. From the setup, it is clear that if the painters congregate at one end, the support force must be greater there.

Another application is the diving board.

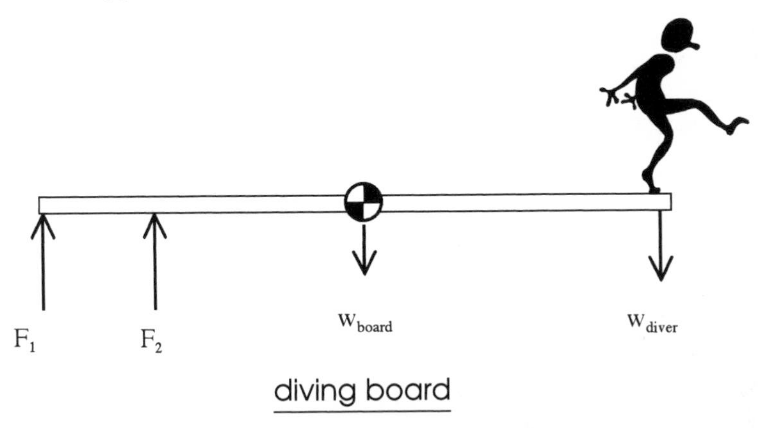

diving board

Taking torques about the leftmost support reveals that F_2 must be quite large, as you would expect, but torques around the other support reveal something else: F_1 is negative. The beauty of free body diagrams is that if your initial equations contain an incorrect direction for a force, it will show up in the final answer as negative. Thus, force F_1 points downward. If you have any familiarity with diving boards, you'll know this is the case, but it's nice to see physics principles match what you already know.

Another classic example is provided by the ladder. If you stand a ladder against a house and begin to climb, thoughts about equilibrium are undoubtedly dancing through your head. The part of the ladder leaning against the house has little friction, so all your hope rests on the friction between the ladder's base and the ground. That's why many ladders have special high-friction feet. The usual failure mode of ladders occurs when friction with the ground is too small and the top of the ladder slides down the house, depositing you on the ground.

STABILITY AND BALANCE

Not all forms of equilibrium are equal. If you balance a stationary basketball on your finger, the tiniest disturbance will create a torque that will destroy the equilibrium, and the basketball will rotate and fall. This is an example of *unstable equilibrium*. Spinning the basketball gives it angular momentum and leads to the gyroscopic effect as discussed in chapter 8. On the other hand, a marble in an empty teacup would return to the center if it was displaced. That is *stable equilibrium*. A cone sitting on its side on a flat surface would still be sitting on its side if you disturbed it, so that would be an example of *neutral equilibrium*. With large bodies that have a substantial base of support, the situation becomes a little more complex. If the body's center of mass is *below* the support base, then the equilibrium is stable. An example of this is a tightrope walker carrying a long pole that droops at the ends. This makes the center of mass of the walker-pole combination below the wire. As long as the walker's feet stay on the wire, the system is stable—any rotational disturbance creates a torque that counteracts the disturbance. If a body's center of mass is *above* the support base, and the body is disturbed to the point that the center of mass goes beyond the vertical line—through the edge of the support point—the body becomes unstable. This is referred to as the *tipping point*—another physics term that has crept into the modern idiom.

Experimentino 6

To illustrate the tipping point, we'll construct a body and support it just below its center of mass, then push it just beyond. You'll need three CDs and some tape. Stack the CDs so that the second one's edge just touches the first one's center, and the third one's edge just touches the second one's center. Tape them lightly at the center/edge contact points to hold them together. Place this composite body on a support such that the edge just matches the edge of

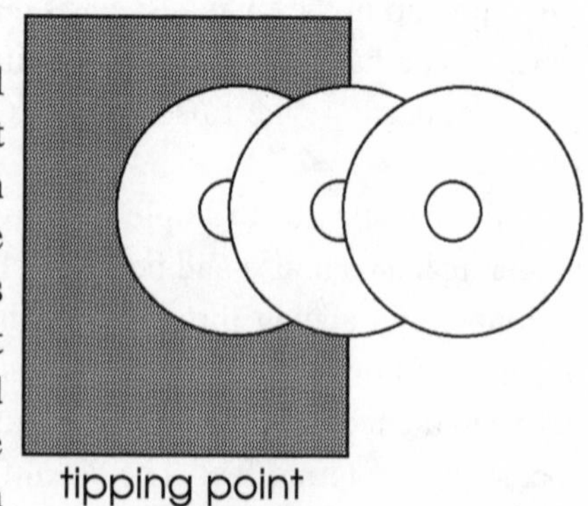

the bottom CD. Note that the top CD extends completely past the support. The center of mass of the three-CD set is in the middle of the composite body, just above the edge of the support. Now push the body slightly and notice the tipping action.

Physics in the Culture—Art and Architecture

To make something that lasts, artistic creations must be designed to retain their structural integrity under anticipated loads and maintain their balance. From fragile-looking figures by Alberto Giacometti and mobiles by Alexander Calder to more substantial works by Henry Moore, Auguste Rodin, and the oversize work of Mark DiSuvero, balance is a key ingredient to a successful sculpture. At a more mundane and personal level, an oversized bird sculpture in my backyard tipped over many times because of torque caused by the wind, but garden stakes now hold its feet down very nicely. Real birds' claws undoubtedly help their equilibrium.

At a more practical level, doorways were needed to move people and goods through holes in walls, but the structural integrity of the wall had to be maintained. Making a doorway by stacking a horizontal beam on top of two vertical ones worked but had built-in limitations. A doorway could be only as large as the single piece of wood, stone, or brick that formed the horizontal piece at the top. This problem was confronted long before free body diagrams, so various configurations were tried, and the successful ones were kept (Darwinian doorway design?). As a result, doorways and arches were constructed in many different shapes in various parts of the world.

Roof supports evolved similarly. First, columns were used, but the interior space was limited by the resulting forest of columns. Eventually, someone realized that a building could be covered by a dome, which is like a three-dimensional version of an arch. Although many different kinds were built, let's analyze the difference between two domes to illustrate the principles at work: the round (Roman) and the pointed (Gothic). The following free body diagram shows half the dome for the sake of simplicity.

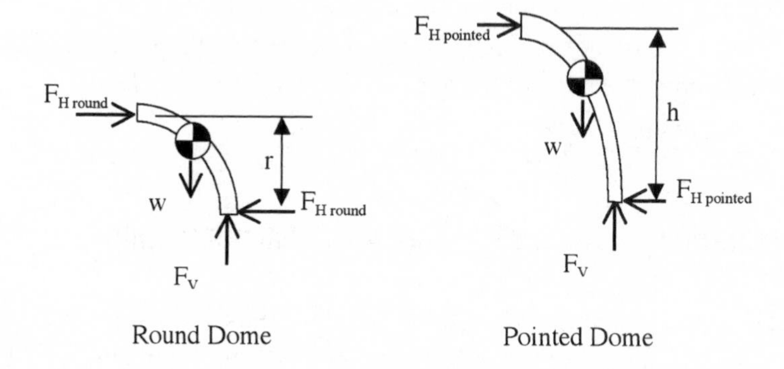

domes free body diagrams

If both domes have the same weight, the vertical support forces must balance the weight ($F_V = w$). The horizontal forces at the base of each dome must be supplied by supports referred to as *buttresses*. To find the amount of horizontal force needed, torques taken around the top point reveal that the pointed dome requires a horizontal support force that is (r/h) times the round dome support force. This means that the buttress requirements for tall, pointed domes are much less than for shorter, round domes. Some buildings with high domes, like the Notre Dame Cathedral in Paris, require smaller horizontal supporting structures called *flying buttresses*. More recent domes have different shapes, including paraboloids and catenaries. Perhaps the most successful has been the *geodesic dome* designed by R. Buckminster Fuller. This dome encloses a maximum volume for its weight.

Architecture must balance practical physics with the aesthetic concerns. In many cases, scratch an architect, and you find an artist.[1] Architects with clear artistic vision and associated Web sites are listed in the endnotes. Frank Lloyd Wright (1868–1959) expressed an interesting view of his profession: "A doctor can bury his mistakes but an architect can only advise his clients to plant vines."

Experiment 11: Stretchy Spring

Equipment needed: springs, force scale, metric tape measure

Bodies that have no net force or torque still have one trick left: they stretch—a little. An ideal example is a spring, whose stretch is proportional to the applied force. The purpose of this experiment is to test some springs to find their spring constant, which tells the amount of stretch they undergo according to the relation: $F = kx$. (These constants will also be useful in Experiment 13.)

Procedure

- For spring 1, apply a force (F) of 1 newton, as measured by the force scale, and measure the resulting deflection (x) of the spring in meters.
- Record the force and deflection in the table below.
- Calculate the spring constant, $k = F/x$, (N/m)
- Repeat for a force of 2 newtons.
- Repeat for a force of 4 newtons.
- Calculate the mean value of k for spring 1.
- Repeat for spring 2.

Force (N)	**Deflection (m)**	**Spring constant = k (N/m)**
Sample	0.005	200
Spring 1		
Mean value for spring 1	———	
Spring 2		
Mean value for spring 2	———	

Chapter 10. GETTING BENT OUT OF SHAPE—ELASTICITY

"At some point, when stretched so far, you break. We are not made of elastic."

—Gerry Adams

"Unnatural work produces too much stress."

—Bhagavad Gita

Just when we think we have bodies whipped into shape so they won't move, it turns out they have some tricks left: they expand a little, or they shrink a little, or they bend a little. To explore the physics of this elastic behavior, we'll start with a simpler case: springs.

Ideally, the harder you pull a spring, the more it stretches.

For an ideal spring, the force applied is directly proportional to the amount the spring stretches and a constant that depends on the spring's stiffness.

$F = kx$, where

F = applied force, N

k = spring constant, N/m

x = spring deflection, m

A spring with spring constant of 1 N/m would be a very loose spring, since 1 N is a small force and a 1 m stretch would be quite large. The larger the value of k, the stiffer the spring.

In a practical sense, there is a limit to a spring's proportional behavior. Eventually, it stretches past its *elastic* limit and won't return to its original position. This is referred to as the *plastic* region. If you stretch it further, it will reach its ultimate limit and break. This is illustrated in the following figure:

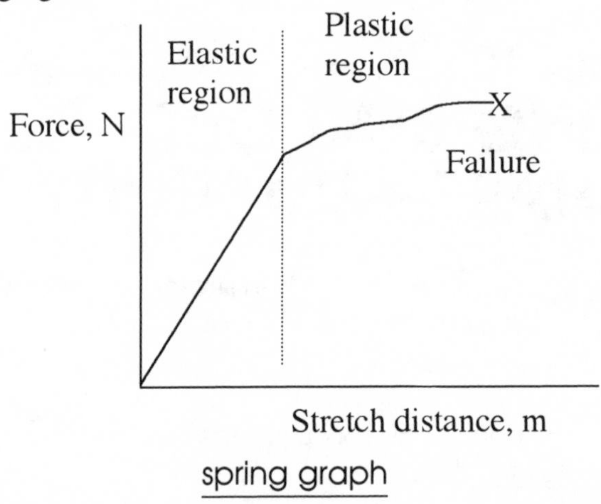

spring graph

Solid matter behaves very much like a spring. At the submicroscopic level, the electrical attraction of one molecule for another is approximately linear, as long as the distances they move remain small. Let's begin with the case where equal and opposite forces pull a solid body in opposite directions, resulting in no motion, but a stretching of the body. Instead of force and stretching distance, there are new quantities:

> *Stress* is the force applied to a body divided by the area over which the force acts. *Strain* is the amount of length increase divided by the original length. The stress applied to a body is directly proportional to the strain and to a modulus, which is a measure of the stiffness of the body.
>
> $$\text{stress} = (Y)\ \text{strain}$$
>
> where
>
> stress = F/A; stress units = N/m^2 = Pa
> Y = Young's modulus, a measure of the body's stiffness;
> Young's modulus units = N/m^2 = Pa
> strain = $\Delta l/l_0$; strain units = m/m = dimensionless

Stress and *strain* are nicely descriptive terms borrowed from ordinary language.

Besides tension, several other stresses and corresponding strains are shown in the figure.

Compression is a bit more complicated than tension because a body in compression can bend and fail suddenly. Further, three-dimensional compression is common in pressure situations, like a body submerged in a fluid.

Shear stress is what you apply to paper or fabric with . . . shears. (Did you think it would be a scissors?)

Architecturally, *bending* stress occurs when some unsupported beam juts out from a wall, like a second-floor deck. This is also known as a *cantilever*.

Torsion occurs when a body undergoes opposing torques, as when you wring out a sponge.

Hoop stress occurs in the metal rings (hoops) that keep barrel staves in place. The outward pressure tries to stretch the hoop (an engineering professor once likened this to the stress in your belt after Thanksgiving dinner).

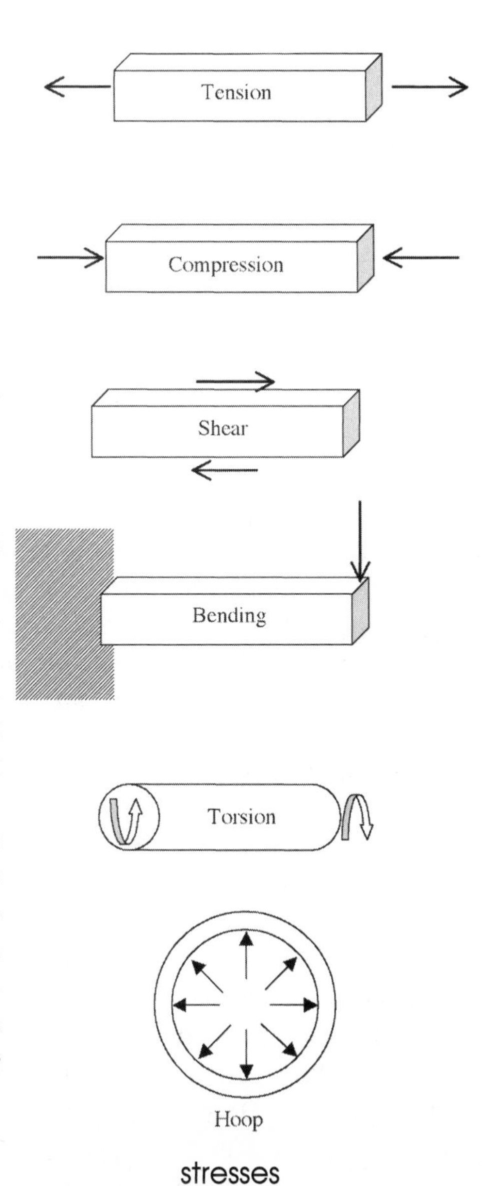

stresses

Experimentino 7

Using a flexible foam ball, or a foam paintbrush, or even a sponge (not too dry or too wet), apply the forces and torques shown for the first five stresses shown in the preceding diagram. Hoop stress can be demonstrated by tightening your belt.

Another type of stress is called *fatigue* stress. This happens after many cycles of a repetitive application of any type of stress. If you bend (bending stress) the pull tab on a soda can back and forth repeatedly, it breaks off quickly, whereas if you tried to pull it off directly (tension stress), it would be very difficult. Many failures in automobiles and airplanes can be traced to fatigue stress. There is even an example of fatigue stress in the human body to be explored later.

PHYSICS IN THE CULTURE

Stress analysis plays a large role in modern culture, since many of our gadgets must withstand stresses in order to function as they were designed. Before analytical methods were developed, designers and builders had to rely on intuition. Often, the result was overbuilding, particularly true in large public buildings, some of which lasted many centuries. From a modern standpoint, if strong enough materials are used, the loads don't cause the stress to exceed the elastic limit, and the body simply returns to its normal state when the load is released. A good example of successful stress analysis is the airplane. The structure of an aircraft needs to be strong enough to handle anticipated loads, but it can't have extra weight without detracting from the plane's performance. An unsuccessful stress analysis example involves early airliners. Passengers walking on an aircraft floor created compressive stress. Since stress is force/area, a large stress can come from a large force or a small area. High-heel shoes have a small heel area, so passengers in early airliners who wore high heels left dents in the floors. Once the designers realized what was happening, they made airliner floors thicker. Another failure of stress analysis was the bridge across the Tacoma Narrows, which will be discussed shortly.

Experiment 12: Gaze Intently at the Swinging Yo-Yo

Equipment needed: yo-yo, metric tape measure, timer

A swinging yo-yo is an example of a body in simple harmonic motion. The time required for one complete cycle of motion is called the *period*, and it depends on the length of the pendulum only.

period $= 2\pi \sqrt{\frac{L}{g}}$, where L is the pendulum length in meters and $g = 9.80\ m/s^2$

Procedure

- Measure the distance from the top of the string to the center of the yo-yo.
- Allow the yo-yo to swing back and forth, and time its period with the stopwatch.
- Repeat for a total of ten swings.
- Find the mean period.
- Compare the mean period to the predicted $\sqrt{\frac{L}{g}}$.
- Allow the yo-yo to swing with a smaller amplitude and measure its period.
- Compare this period to the mean period found earlier.
- Attach the metric tape measure to the yo-yo, allow it to swing, and measure its period.
- Compare this period to the mean period found earlier.

If the period depends only on the length of the pendulum and not on the amplitude or mass, all the measured periods should be the same. Galileo noticed the independence of a pendulum's period by observing the hanging lights in church, which swung back and forth. He timed them by using his pulse.

Distance from string top to yo-yo center = (sample = 1.03 m ____)

Trial	**Period (s)**
Sample	2.00
1	
2	
3	
4	
5	
6	
7	
8	
9	
10	
Mean	
Predicted period (sample)	2.04
Predicted period	
Smaller amplitude swing	
Yo-yo with additional mass	

Experiment 13: Incredibly Repetitive

Equipment needed: springs, mass, stopwatch

A mass oscillating on a spring is another example of a body in simple harmonic motion. Assuming the spring has a much smaller mass than the mass hanging from it, you will see that the period of oscillation of such a system is dependent only on the mass and the spring constant of the spring. This is according to the equation

$$\text{period} = 2\pi \sqrt{\frac{m}{k}}$$

where m is the mass (kg) and k is the spring constant (N/m).

Procedure

- Attach a (heavy) mass from Experiment 3 to one of the springs from Experiment 11.
- Allow the mass to oscillate and measure the period of oscillation with the stopwatch.
- Record the period in the attached table.
- Repeat for a total of ten trials.
- Find the mean period.
- Calculate the predicted period $= 2\pi \sqrt{\frac{m}{k}}$
- Repeat for the other spring from Experiment 11.

Trial	**Period (s)**
Sample	2.00
Spring 1 #1	
2	
3	
4	
5	
6	
7	
8	
9	
10	
Mean	
Predicted period (sample)	(2.04)
Spring 2 #1	
2	
3	
4	
5	
6	
7	
8	
9	
10	
Mean	
Predicted period	

Chapter 11.
BACK AND FORTH, BACK AND FORTH . . . SIMPLE HARMONIC MOTION AND WAVES

"I'm pickin' up good vibrations."
—from "Good Vibrations" by the Beach Boys

"Kenneth, what's the frequency?"
—Dan Rather's assailant

Here's a variation on the motion theme: How about a body that is in almost constant motion yet never really goes anywhere? You may recall the discussion from chapter 9 about the marble in the teacup being an example of dynamic equilibrium. As it rolls up one side and back down and up the other side, the average position of the marble is the center of the teacup. The term that describes this motion is *oscillation*, and the first kind we'll look at is called *simple harmonic motion*.

To make it really simple, think about sitting on a rubber raft in a lake. You and your raft bob up and down, but your average position is the undisturbed water surface. If you plot a graph of your position as a function of time, it would look like this:

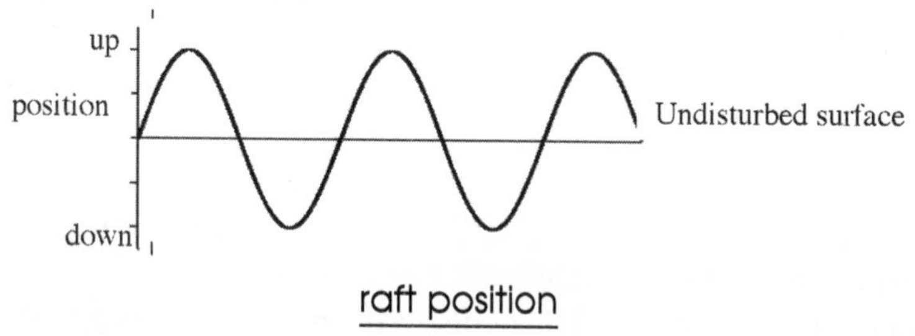

raft position

The time required to make one complete cycle of the motion (down to up to down, say) is called the *period* of the motion and is given the symbol t. Since this is the time for one cycle of the motion, it is measured in seconds/cycle. The inverse of the period is called the *frequency* and is given the symbol $f = 1/t$. The frequency's unit is cycles/second, which has been renamed the hertz, abbreviated Hz. As you saw in Experiments 12 and 13, the period of oscillating bodies depends only on certain characteristics of the bodies themselves.

From another point of view, think about what's causing you to bob up and down. It's the motion of the water. You're just along for the ride. The water moves because there has been some disturbance that propagates across the water surface. Maybe it's a moving boat, or wind, or something that fell into the water and made a big splash. The point is, the disturbance traveled. Let's take a cross section through the water surface at a single instant of time:

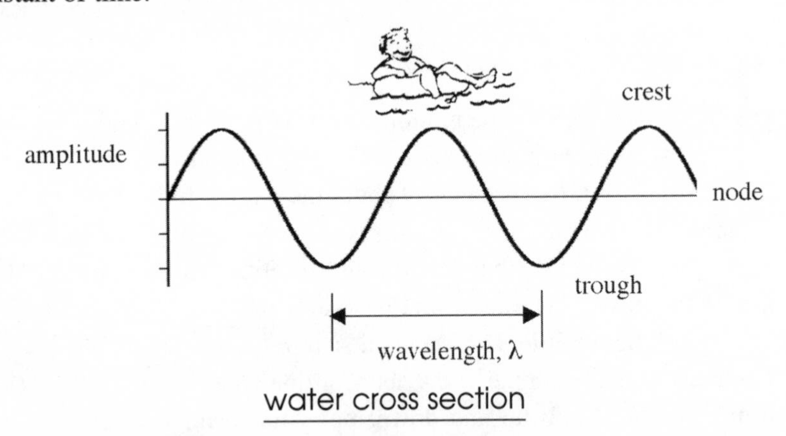

water cross section

The top of the wave is called the *crest*, the bottom is the *trough*, and the part that is even with the undisturbed surface is called the *node*. The height of the wave above the undisturbed surface is the *amplitude*. The length of one complete wave is called the *wavelength* and is given the Greek symbol lambda, λ. Before long, you and your raft move downward, and the wave moves on. In fact, moving is essential to a wave. A wave is defined as *a traveling periodic disturbance*. Most waves have a medium in which they travel, and there are two types of waves, depending on the relationship between the direction the wave moves and the direction of motion of the medium. With a *transverse* wave, the medium moves perpendicular to the wave, the way that you and your raft move up and down while the wave moves horizontally across the water surface. The medium moves in the same direction as the wave for a *longitudinal* wave. Chapter 12 will explore the most familiar of the longitudinal waves—sound. Examples of transverse waves, however, include water surface waves, waves on a stretched string, and light waves. Experiment 14 will demonstrate both kinds of waves.

All waves obey a simple relationship based on their velocity. Let's return to your raft. If a wave came by, you could estimate the wavelength visually and time the period with your watch (waterproof, I hope). You could then determine the wave's velocity by dividing the wavelength by the elapsed time. Since 1/period is the frequency, the relationship between wave speed, wavelength, and frequency is:

$$\text{speed} = (\text{frequency})(\text{wavelength}), \text{ or } v = f\lambda$$

When disturbances happen to the Earth (earthquakes), four different seismic waves are generated: P-waves (primary), which travel through the body of the Earth and are longitudinal; S-waves (shear or secondary), which travel through only the solid parts of Earth, and are transverse; L-waves (Love, for their discoverer), which are surface waves that shake the surface side to side; and R-waves (Rayleigh, for their discoverer), which roll the surface up and down, like water waves. Because all these waves have their own unique characteristics, when they are recorded on seismographs around the world, the location and intensity of the earthquake can be determined.

Waves have interesting properties, some of which make them quite different from matter.

Characteristic Speed

Both longitudinal and transverse waves have a speed that depends on properties of the medium. For example, waves that travel on a stretched string depend on the tension in the string, the string's length, and the string's mass according to:

The speed of a wave on a stretched string is directly proportional to the square root of the tension in the string and the string's length, and inversely proportional to the string's mass.

$$v = \sqrt{\frac{T}{\left(\frac{m}{L}\right)}}$$

where T = tension, N
L = string length, m
and m = string mass, kg

Interesting things happen to waves when they encounter an abrupt change in medium.

Reflection: At the boundary between the new and old media, a portion of the wave is sent back in the direction from which it came, sometimes with an inversion (crest becomes trough and vice-versa).

Absorption: In the new medium, some of the wave is converted to energy of molecular motion, so the wave is correspondingly diminished.

Transmission: A portion of the wave simply continues moving in the new medium from the old medium.

Refraction: If the wave strikes the boundary of the medium change at an angle, the part of the wave that is transmitted into the new medium moves at a different angle.

Diffraction: When a straight wave encounters a small opening in a barrier, the wave that is transmitted through the opening is circular rather than straight.

Interference: When two or more waves occupy the same spot at the same time, the amplitudes of the waves are simply added to each other. This is called the *principle of superposition*. Two crests make a doubly deep trough. This is called *constructive interference*. When a crest meets a trough of the same amplitude, there is a temporary cancellation of the wave, called *destructive interference*. Notice how this differs from matter. Two bodies cannot occupy the same spot at the same time, but waves can do it because they are disturbances, not molecules. In some sports, two players who attempt to occupy the same spot at the same time are penalized for . . . interference.

Returning to the vibrating masses on springs and oscillating pendulums, we need to add an important ingredient to the picture: friction. If the motion of harmonically oscillating bodies is impeded by friction, this is called *damping*. With no damping (*undamped motion*), the position of the mass on the spring was described graphically by a sine wave with constant amplitude. If there is a small amount of damping (*lightly damped motion*), the pattern looks like a sine wave with decreasing amplitude. If there is way too much damping (*overdamped motion*), the body never even completes a single oscillation and approaches zero position slowly. The borderline case between damped and overdamped motion is called *critically damped motion*, and the body approaches zero as quickly as possible but does not overshoot and begin to oscillate.

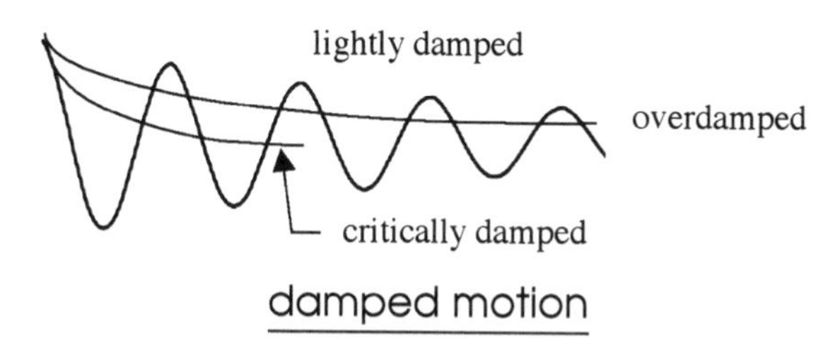

damped motion

One way to envision frictional damping would be to bring up a container of molasses under the mass vibrating on a string. (Experimentino? No, too messy.) Perhaps a better example of damping is provided by a door-closer. Old-fashioned storm/screen doors had a cylinder that contained a piston and some fluid. An adjustment screw at the end allowed the piston's size to be changed, making more or less fluid friction. If there was too little friction, the spring would return the door to its closed position quickly, and there would be a loud bang as the door tried to oscillate through the door jamb. If there was too much friction, the door would close very slowly, if at all. The goal was to adjust the friction to allow critical damping, closing the door in the minimal time, with no bang.

Another example of vibration with damping is provided by an automobile. Cars have both springs and shock absorbers. The springs distribute road disturbances throughout the whole car, rather than transmitting them directly to the passengers. (Ever ride in a sports car with tight suspension?) Shock absorbers apply damping so that the oscillations last only a few cycles before their amplitude diminishes to the point where they aren't noticeable. Periodically, TV commercials show cars that need new shocks as they bounce down the road. A simple way to demonstrate this yourself involves a technique often used in buying a used car. Stand on the front bumper then jump off suddenly. The car's springs will cause the car to oscillate, but if the shocks are in good shape, the oscillations should diminish quickly. If the oscillations continue for a long time, it means the shocks are ineffective. (Of course, there are a few more things to check before buying a used car.)

So a body's frequency of oscillation depends on its mass and spring constant. This is called the body's *natural frequency*. What happens if a forcing function is applied to a body in an attempt to cause vibration? As you would surmise, it depends on the relationship between the frequency of the forcing function and the natural frequency of the body. If the frequency of the forcing function is way lower or way higher than the body's natural frequency, not much happens. (Think about pushing a child on a swing at too high a frequency—"Swing fast, it's almost time to go.") But if the frequency of the forcing function is about the same as the body's natural frequency, a lot happens, and it's called *resonance*. Just as constructive interference makes bigger wave amplitudes, resonance produces larger amplitudes of oscillation, with some interesting results. You've undoubtedly seen the ads where a singer sounds a particular note (frequency) and a glass shatters. The singer's note matched the glass's natural frequency, so the resulting resonance produced such large amplitude waves that the glass broke.

Experimentino 8

You can find the natural frequency of a glass by simply rubbing the rim with a moistened finger. The friction caused by the motion of your finger (moisture cuts down the lubricating effect of skin oil) creates waves, and those with the same frequency as the natural frequency of the glass will resonate enough to create a wave that is audible. If there is liquid in the glass and an air column above the liquid, the tone will be affected. Singing the same note to shatter the glass isn't something I'd advise, unless you're a budding opera singer who wears safety glasses.

On a bigger scale, the spectacular collapse of the Tacoma Narrows bridge in 1940 was an example of both transverse waves and torsional vibrations working together, *plus* fatigue stress. How much resonance was involved is not clear, but videos of the disaster are available for viewing on the Web (see endnotes).[1]

A smaller example shows up in your car. Sometimes driving at a particular speed causes a resonance in some part located deep in the innards of your dashboard. Driving faster or slower makes the rattle quit, but law enforcement officers don't always appreciate the physics of the situation.

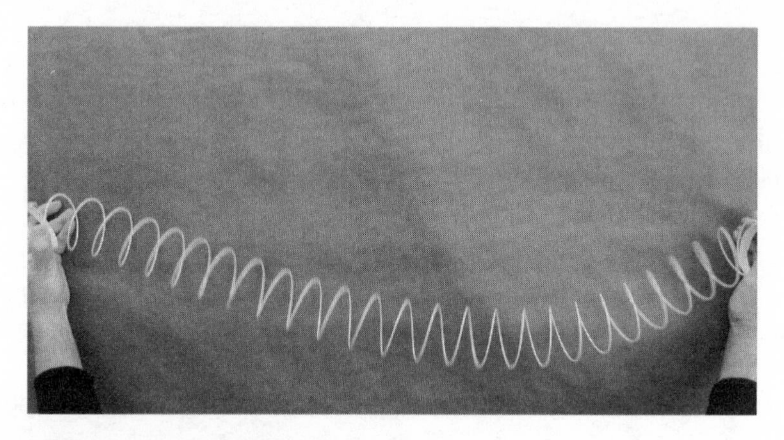

Experiment 14: Magic Spring Tricks

Equipment needed: magic spring

Extend the magic spring (sometimes called a Slinky) to a distance of about 1 meter (3 feet) and hold one end in each hand in front of you. No measurements will be made, but three different waves will be demonstrated.

PROCEDURE

- Jiggle the right end of the spring upward and observe the resulting movement. You should see a crest propagate in front of your eyes. Repeat until you see it clearly. This is a *transverse* wave.
- Make a sharp movement to the left with the right end of the spring and observe the resulting movement. The disturbance should propagate across the wave. Repeat. This is a *longitudinal* wave.
- Holding the left end of the spring steady, jiggle the right end up and down at such a rate that the spring appears to move up and down in halves. If you jiggle the spring faster, you might get it to move up and down in thirds or even quarters. These are all part of a series called *standing* waves.

Chapter 12.
HEY, LISTEN, WHAT'S THAT SOUND?

"When out on the lawn there arose such a clatter, I sprang from the bed to see what was the matter."
—from "The Night Before Christmas" by Clement Clarke Moore

Sound is undoubtedly the most familiar longitudinal wave. The air is full of sounds, sometimes too many. But if you try to escape by going underwater, that won't work because there are sounds there, too. Even worse, if you become exhausted from all this sound and try to take a snooze by resting your head on a solid surface such as a table or a desk, you're still not safe, because sound travels through solids, too. Sound travels through solids, liquids, and gases, but there is a difference in the speed. Sound travels fastest through solids and slowest through gases. That isn't too difficult to understand when you think about how waves propagate and how the architecture of matter is arranged. A longitudinal wave starts when the medium is disturbed, then the wave propagates in the same direction as the original disturbance. The medium consists of molecules, so a disturbance pushes a molecule in the direction of its neighbor molecule until their electron clouds overlap. Then the neighbor molecule is pushed toward its neighbor, and so on, propagating the sound wave.

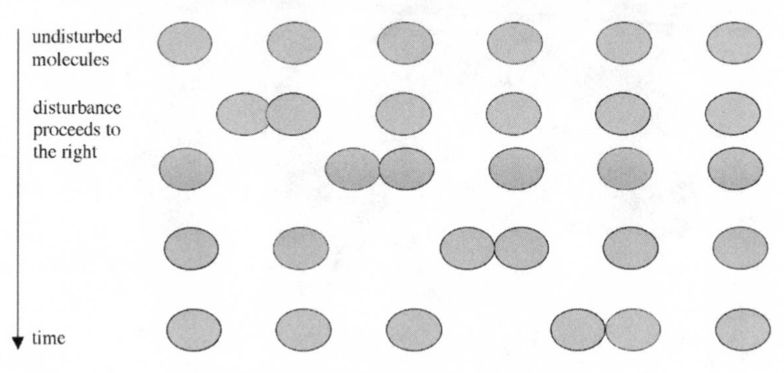

sound wave propagation

If the molecules in the diagram represent a gas, the overall speed of the wave would be governed by how fast the molecules move in the direction of their neighbors. In a liquid, molecules are closer, so the wave speed is greater. A solid's molecules are closest, so the wave speed is greatest of all. The speed of sound in air is around 300 m/s; liquid water has a sound speed of about 1,500 m/s; and steel allows sound to travel at about 5,000 m/s.

Our familiarity with sound is based at least partly on our extremely effective built-in sensors. Sound waves are routed through gases, liquids, and solids before the auditory signals are turned into messages sent to the brain. Details of this intricate system are available in biology texts, but let us explore at least some of the physics implications of the human hearing system. A perfect human ear can hear a frequency range from about 20 Hz to 20,000 Hz. This wide frequency spectrum enables us to sense most of the significant sounds of our environment and alerts us to danger or opportunities for finding food. Normal aging, however, changes this range. Just before the signals are sent to the brain, sound waves arrive in the part of the human ear known as the *cochlea*. Within the cochlea, many hairs of different thicknesses and lengths respond to the incoming sound waves by resonating when the incoming frequency matches their natural frequency. Short, thick hairs resonate at low frequencies and long, fine hairs resonate at high frequencies by waving back and forth very quickly. After these fine hairs have vibrated for a long time (years), fatigue stress sets in, and the hairs break, leaving the high end of the frequency spectrum without any sensors. Thus, older people have trouble hearing high-frequency sounds because their cochlea lack the long, thin hairs. (This provides a way of judging people's age, just as the age of horses is directly related to the length of their teeth.) Sound waves with frequencies above about 20,000 Hz are called *ultrasound*, and they are used in medicine and sonar. Medical use involves sending sound waves (often 1–5 MHz) into human bodies and constructing images based on sensing the reflections. Few side effects have been documented, so ultrasonic imaging is far less damaging and intrusive than electromagnetic radiation. Sonar is used in underwater applications in either an active or a passive mode. In a passive mode, sonar detectors listen for sounds made by ships, submarines, or fish. Active sonar sends out signals (*pings*) anywhere from 50 kHz to 200 kHz and identifies objects by the reflections of their pings. Sound waves of low frequency (below 20 Hz) are known as *infrasound* and have some interesting properties. Although infrasound is too low in pitch to be detected by human ears, it may still be sensed crudely by humans. These low-frequency disturbances cause internal body parts to resonate, possibly generating "strange" feelings that have no traceable source. Some animals' hearing extends to this region, so there is a possibility that animals can sense extreme weather conditions such as hurricanes or tsunamis, or even geological events such as earthquakes, both of which are possible generators of infrasound waves.

Besides frequency, which is often referred to as *pitch*, the human ear is capable of hearing a wide range of sound *intensities*, more commonly called *loudness*. In physics terms, a sound's intensity is measured in units of watts/meter^2 (W/m^2). The human ear is capable of hearing an incredible range of intensities, from 10^{-12} W/m^2, corresponding to the proverbial pin drop, to about 1 W/m^2, which is the threshold of pain. Because of this wide range of intensities, all with negative exponents, the W/m^2 unit is cumbersome. So a different scale was devised, the decibel.

$$\text{decibel} = \text{dB} = 10\log_{10}\left(\frac{I}{I_0}\right),$$ where I_0 is the lowest audible sound, 10^{-12} W/m^2

In terms of dB, human hearing ranges from 0 dB to 120 dB, a more convenient scale.

Intensity level, dB	Typical sound
0	Threshold of hearing
10	Rustle of leaves on a quiet day
20	Whisper, ticking watch
30	Library, broadcast studio
40	Quiet TV or radio background
50	Mild traffic, rainfall
60	Conversation, ventilation fan
70	Noisy restaurant, freeway traffic
80	Buses, garbage disposal
90	Siren, heavy truck nearby
100	Garbage truck, factory machinery
110	Construction pile driver
120	Loud rock concert, threshold of pain

The human ear's sensitivity is not uniform. It's harder to hear both high and low frequencies, but the midrange from about 1,000 Hz to 5,000 Hz is easiest to hear. This helps to explain why the telephone was designed to pick up, transmit, and receive the frequency range from 400 Hz to 3,400 Hz.

Many sounds in our environment have little or no discernable pattern and would best be characterized as noise. But musical sounds are generated to please the ear, and musicians accomplish their goal wonderfully. There are a huge number of musical instruments, and it would require too much space to investigate the physics of all of them. So let us discuss just a few categories and examples of each.

Ways to Generate Musical Sound

Vibrating Strings

Recall from chapter 11, a stretched string allowed disturbances to move at a speed determined by the string's tension, length, and mass according to

$$v = \sqrt{\frac{T}{\left(\frac{m}{L}\right)}}$$

But waves generated in a string reach the string's end and are reflected. If the ends are fixed in place, as they would have to be in order to support any tension, waves undergo reflection at the ends. Further, not only are the waves reflected, but they are inverted, so a crest reflected from a fixed end goes back the other way as a trough, and vice versa. So a vibrating string has waves going back and forth. This creates a pattern that tends to cancel most of the waves, but there are a few exceptions. Waves of just the right frequency will "fit" within the string's length and cause the string to oscillate up and down all at once, or in two halves, or three thirds, and so on. These are called *standing waves.*

The longest wave that would just fit and make the string vibrate up and down as a whole is called the *fundamental.* The wavelength of this wave is twice the string length, and the speed of the wave is v, so the frequency of the fundamental is

$$f_0 = \frac{v}{2L}$$

The next-longest wave is called the *first overtone*, and its wavelength is exactly L, so its frequency is

$$f_1 = \frac{v}{L} = 2\left(\frac{v}{2L}\right) = 2f_0$$

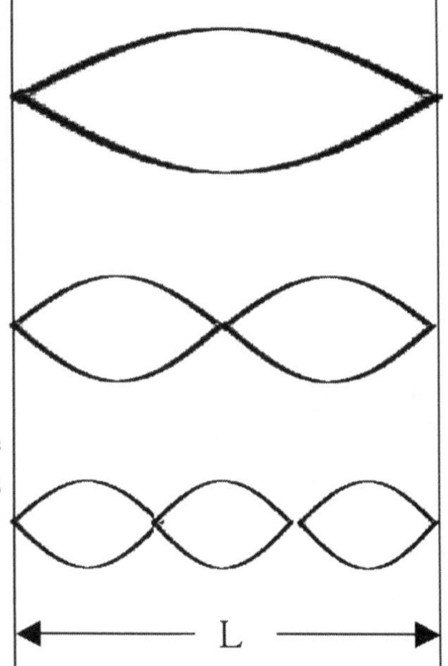

standing waves on a string

With the first overtone, the string vibrates up and down in halves. This was demonstrated in Experiment 14. Moving on to the next possible wave, the *second overtone*, its frequency turns out to be $f_2 = 3f_0$, and the string vibrates up and down in thirds. Remarkably, there is a whole series of waves that "fit" into the string's length, and their frequencies are all whole-number multiples of the fundamental frequency. This is called the *harmonic series*, and forms the basis for most music. Of course, waves on the string move too fast for us to see, but since strings move air molecules, our ears can hear the results of these harmonic waves. Here's a little test.

Experimentino 9

Since the frequency of the fundamental wave depends on the velocity of the wave on the string, and the velocity in turn depends on the tension in the string, the greater the tension, the higher the frequency. Take the yo-yo used in Experiment 5, and fix one end on the floor by stepping on it. Go ahead, you won't hurt its feelings. Pull the other end tight and pluck the string. As you pluck it, increase the tension. Do you hear the pitch (frequency) rise? Now change the length of the string and pluck it again. If you made the string shorter, do you hear the frequency rise? Do you feel like a bass player?

The guitar, violin, piano, mandolin, bass, viola, contrabass, harp, ukulele, cello, and harpsichord all rely on vibrating strings that are bowed, plucked, or struck, but they all produce sounds that consist of a fundamental frequency and overtones that are integer multiples of that fundamental.

Vibrating Air Columns

Columns of air vibrate similar to the way stretched strings do, but there are some interesting variations. In air columns, the waves are longitudinal and their fixed speed is the speed of sound. Strings have transverse waves, and their speed is adjustable by changing the tension. Also, reflections from the ends of air columns have different possibilities than strings that have only fixed ends. If a tube containing air is open at both ends and a disturbance is created in the tube, the sound wave reflects from the open ends, but without being inverted. As a result, the open ends are not nodes as they were with strings, but antinodes. Here's a diagram:

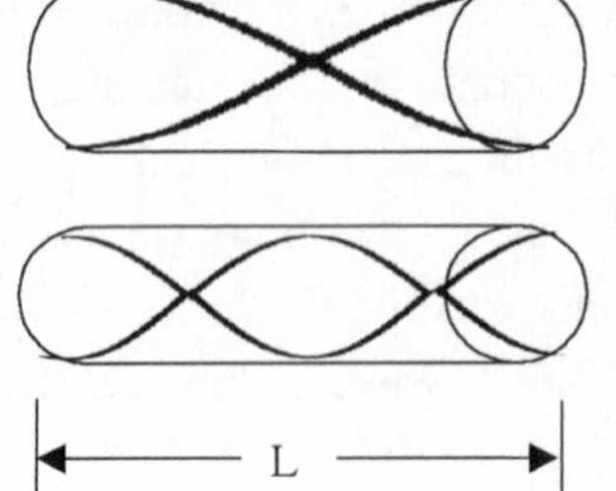

The longest wavelength that "fits" is again called the *fundamental*. Its wavelength is twice the tube length and the corresponding frequency is

$$f_0 = \frac{v}{2L}$$

standing waves in a vibrating air column

where v is the speed of sound and L is the length of the tube. The next-largest wavelength, which generates the first overtone, has a wavelength equal to the tube length, so its frequency is double the fundamental. Similarly, the second overtone has a frequency of three times the fundamental. In fact, the complete harmonic series is generated, with each overtone frequency being an integer multiple of the fundamental. An interesting variation is provided by a tube that is open at one end and closed at the other. The longest wave that will fit and satisfy the boundary conditions of a node at one end and an antinode at the other is a quarter-wavelength; so in this case, the fundamental frequency is given by

$$f_0 = \frac{v}{4L}$$

For the first overtone, the wave fills three-quarters of the tube, so the first overtone frequency is

$$f_1 = \frac{3v}{4L} = 3f_0$$

The next overtone has a frequency $5f_0$, and the rest of the harmonic series contains only *odd* integer multiples of the fundamental.

In the vibrating string, the source of waves was a physical disturbance to the string itself. It was bowed, plucked, or struck by the instrument player, and the only waves that survived were ones that "fit" the instrument. With vibrating air columns, again only "fitting" waves survive all the interference. But the original source of the sound waves depends on the instrument.

Blowing across the top of the air column: The pipe organ, piccolo, flute, jug, and panpipes are all instruments that feature an air jet blown across an air column to initiate the sound. A car's open sunroof can also exhibit annoying resonances.

Blowing through a reed or reeds: The saxophone, clarinet, and some bagpipes all work by having the player blow through a single reed, which vibrates to create the initial sound. The oboe, bassoon, and other bagpipes utilize double reeds to start the waves.

Blowing through vibrating lips: The bugle, cornet, euphonium, French horn, trombone, trumpet, and tuba all require the players to blow through their lips; these vibrations provide the initial disturbance that starts the wave.

Experimentino 10

Designing and building a huge variety of musical instruments has demonstrated people's creativity for centuries. Here's a chance to participate.

1. Using the plastic soda bottle from Experiment 15, fill it about half full of water (half empty for pessimists?) and blow across the top. You should hear a strong resonance at a particular frequency. Now empty some of the water, making a longer air column. Blow again and listen for a lower-frequency note. Some creative folks have "tuned" several bottles to particular notes and blown across them in the proper sequence and rhythm to produce a recognizable tune.
2. Using the straw from Experiment 15, use a scissors to cut two pieces off one end to make a V. Place the cut end in your mouth, close your lips, and clamp down a little on the straw, just beyond the V, and blow. This may take some adjustment before you get it, but you should eventually be rewarded with a sound that sounds somewhat like a duck call. This is actually a double-reed action similar to that of a bassoon.

Vibrating Objects

Music is also produced when objects are struck (*percussion*). The object may vibrate in such a way that the body of the instrument itself vibrates (triangle, cymbal, gong, xylophone, steel drum, or marimba) or a membrane attached to the instrument does the vibrating (timpani, snare drum, bass drum, or bongo drum). The resulting wave pattern is often quite complex. Percussion instruments do *not* form a harmonic series of frequencies like stringed or wind instruments.

Electronically Driven Vibrations

Electronic oscillators attached to speakers and controlled by keyboards or more exotic input devices make music with fascinating sounds, some of which don't exist in nature. They include the theremin, the synthesizer, and the electric organ. For more information about electronic music and its history, look up any biography of Robert Moog (rhymes with *vogue*), whose degrees in physics were put to good use in inventing and building electronic musical instruments.

Knowing that stringed and wind instruments both rely on harmonic series in which overtone frequencies are integer multiples of the fundamental, you might wonder how the same note played on two instruments could be distinguished from each other. To be specific, suppose you played A above middle C on a piano and a French horn. The fundamental frequency, 440 Hz on an even-tempered musical scale (that's another story, too big for this chapter), would be the lowest frequency for each note, and then there would be overtones of 880 Hz, 1320 Hz, 1760 Hz, and so on, up to the highest audible frequency. Both instruments would contain exactly the same set of frequencies. Yet your ear would be able to tell them apart. How is this possible? The answer lies in the amplitude of each of the overtones. If the sounds for each instrument were analyzed on the basis of the frequencies involved (this is called *Fourier analysis*), the same frequencies would show up for both. But the overtones for one instrument would have a different amplitude pattern than the other. The overtones are emphasized or minimized for a particular instrument because of the sounding board, the details of the instrument's finishing materials, the type of wood or metal, and so on. (*Not an experimentino—* If you have access to a digital keyboard, a quick way to hear the differences in overtone amplitudes is to play the same note with one hand and dial different musical voices with the other hand.) Even knowing how the sounds differ, you could still wonder how your ear can tell which instrument made which sound. The answer lies in your brain. Among the souvenirs stored in your brain's attic are frequency-analyzed sounds for all the musical instruments you have heard. Within tiny fractions of a second, you can hear a sound, compare it to the sounds in your mental library, and identify the instrument that played it. As if auditory memory alone isn't enough, recent active brain-scan research at Dartmouth College links musical memory to emotional response and control. Impressive things go on inside your head.

We've seen that stringed instruments can be tuned by adjusting the tension in the strings. You may wonder how tuning is accomplished for the piano, which has eighty-eight keys to be adjusted—some of which strike more than one string. The answer has to do with a phenomenon known as *beats*. If two different notes are sounded at the same time, their interference pattern produces sounds at frequencies equal to the sum and difference of the two frequencies involved. Suppose you are trying to make sure that A above middle C is exactly 440 Hz. If you have a tuning fork that puts out exactly 440 Hz, first excite the tuning fork and then strike the piano key. (If the key strikes three strings, the two outer strings must be muted.) If the frequencies don't match exactly, you will hear a beat frequency equal to the difference between the piano note and the tuning fork. Beats make a rising and falling sound superimposed on the normal note. To tune the piano, you then adjust the tension until the beat vanishes. In the modern version of the tuning process, the tuning fork is replaced by an electronic oscillator with adjustable frequencies, but you still have to listen for the beat frequency.

Another interesting thing happens if the source of the sound is moving. It is called the *Doppler effect*. Fundamentally, what happens is that the sound waves emitted later begin to catch up with the sound waves emitted earlier.

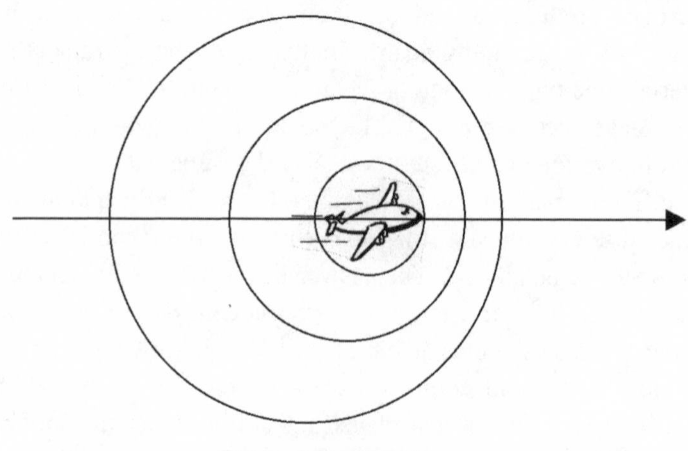

Doppler effect

If the circles represent crests of sound waves emitted by a moving body, you can see that an observer ahead of the moving body will hear more crests per unit time, therefore a higher frequency. This is as opposed to an observer behind, who will hear fewer crests and a lower frequency. In a similar way, if the sound source is stationary and the receiver is moving toward the source, a higher frequency is heard, while movement away from the source yields a lower frequency.

This is summarized by the Doppler equation: $f_{heard} = f_{emitted}\left(\dfrac{1 \pm \dfrac{V_{receiver}}{V_{sound}}}{1 \mp \dfrac{V_{source}}{V_{sound}}}\right)$

where the appropriate signs are used to make approaching frequencies higher and receding frequencies lower. There is a Doppler effect for light also, which shifts light frequency higher (blueshift) or lower (redshift), depending on the motion of the light source either toward or away from the observer.

If the moving body travels faster than the speed of sound, things get even more interesting. The sound waves add up in such a way that a shock wave is formed.

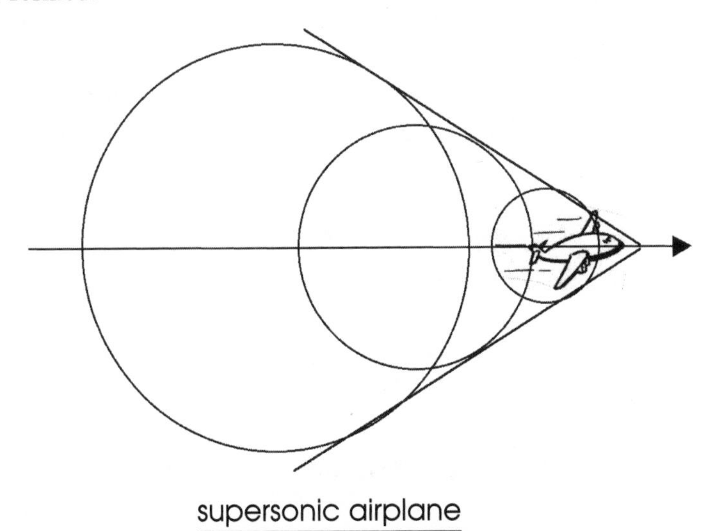

supersonic airplane

The ratio of the speed of the body to the speed of sound is called the *Mach number*, and the cone of sound waves produced by this body is called the

Mach cone. Clearly, the greater the speed of the body, the smaller the angle of the Mach cone. An aircraft speed of Mach 1 means the plane is traveling at exactly the speed of sound. Supersonic speeds refer to speeds beyond sound speeds, and hypersonic denotes even higher speeds—usually Mach 6 and above. As the body moves, it drags the Mach cone along with it, so the consequent disturbances—where the Mach cone intersects the ground—depend on how high the plane is flying. The point of aviation regulations on the altitude of supersonic flights is to keep ground-level sonic disturbances to an acceptable level. If a plane is going fast enough, it may be out of sight when the Mach cone disturbances reach an observer on the ground. So you may occasionally hear a sonic boom that seems to have no cause.

Physics in the Culture

The house lights dim, a spotlight shines on the stage, and the main character enters and launches into the opening statement, which sets the tone for the whole play. "@#$% ^&^% **&$." "What did she say? I couldn't hear a word." In a well-designed theater, concert hall, or auditorium, sounds must be delivered faithfully from the performers in front to the audience throughout the room. But funny things happen along the way: reflection, refraction, absorption, diffraction, and interference of the sound waves. To investigate the way sound waves propagate through a room, let's examine two limiting cases.

Perfectly *absorbing* walls, floor, and ceiling

There would be no echoes in such a room, called an *anechoic chamber*. Some of these exist in research and measurement labs, but are truly impractical for concert venues, because listeners perceive the sound in such a room as very unpleasant or "cold."

Perfectly *reflecting* walls, floor, and ceiling

This kind of room would have many echoes. A listener at a given location in the room would hear the sound projected directly from the stage, plus the sound bounced off one wall, two walls, three walls, and so on. These multipath reflections would lengthen the duration of sound or make multiple echoes at the listener's location. Further, dependent on the listener's position and the frequencies of the sound from the stage, there could be constructive interference (loud spots) or destructive interference (dead spots).

Experimentino 11

If you have access to a large, empty room (or some scenic spot famous for echoes), go in when you won't disturb anyone (I've done this at an indoor tennis facility at 7 AM), and shout something at the top of your lungs. "Hello" is good, but if you're not disturbing anyone, be creative. Your word will reflect off the walls and return to you in the form of an echo. Listen for how many echoes of your word are audible. If you can time the interval from the first shout to the last audible echo, do so. This is related to the *reverberation time,* discussed below.

One quantity that assists in acoustic design is called the *reverberation time*, defined to be the time it takes for the intensity of the original sound to decrease by 60 dB. In other words, how long it takes for the sound to fade. An anechoic chamber would have a reverberation time of zero, and a room with perfectly reflecting walls would have a reverberation time of perhaps six seconds or more. A long reverberation time would produce a "bright, lively" musical sound, but speech would be "muddy." A short reverberation time would make speech clear, but music would be less "rich." To paraphrase Goldilocks, reverberation time shouldn't be too long or too short, it should be just right. To keep reverberation times in the appropriate range and reduce unwanted interference, various architectural/decorative elements are used: angled walls, curtains, draperies, minimal obstructions, fluted walls, acoustic ceiling tiles, and irregular, protruding wall designs (bas-relief). Top concert halls have reverberation times in the 1.5- to 2.5-second range, but there is a lot more to acoustical design than just reverberation time. After all, beautiful sound is in the ear of the beholder.

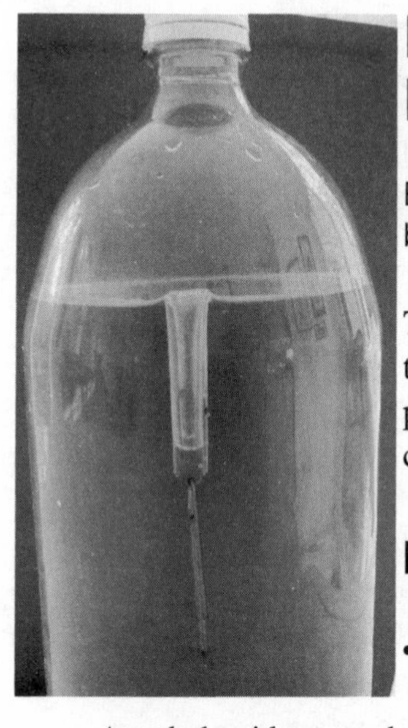

Experiment 15: Dive, Dive

Equipment needed: 2-liter plastic soda bottle, straw, paper clips, glass, water

This is a classic experiment, usually called the Cartesian diver. The idea is to see how pressure and density can work together to create a force that accelerates a body.

PROCEDURE

- Take about half a plastic straw and bend it so the two open ends are side to side.
- Attach the sides to each other using two paper clips. This is the diver.
- Fill the diver about halfway with water.
- Fill the glass with water and place the diver in the water, paper clip side down.
- The goal is to have the diver just barely float, with only a tiny amount above the water. (If this happens right away, go on to the next step. If not, add or subtract water from the diver until this critical balance is achieved.)
- Fill the 2-liter bottle with water and transfer the balanced diver to the bottle without spilling any water from the diver, which would disturb its balance.
- Pour some of the water out of the bottle, allowing a little air space at the top, and cap the bottle tightly.
- Squeeze the bottle anywhere and observe the diver carefully. You should see the water level within the diver rise, and the entire diver should dive.
- If the diver doesn't dive, you must remove it and rebalance it with more internal water.

Chapter 13.
GOOEY AND GASSY—FLUIDS AT THEIR FINEST

"There is no philosophy which is not founded upon knowledge of the phenomena, but to get any profit from this knowledge it is absolutely necessary to be a mathematician."

—physicist Daniel Bernoulli

"That we have written an equation does not remove from the flow of fluids its charm or mystery or its surprise."

—Richard Feynman (1964)

So far, most of our attention has been focused on solids. Now, it's time to admit that there's a bit more to matter. Let us broaden our field of view to include liquids and gases. We'll discuss them together here, in terms of their common properties, in which they are called fluids. First, we'll analyze fluids at rest, *fluid statics*, then fluids in motion—*fluid dynamics*. In the next chapter, gases will finally get their moment of glory as we explore them all by themselves.

Fluid Statics

Just as solids at rest constitute an interesting case, so do fluids at rest, although we'll have to get down to the molecular level to understand a few aspects of their properties. Let's start by drawing a free body diagram for a cube of fluid in the middle of a container of fluid at rest.

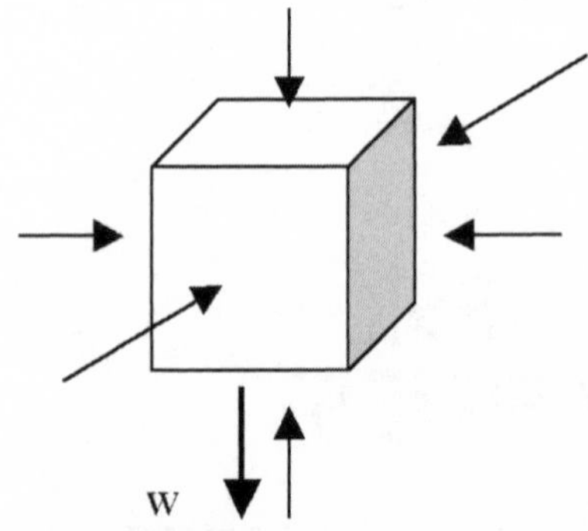

fluid cube free body diagram

There are normal forces on every face of the cube because of collisions with adjacent molecules. But since the cube is not moving, many of the forces cancel out, namely, the ones from left and right and front and back. The up and down forces, however, have a complication—the weight. The force on the bottom of the cube must balance the force on the top plus the cube's weight.

In equation form this is expressed as: $F_{bottom} = F_{top} + mg$. Now for a little mathematical exercise. Divide each term by the area of one face of the cube, A. The equation becomes:

$$\frac{F_{bottom}}{A} = \frac{F_{top}}{A} + \frac{mg}{A}$$

We saw the quantity $\frac{F}{A}$ before, in chapter 11, where it was called *stress*. In this context, the quantity is known as *pressure,* p, and its unit is N/m^2 = Pa. With one more manipulation, the equation will be complete. Multiply and divide the last term by h, the length of the cube. The whole equation becomes:

$$p_{bottom} = p_{top} + \frac{mgh}{Ah}$$

Ah is just the volume of the cube, V, and the ratio $\frac{m}{V}$ = density = ρ, with units kg/m^3. Density is an extremely useful quantity, which we will see again shortly. For now, let's put this back into the equation to get the final result, known as Pascal's principle:

Pressure anywhere in a fluid is the same as the pressure at the top of the fluid plus an amount that increases with the fluid density and the depth below the surface.

$p_{below} = p_{above} + \rho gh$, where the symbols are defined above

This relationship carries no surprise for anyone who has spent time underwater. The deeper you dive, the higher the pressure. Submarines and deep-sea explorers are well aware of this property.

One of the intriguing features of *density* is its dependence on the kind of material but not its amount. Here's a way to approach the understanding of density. A thought experiment—not as good as an experimentino, but easier to perform. Suppose you have a large number of identical rock cubes. Each one is 1 m on a side and has a mass of 2,000 kg. (See why it's a thought experiment?) The density of one rock cube would be

$$\rho = \frac{m}{V} = \frac{2{,}000\text{kg}}{1\text{m}^3} = 2{,}000\text{kg/m}^3$$

Stack a second cube on top of the first (ugh, heavy). Now what's the density of the double cube?

$$\rho = \frac{m}{V} = \frac{4{,}000\text{kg}}{2\text{m}^3} = 2{,}000\text{kg/m}^3$$

How about three cubes? 2,000 kg/m^3. How about 672 cubes? You get the point. The density of water is about 1,000 kg/m^3 and the density of air at sea level is about 1.3 kg/m^3. The following table shows densities of common substances:

Substance	**Density, kg/m^3**
Air	1.3
Ice	920
Water	1,000
Sea water	1,030
Aluminum	2,700
Steel	7,800
Gold	19,300

If our original water cube is replaced by something with a different weight than water, let's return to the free body diagram and see what changes. The weight is different, but all the normal forces are due to collisions with fluid molecules and they aren't aware that another body has been substituted, so they stay the same. The sets of forces left/right and front/back cancel. The forces up and down don't cancel, however, since they just balanced the weight of the fluid that used to be there. Calling the net upward force a *buoyant force*, if the cube is fluid, the buoyant force exactly balances the fluid cube's weight. What this means is that the buoyant force is the same as the weight of the fluid that would have been there if the new body wasn't there. Think about this for a minute. When Archimedes first realized this, it amazed him so much that he ran, at least in myth, through the streets of Athens, shouting, Eureka! In equation terms, the net force on an object immersed in a fluid becomes:

$$F_{net} = w_{object} - F_{buoyant} = (\rho_{object} - \rho_{fluid})V_{displaced}g$$

This tells us what we already knew, that an object with a greater density than a fluid sinks, and if the object's density is less than that of the fluid, it floats.

Let's apply these ideas to Experiment 15. The diver starts out just floating, so its density must be close to the density of water. But when you squeeze the bottle, a funny thing happens. Watch the water level inside the diver. More water flows in. The reason is that the pressure at the top is increased, so the pressure throughout the water increases, including the water pressure inside the diver. The air trapped inside the diver must also have its pressure increase, so this causes its volume to decrease. If there is less air inside the diver, the water rises, as you can see. This increases the mass inside the diver, so the diver's density increases, and it sinks. Release the pressure and it floats to the top.

The top layer is quite different from the body of a liquid. Instead of having attractive forces pulling molecules evenly in all directions, the very top has no molecules above, so this layer is pulled down, making a slightly more dense thin film. As a result, as long as the layer isn't penetrated, more dense objects may rest on the surface. This is referred to as *surface tension*.

Experimentino 12

Fill a wide-mouth container (I used a cat's water dish) with water. Then take a piece of aluminum foil about 3 cm by 3 cm (slightly over 1 in by 1 in) and place it on the surface of the water. It should float. The pressure (force/area) is not sufficient to penetrate the surface layer. Fold the foil piece in half and try again. Keep folding until it finally sinks. Now the pressure is more than the surface tension can withstand.

FLUID DYNAMICS

Fluids in motion are much trickier than fluids at rest, but a few restrictive assumptions yield relationships that give some insights into fluid flow. The first relationship is called *continuity*. In application, it is quite familiar. Here's another thought experiment: Suppose you're watering your garden with a hose that has no nozzle on the end. Further, let's say you've stretched the hose to its limit and are watering away happily. Suddenly you notice a thirsty plant a few meters (or feet) beyond your reach. Here are your choices: Put down the hose and go back to the shed, get more hose or a nozzle, turn off the water, connect the hose or nozzle, turn on the water, and water the plant. Or, you can do something right on the spot that doesn't involve moving at all. Your choice? If you're a lazy gardener like me, you'll go for the nonmoving solution. Put your thumb over the end of the hose. The water sprays out faster, so you can reach that faraway plant. You may also notice a similar effect in the flow of water in a river. When there are rapids (for you white-water enthusiasts), the river is either narrower or shallower or both. It must speed up to avoid bunching up behind the smaller section.

In terms of physics relationships, this translates into a requirement that the fluid doesn't pile up anywhere. This is referred to as *continuity*.

> The same amount of mass flows through any cross section of fluid flow.
>
> $\rho_1 v_1 A_1 = \rho_2 v_2 A_2$, where ρ is the fluid density,
> v is the fluid velocity,
> and A is the cross-sectional area

Wouldn't it be nice if traffic worked that way? When a freeway has fewer than normal lanes open, the traffic backs up behind the narrow spot. If traffic operated like a fluid, it wouldn't accumulate. Of course, the speed would increase through the narrow place. Lane restrictions are often due to construction, an accident, or some other cause that is hardly compatible with higher traffic speeds. Oh, well.

Another relationship that applies to fluid flow is an extension of energy conservation, first published by Daniel Bernoulli in his 1738 book *Hydrodynamica*.

Bernoulli's equation is very handy for explaining many different fluid flow phenomena, but before we explore the relationship and its applications, let's look at the circumstances of its formulation.

Daniel Bernoulli Mini-Biography[1]

(Note: The subject of this biography, Daniel Bernoulli, will be referred to as Daniel, since 90 percent of the other people discussed have the same last name. Clarity only is intended, not undue familiarity.)

Daniel Bernoulli

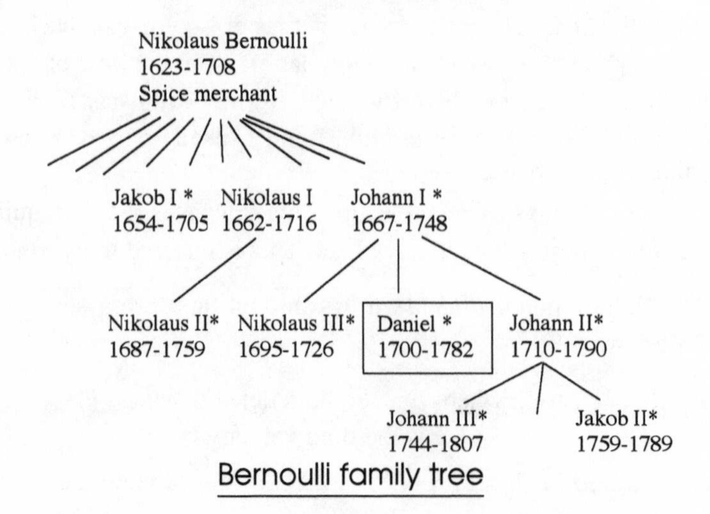

Bernoulli family tree

Originally, the Bernoulli family fled the Netherlands to avoid religious persecution for their Calvinist beliefs. Daniel Bernoulli's grandfather, Nikolaus, married into a family of wealthy spice merchants in Basel, Switzerland, and raised a family of ten children. Daniel Bernoulli's father, Johann, was the tenth child, and very strong-willed. Though Johann's father wanted him to go into the family business, Johann failed at his apprenticeship. The father was displeased but finally allowed Johann to study medicine as a compromise. Johann entered the University of Basel, where his older brother Jakob (child #5) taught mathematics. Ostensibly studying medicine, Johann got Jakob to teach him the latest mathematics—calculus—based on the work of Leibniz. He was a quick study and soon bragged about how he knew more mathematics than his brother, the professor. Jakob wasn't pleased, and the two of them quarreled, often in public. After a several-year tour of the Continent, including more than a year tutoring Guillame Marquis de L'Hôpital, Johann decided to finish his doctorate and start a family, so he needed a steady job. Since Jakob blocked him from any faculty position in Basel, he took a teaching job in Groningen, the Netherlands. His first son, Nikolaus, was born several months before the family traveled to Groningen, which turned out to be a difficult journey. Things didn't improve after they arrived. In France, de L'Hôpital published the first calculus textbook. Johann received scant mention but swore the book was based on notes from his tutorial sessions in France. In Groningen, Johann introduced experiments into his physics lectures, which offended both Calvinists—who thought sensory data should be used to ascertain God's plan—and Cartesians—who saw sensory information as greatly inferior to reason. Johann's next child, a daughter, lived only a few weeks, which saddened Johann greatly. Several student disputes of a religious or philosophical nature also occurred, as well as the ongoing controversy with his brother Jakob. In 1700 Johann's second son and the subject of this mini-biography, Daniel, arrived into this unhappy brood. It wasn't long before the Bernoullis left Groningen. Johann, his wife, their firstborn son Nikolaus, Daniel, and Johann's nephew Nikolaus (who had come to study math with his uncle) all set out for Basel. Johann had finally been offered an academic position—teaching Greek. During the journey, they learned that Jakob had died, so when Johann arrived, he lobbied hard and was rewarded with his brother's teaching position in mathematics. You might think this would be a good time to start playing the "happily ever after" music, but it didn't turn out that way. After a few years, Johann's third son, also named Johann, arrived. Johann busied himself with his mathematics and spent a lot of time writing letters supporting the Leibniz side of the Newton/Leibniz feud. Soon, it became time to guide his sons' choices for their life's work. The first and favorite son, Nikolaus, was already studying math, so Daniel was sent to the University of Basel to study logic and philosophy. He obtained a master's degree at sixteen but spent a lot of time studying mathematics with his older brother. Johann was determined that Daniel would become a merchant, not a mathematician, so he attempted to place him in an apprenticeship. But Daniel failed, preferring to continue his study of mathematics. Johann said there was no money in math but finally bent slightly and demanded Daniel study medicine instead. Daniel agreed, and studied in several places in Europe before completing his doctor's degree in Basel in 1720. His thesis dealt with the mechanics of breathing, based partially on his father's ideas about energy conservation. (Does this sound extremely familiar? Genetics strikes again.)

After failing to obtain an academic post at the University of Basel, Daniel journeyed to Venice, on his way to Padua to further his medical training. But he became ill in Venice and stayed there until he recuperated, working on mathematics and publishing a mathematics book. He also designed an hourglass that would continue to function on ships during heavy seas, and this won the Paris Academy Prize in 1725. Having acquired a measure of fame, he was offered an academic job. Both Daniel and his older brother Nikolaus were offered faculty positions in mathematics at a new academy at St. Petersburg, which they accepted. Unfortunately, Nikolaus died within a year after their arrival. Sadness at his brother's death and the harshness of the climate made life difficult for Daniel, but a very interesting thing happened, based on earlier developments back home. Several years prior, Daniel's father, Johann, did something entirely out of character. His old college roommate, Paul Euler (pronounced Oiler) had a son, Leonhard, who wanted to study mathematics rather than become a minister like his father. Johann convinced Paul Euler to let his son study mathematics. Just prior to Nikolaus's death, Leonhard Euler finished his doctorate at Basel. When he learned of Nikolaus's death, Johann recommended that Leonhard Euler should fill Nikolaus's post at St. Petersburg. Leonhard took the job, and Daniel was delighted. Daniel offered him a place to live in St. Petersburg, but, in exchange, Leonhard had to bring brandy, tea, coffee, and other delicacies from home. Thus began an incredibly fruitful collaboration for both of them. Leonhard's great analytical skills combined with Daniel's superior physical insights produced top-quality work in a wide variety of scientific and purely mathematical areas, including the harmonic vibrations of a stretched string and odd harmonics of an open/closed air column (chapter 12), hydrodynamics (chapter 13), and a remarkable anticipation (by one hundred years) of the kinetic theory of gases (chapter 14).

In spite of the productive collaboration with Leonhard Euler, Daniel was homesick and ready to leave St. Petersburg. In 1733 Daniel and his brother Johann, who had joined him in St. Petersburg, toured the continent as they journeyed back to Basel. Daniel finally landed an academic position there—teaching botany (well, better botany in Basel than blizzards in St. Petersburg, right?). Before leaving St. Petersburg, Daniel had submitted an entry into the Paris Academy competition that applied some of his ideas to astronomy. He arrived in Basel to find that he had won the prize. The only problem was that it was awarded jointly to his father, who had also entered. Johann was furious that his son could be considered an equal, so he wouldn't let Daniel live at home (the worst is yet to come). Daniel had left a copy of his major work, *Hydrodynamica,* at the printer's in St. Petersburg, but he continued to polish it for publication, finally getting it issued in 1738. His father also had a book published, *Hydraulica*, and although it didn't appear until 1739, he had it postdated and claimed Daniel had stolen the work from him. Ironically, the frontispiece of Daniel's work is signed "Daniel Bernoulli, son of Johann." The plagiarism was noticed before long and Daniel received appropriate credit for the work. After sixteen years of teaching botany and physiology, Daniel was finally appointed to the chair of physics in 1750. He was extremely popular and incorporated many experiments in his lectures. He taught until 1776 and received many honors from eminent scientific societies.

The Bernoulli family genes must be quite strong. In 2004 the chairman of the earth sciences department of the University of Basel was Professor Dr. Daniel Bernoulli.

Applying energy conservation to fluid flow, Daniel Bernoulli and Leonhard Euler derived very general equations and made simplifying assumptions to arrive at a result that provides fascinating insights.

For the smooth, steady flow of a frictionless, incompressible fluid, the total energy along a streamline of the flow is constant.

$p + \frac{1}{2}\rho v^2 + \rho gh = \text{constant}$, where p is the pressure, a measure of random kinetic energy;

$\frac{1}{2}\rho v^2$ represents normal kinetic energy;

and ρgh is gravitational potential energy

Because of the restrictive nature of the assumptions needed to solve the original equations, the Bernoulli equation doesn't apply to unsteady flow, turbulent flow, compressible fluids, or frictional cases, but it does provide interesting insights into a wide range of physical phenomena.

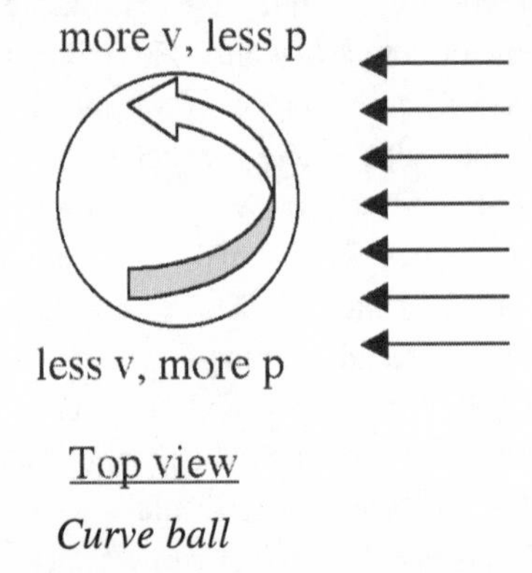

Top view

Curve ball

In the diagram, a ball moving from left to right has air flowing across it. If the ball spins counterclockwise as viewed from the top, the ball drags air with it on the left side, increasing its velocity. According to the Bernoulli equation, an increase in velocity must be accompanied by a decrease in pressure, so the pressure on the left side of the ball is smaller. The right side slows the air's velocity, so the pressure on the right side is greater. This imbalance in pressures leads to a net force from right to left, so the ball's path curves to the left. Baseball fans would note that this corresponds to the normal curveball thrown by a right-handed pitcher. If the batter is also right-handed, the ball curves away from him, possibly causing him to swing where the ball isn't. Strike. Work out what would happen for left-handers, and check TV slow-motion camera shots for examples. Curveballs also show up in soccer and whiffle ball, as you can see from Web sites listed in the endnotes.[2]

Backspin

Look at the previous diagram and imagine it showed a view from the right side of the gallery watching a golfer hit a shot toward the green. The counterclockwise spin causes the golf ball to rise higher than normal, and when it lands, the ball would roll forward very little and might even roll backward.

Topspin

In tennis, the goal is often to bring the ball down quickly so that it stays inbounds. To accomplish this goal, the spin direction must be reversed from the diagram, so the path of the ball would dip downward. To apply topspin, tennis players "brush up" the back of the ball as they hit it.

Airplane Wing

In normal flight, an airplane's orientation is such that it looks like it is going uphill. This is called an *angle of attack*. From the Bernoulli equation, if the air must go farther across the top of the wing than the bottom, it must also go faster across the top, leading to less pressure above the wing and more pressure below. The net force is known as *lift*. Actually, the situation is much more complicated, because the flow is not always smooth (laminar) and the turbulence affects flight, as does the net motion of air downward behind the wing and air flowing from below the wing to above, creating wingtip vortices.

Sail Boat

The sail on a sailboat functions like a wing, and the force generated helps to propel the boat forward. Another component of the force from the sail pushes sideways, and since the sail is high above the boat deck, this other component creates a torque that tends to rotate the boat. To counteract this torque, boats have daggerboards, centerboards, or keels, which generate a torque in the opposite sense. The other possibility for counteracting this torque is called hiking out. Sailboat passengers or crew hang their bodies out of the boat in the opposite direction to the boat's tilt.

Water Tower

Many communities have water towers from which water is pumped so that it can be distributed later with minimal pumping action. From the standpoint of the Bernoulli equation, this is an example of trading gravitational potential energy for kinetic energy. Pumping the water into the tower often occurs late at night, when there is excess electrical energy available because of minimal demand.

Beach Ball in Blower Stream

You may have seen a demonstration where a blower is aimed upward and a beach ball is suspended in the airstream. This exercise is probably designed to show the power of the blower, but it also reveals some interesting physics. The ball is remarkably stable. You can try to push it out of the airstream, but it just returns to the center. From the Bernoulli principle, air in the stream is moving faster than air outside the stream, so the higher pressure outside pushes the ball back.

Chimneys

When winter winds howl, smoke from the fire in the fireplace moves up the chimney and out into the air quite smartly. The wind causes reduced pressure above, pulling the smoke out.

Experimentino 13

Hold two sheets of paper vertically in front of your face. Blow between the sheets and watch to see if they move toward or away from each other. As the air velocity increases between the sheets, the pressure decreases, so they should move together.

FLUID FRICTION

It's time to confess. Fluids *do* have friction, and this friction is often described by the term *viscosity*. Viscous fluids seem thick and flow slowly (think molasses in January), whereas fluids that are nonviscous are thin or runny and flow quite easily (think water). There are equations that define viscosity, but their applicability is limited to fairly narrow ranges, so they will not be explored here.

Fluid flowing in a tube does not have a constant velocity across the whole cross-sectional area of the tube. At the boundaries, the velocity diminishes to zero, then increases over a small layer until it reaches the amount given by the Bernoulli equation. This is called the *boundary layer* and may have either laminar or turbulent characteristics. Smooth, laminar flow may be analyzed fairly easily, but turbulent flow is chaotic and requires statistical treatment. The greater the resistance at the walls, the more pressure is required to pump a fluid. For the human circulatory system, the clear implication is that smooth flow is preferable, but arterial deposits create more friction, so blood pressure rises and the risk of rupturing the arterial walls increases.

Bodies moving in the atmosphere also experience fluid friction, called *aerodynamic drag*.

> The drag on a body moving through the air is directly proportional to the air density, the body's velocity squared, the cross-sectional area of the body, and a dimensionless coefficient that is determined by the body's shape.
>
> $$\text{drag} = \frac{1}{2} \rho v^2 C_D A,$$ where ρ is the air density,
>
> v is the body's velocity,
> C_D is the coefficient of drag,
> and A is the body's cross-sectional area

The drag coefficient depends on the body's speed. It is close to constant at low speeds, rises to a maximum at Mach 1, then decreases at supersonic and hypersonic speeds. Theoretical values for drag coefficient are limited at best and often must be determined experimentally in wind tunnels. Some automakers advertise their car's drag coefficient, since aerodynamic loads account for most of the frictional resistance of cars at high speeds.

Skydivers know a lot about drag, because the drag force matches the weight of the diver when the speed has increased enough to reach what is called the *terminal velocity*. It is called terminal because the velocity doesn't increase beyond this value. Often, a diver assumes a spread-eagle position, which increases both the drag coefficient and the cross-sectional area. This makes for the slowest terminal velocity, enabling the diver to enjoy the ride longer than if he curled up into a ball and fell faster. Opening the parachute increases the cross-sectional area greatly, which reduces the velocity to a level that diminishes the impulse felt at impact. Note that divers crouch and roll at impact in order to maximize impact time, thus minimizing impact force.

You may wonder where the frictional work shows up in fluid friction. The work becomes the internal energy of the air and the body, increasing their temperatures. At low speeds, the effect is minimal, but at high speeds the effect can be very substantial, as evidenced by the thermal protection needed for reentry vehicles.

Experiment 16: Soda Bottle Crush[1]

Equipment needed: 2-liter plastic soda bottle, hot water, cold water

The atmosphere of gases that surrounds and sustains us is so ubiquitous that we often ignore it until a gust hits us in the face. This experiment is designed to illustrate the pressure normally exerted by the atmosphere, even without any gusts.

Procedure

- Prepare a container of cold water (ice cubes would be appropriate) large enough to submerge a plastic soda bottle.
- Fill a plastic soda bottle (2-liter, if possible) with water as hot as possible.
- Let the hot water stand in the bottle a while, then empty it quickly and cap the bottle tightly.
- Quickly submerge the bottle in the cold water and observe what happens to the bottle.

Chapter 14.
EVEN PERFECTION HAS ITS FLAWS[1]— THE IDEAL GAS

"The true ideal is not opposed to the real but lies in it."
—James Russell Lowell

"The ideal mother, like the ideal marriage, is a fiction."
—Milton R. Sapirstein

Gases have fascinated scientists through the ages. Even Newton took a shot at explaining gases, theorizing they were a collection of particles that repelled each other. He even derived the mathematical consequences of that idea. Even though he was wrong, his conjecture illustrates an appreciation that something was happening at a deeper level. The fact that gases' inner workings couldn't be observed directly made the situation particularly difficult. From a modern perspective, gases can be approached either from a macroscopic or a submicroscopic point of view. Historically, the macroscopic view was first, so that's where we'll start. There were early conjectures about the submicroscopic view, but they took some time to be developed. Linking the two points of view was eventually accomplished through statistical mechanics.

MACROSCOPIC PERSPECTIVE

Chemists were especially interested in making measurements to determine the macroscopic properties of gases. But there are many different gases, and it was unclear how similar they might or might not be, so many experimental studies had to be made before any general conclusions could be drawn.

Thermometer

Galileo invented a water-based thermometer in 1593, but it was Daniel Gabriel Fahrenheit who perfected the mercury-in-glass thermometer in 1714 that simplified temperature measurements. Many different temperature scales were used, including one devised by Fahrenheit himself. There are several temperature scales in modern use, all of which require two fixed points. The two scientific scales are the centigrade or Celsius (°C) scale and the kelvin (K) scale. On the centigrade scale, water freezes at 0 °C and boils at 100 °C. The kelvin scale simply adds 273 to the centigrade reading, so water freezes at 273 K and boils at 373 K. Another scale in common use, the Fahrenheit (°F) scale, has fixed points of 0 °F for the freezing point of a saturated solution of salt (NaCl) in water, and 100 °F as slightly above normal internal human body temperature. The method used by Fahrenheit to arrive at these fixed points is not clear. Converting from Fahrenheit to centigrade uses the formula °F = 9/5 °C + 32.

Barometer

In 1644 Evangelista Torricelli invented the barometer by inverting a mercury-filled tube into a dish of mercury open to the atmosphere. Atmospheric pressure was measured by the height of the mercury column in the tube. This device has formed the basis for other pressure-measuring instruments. Many different pressure units have been used, beginning with the height of the mercury column. The pressure exerted by the atmosphere at sea level is given by:

> 1 atmosphere = 760 mm of mercury = 29.92 in of mercury = 101 kPa = 14.7 lbs/in^2 (psi) = 760 torr = 2120 lbs/ft^2

Sometimes, pressures are stated in terms of gauge pressure, which is the amount of pressure above atmospheric pressure, since many gauges are open to the atmosphere. Absolute pressure is obtained by adding atmospheric pressure to gauge pressure. For example, if the pressure in your tires reads as 32 psi, that is a gauge pressure, and the absolute pressure is 46.7 psi. Pressure is exerted by the atmosphere in many situations that we seldom notice. For example, in sucking liquid through a straw, the sucking action isn't what forces the liquid to rise; it's atmospheric pressure. If you would like to conduct an experimentino to illustrate this, try sucking liquid directly from a soda bottle (but don't tell anyone it is a physics experiment, because it is doomed to failure).

To discover the macroscopic properties of gases, measurements were carried out from the late 1600s to the 1800s.

Constant Temperature

Robert Boyle (1627–1691) was an Irish natural philosopher who conducted experiments (with the help of Robert Hooke) that showed the pressure of any gas is inversely proportional to its volume, as long as the temperature and amount of gas is unchanged.

> If a given amount of gas is held at constant temperature, the product of the gas's pressure and the volume it occupies is a constant.
>
> Boyle's law
>
> pV = constant for constant amount and temperature,
> where p = pressure and V = volume

Boyle's law and Pascal's principle from chapter 13 complete the explanation of Experiment 15. Squeezing the bottle causes additional pressure throughout the water. Within the diver, the water is in contact with trapped air in the top of the diver. Increased water pressure also increases the air pressure. According to Boyle's law, higher pressure reduces the volume of air trapped in the diver. The volume of water inside the diver thus increases, increasing the diver's mass. This, in turn, increases the diver's density, so the diver sinks when its density surpasses the water density.

Constant Pressure

Jacques Charles (1746–1823) was a French balloonist (hydrogen) and inventor who conducted experiments in 1787 that showed that the ratio of any gas's volume to its temperature remains constant as long as the amount of gas and its pressure are held constant. Charles didn't publish his result—possibly he was too busy ballooning. His first ascent was only the second hydrogen balloon ascent ever made. The Montgolfier brothers had made numerous hot air balloon flights just prior to this hydrogen effort.

> If a given amount of gas is held at constant pressure, the quotient of the gas's volume and its temperature is a constant.
>
> Charles's law
>
> $\frac{V}{T}$ = constant for constant amount and pressure,
> where V = volume and T = temperature

Constant Volume

In 1802 another French balloonist, Joseph-Louis Gay-Lussac (1778–1850), published Charles's law, giving Charles full credit. Gay-Lussac later added a gas law of his own discovery, namely, that the pressure of a gas is directly proportional to its temperature as long as a fixed amount is held at constant volume. His ballooning allowed him to study the atmosphere directly and obtain samples at various altitudes. Ascending in a balloon left over from Napoleon's Egyptian military campaign, Gay-Lussac and physicist Jean Baptiste Biot rose to a height of 7,000 meters (23,000 feet), establishing an altitude record that stood for fifty years. Among his many activities, Gay-Lussac helped to discover boron in 1808 and, the following year, married a young draper's assistant he had caught reading a chemistry textbook under the counter.

If a given amount of gas is held at constant volume, the quotient of the gas's pressure and temperature is a constant.

Gay-Lussac's law

$\frac{p}{T}$ = constant for constant amount and volume, where p = pressure and T = temperature

Putting all these laws together yields a single combined gas law:

combined gas law

For a fixed amount of gas, $\frac{pV}{T}$ = constant

Italian chemist Amadeo Avogadro (1776–1856) added an important piece to the puzzle in 1811 with the hypothesis that, at the same temperature and pressure, equal volumes of gases contain the same number of molecules, regardless of other chemical or physical properties. Avogadro's work was mostly ignored. The idea of a molecule, much less an atom, was not well accepted by the scientific community at this time.

Submicroscopic Perspective

Let's start with a simple-looking case and build our way up to a complete gas. Suppose there is a particle of mass m, moving back and forth between two walls. At the moment we look at it, the ball is moving with velocity v and is headed toward the right.

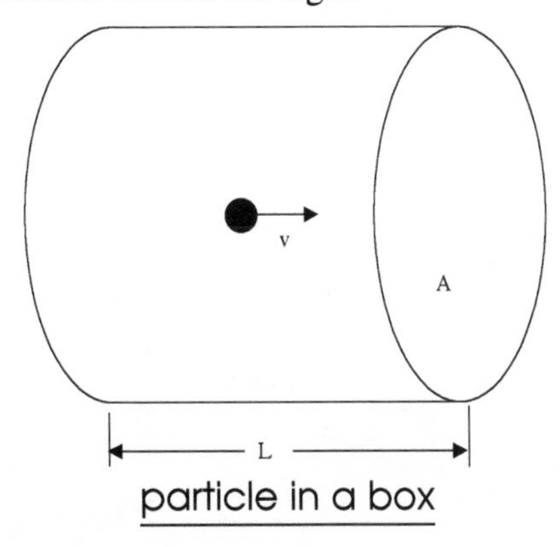

particle in a box

When this particle arrives at the right-hand wall, there is a collision. The particle accelerates backward because it experiences a force due to contact with the wall. The relationship between the force and acceleration is Newton's second law: F = ma. If we assume the collision is perfectly elastic, then after the collision, the particle will be moving with velocity v to the left. This allows us to find the acceleration:

$$a = \frac{\Delta v}{\Delta t} = \frac{v_{left} - v_{right}}{\Delta t_{collision}} = \frac{2v}{\Delta t_{collision}}$$

(Don't make the mistake of thinking $\Delta v = 0$. The particle doesn't go through the wall; it bounces back.) So the wall exerts a force on the particle and the particle exerts a force on the wall (Newton's third law).

$$\text{The force is } F = \frac{2mv}{\Delta t_{collision}}$$

If the particle then moves to the left, collides with the left wall in a perfectly elastic collision, then heads back toward the right, it won't be long before the right wall experiences another force. Graphically, this periodic force would look like this:

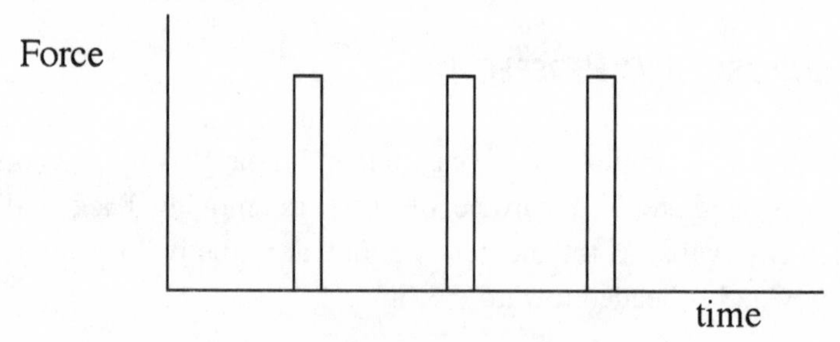

force exerted by particle

The time between force spikes would be the time for the particle to make a trip from one wall to the other and back: 2L, divided by its constant speed, v. To find the average force exerted on the wall, let's use the average time between collisions rather than the impact time.

$$\text{This yields } F_{average} = \frac{2mv^2}{2L}$$

Now for a little mathematical manipulation. Divide both sides by the wall's cross-sectional area, A.

$$\frac{F_{average}}{A} = \frac{mv^2}{LA}$$

The left side is the average pressure, and the denominator of the right side is the volume, so this equation becomes $p_{average}V = mv^2$. If we now generalize by allowing N particles to be within the box, and assume

$\frac{N}{3}$ move left and right, $\frac{N}{3}$ move up and down, and $\frac{N}{3}$ move forward and back

and further assume that the temperature of a gas is proportional to the average kinetic energy of its molecules, the equation becomes

ideal gas law: $pV = NkT$, where p, V, and T have their usual meaning,
N is the number of molecules, and
k is the Boltzmann constant

Assumptions made along the way were that the particles (molecules) occupy zero volume, are moving randomly, interact with each other only through perfectly elastic collisions, and have an average kinetic energy proportional to the temperature. See why it's called ideal? This *kinetic theory of gases* is an extension of the ideas of Daniel Bernoulli from the 1730s. James Clerk Maxwell, of electricity fame, and Ludwig Boltzmann revisited the problem of the inner structure and dynamics of gases in the mid-1800s. Boltzmann's statistical mechanics bridged the submicroscopic world of molecules and atoms and the macroscopic ideas of pressure, temperature, and volume, but these ideas were not appreciated immediately because atoms and molecules were still not accepted by mainstream scientists.

Chemistry often uses another form of the ideal gas law:

ideal gas law: $pV = nRT$

The quantities p, V, and T have the meanings discussed earlier, R is a universal gas constant needed to make the units balance, but n is unusual. It is the number of *moles* of the gas. Moles are found by taking the mass of the gas in grams and dividing by the molecular mass. For example, in a sample of 88 g of carbon dioxide (CO_2) there would be 88/44 = 2 moles of the gas. Each mole consists of a particular number of molecules of the substance, called *Avogadro's number* = 6.02×10^{23}. The huge size of this number tells us that even a small sample of gas would involve so many molecules that treating them statistically would be a valid approximation.

Let us discuss moles. The concept is quite useful, but the name has puzzled students for years.[2] It gets confused with molecules, skin eruptions, furry animals that burrow under your lawn, and spies under deep cover. Besides, moles were introduced by the chemist Wilhelm Ostwald, who led a vigorous opposition to Ludwig Boltzmann's statistical mechanics because he opposed the atom on philosophical grounds. How can you use moles and not accept atoms? Boltzmann and Ostwald were colleagues at Leipzig in 1902, when Boltzmann attempted suicide.

REAL GASES

If some of the assumptions made in kinetic theory are relaxed, the ideal gas more closely approaches a real gas. For example, assuming that molecules are points that occupy no volume is clearly inaccurate, especially if the gas is dense. In addition, presuming that molecules don't interact with each other is also wrong. Correcting for both of these inaccuracies leads to the van der Waals equation of state for a gas:

$$\left(p+\frac{an^2}{V^2}\right)(V-nb)=nRT$$

where a and b are constants that depend on the specific gas. This equation is a better approximation of the behavior of real gases at high pressures and low temperatures, but it is far more difficult to work with mathematically.

The assumption of perfectly elastic collisions is also unrealistic at low temperatures. The collisions become less violent and the attractive force between molecules becomes dominant, so the molecules stick together. Take the limit of a very low temperature and let electrically sticky molecules collide perfectly *inelastically*, and what have you got? A solid. Electrical bonds hold molecules together and they simply vibrate around fixed positions like so many masses on springs. So if you think of solids as molecules interacting with each other because they are stuck together and gases as molecules that collide so hard that they don't stick together at all, what does that make liquids? They represent a transitional zone between solids and gases. Some of the molecules stick together, giving liquids a fixed volume, and some don't, allowing the liquid to assume the shape of the container.

The situation gets even more fascinating at high temperatures. Collisions between molecules become more violent, and, eventually, parts of the molecules break loose, the way demolition derby cars come apart. The pieces that come off first are the outermost ones—electrons—and what is left are molecules and atoms with electrons missing. These are known as *ions*. So, at high enough temperatures, what used to be a gas now contains ions and free electrons. We'll see more about electrical forces in chapter 16, but you can probably imagine this collection won't behave like an ideal or even a real gas. It is called a *plasma*. This fourth state of matter is present in our environment in the form of plasma TVs, fluorescent lamps, neon signs, and lightning. So there isn't that much plasma around then? Oh, yes. Stars consist of plasma, as do the interplanetary medium, the interstellar medium, and the intergalactic medium. In fact, matter in the universe is mostly plasma. Solids, liquids, and gases account for an extremely small fraction of matter.

Experiment 17: Kitchen Physics

Equipment needed: 2 Styrofoam cups, ice, water, aluminum foil, spring scale

Heat energy can either raise the temperature of matter or change its state. The purpose of this experiment is to illustrate both processes and the conservation of energy at the same time.

PROCEDURE

- Punch a hole near the lip of an insulated cup so its mass can be measured by the spring scale.
- Place ice cubes in the insulated cup to make a mass of approximately 10 grams.
- Allow enough time for the ice cubes to melt slightly. Pour off the liquid water and measure the mass of the ice cubes.
- Multiply the ice cube mass by 10 and measure that mass of water in the other insulated cup.
- Pour the liquid water into the cup with the ice and cover with aluminum foil.
- Wait several minutes and check to see if all the ice has melted.

Chapter 15.
SOME LIKE IT HOT—THERMODYNAMICS

"Heat is in proportion to the want of true knowledge."
—Laurence Sterne

"It ain't the heat; it's the humility."
—Yogi Berra

Heat. A houseful, a roomful—you can't catch a spoonful. Just like gases, the deeper nature of heat is very difficult to analyze because no parts are visible. Further, heat flows from one body to another and is somehow tangled up with fire and temperature, so it has been confusing to sort out these concepts. The first recorded ideas came from Heraclitus (around 500 BCE), who thought that fire was the most important of the three elements: fire, earth, and water. Diogenes wrote of Heraclitus's view: "Fire is the element: all things are an exchange for fire." The confusion was long-lasting. The next recorded thoughts showed up in 1620, when Francis Bacon wrote: "The very essence of heat, or the substantial self of heat, is motion and nothing else." Robert Hooke proposed in 1665 that "heat is a property of a body arising from the vibration of its parts."

The question was, what parts were moving? The idea of atoms was considered "godless" and therefore not easily entertained without serious evidence, which was a long way off. Instead of this kinetic theory, a new hypothetical substance was invented to explain combustion and heat flow. Johann Becher suggested something that had no odor, taste, or weight but that is emitted when substances are burned. He called this hypothetical substance "inflammable earth." Georg Ernst Stahl modified this idea somewhat in 1718 and gave it a much more colorful name, *phlogiston*. One of phlogiston's chief supporters was Joseph Priestley, the full-time minister and part-time dabbler in science who discovered oxygen. The phlogiston theory expired in 1775, when Antoine Lavoisier demonstrated, by extremely accurate weighing of reacting elements, that some substances gain mass when they burn. If burning a substance loses phlogiston, but the substance's mass increases, phlogiston would have to have both positive and negative mass. In place of phlogiston, Lavoisier proposed a "subtle fluid" called *caloric* as a substance of heat whose total amount in the universe is constant and flows from warmer to cooler bodies. Lavoisier conducted substantial research in many different topics in chemistry, and was aided materially by his wife, Marie-Anne, who was thirteen when they married (more about her later).

Near the end of the 1700s, there were two theories of heat: the caloric theory and the kinetic theory. In the absence of strong experimental support for either, both theories were held as being equivalent, although caloric theory was more recent and therefore more modern. This was about to change, and the change came from an unexpected direction.

Benjamin Thompson, Count Rumford Mini-Biography[1]

Benjamin Thompson, Count Rumford

More than a hundred years had elapsed since the Thompson family started farming in Massachusetts. Into such an unremarkable family Benjamin Thompson was born in 1753. Who would have dreamed he would become such a flamboyant character? Benjamin's father, also named Benjamin, died within two years, and Benjamin's mother remarried quickly. Young Benjamin became very self-reliant at a very early age. Although he excelled at scientific studies and expressed himself with great facility, Benjamin didn't progress very far in school. Yet he resolved he would never be a farmer. Thompson went to Salem at age thirteen to become a shopkeeper's apprentice. One of his duties involved working with gunpowder, and a batch of it exploded, injuring Thompson seriously. After he recuperated, he went to Boston to become a doctor's apprentice. That didn't work out, so he decided to try his hand at teaching and sat in on a few scientific lectures at Harvard. Thompson then moved to Rumford, New Hampshire (now Concord), to take a teaching job. In Rumford, he lived with the new school's headmaster, Timothy Walker. Walker was delighted to have this tall, handsome new teacher with such genteel manners. Sarah Walker Rolfe, the minister's recently widowed daughter, was even more taken with Thompson. She was fourteen years his senior and the largest landowner in Rumford. Thompson was also pleased, because this time he hit the jackpot. Within months of his arrival, Thompson was married to Sarah. Moving easily into Sarah's circle of friends, Thompson became almost instantly acquainted with New Hampshire's governor, John Wentworth. The governor was mightily impressed and appointed Thompson a major in the local militia. Soon, Sarah delivered a daughter, Sarah, called Sally. With the colonist/loyalist conflict brewing, Thompson's sympathy for England created great animosity within his militia command. Things finally grew so bad that he abandoned his wife and daughter in Rumford and went to Boston, eventually leaving with the British troops for England. There, he clerked for the secretary of state, Lord Germain, working his way up the ladder to become undersecretary of state. (With the Revolutionary War raging, doesn't it seem strange that a young colonial militia officer would be welcomed into the heart of the British war effort, becoming an assistant to the secretary of state?) In his spare time, Thompson continued his scientific hobbies, particularly experiments with gunpowder, and was elected to the Royal Society in 1779. After Lord Germain's forced resignation—he was the least respected minister of Lord North's unpopular administration—Thompson was reassigned. Given the rank of lieutenant colonel, he was sent back to the colonies for a brief stint at a British garrison in Manhattan. The war concluded soon—in favor of the colonials—and Thompson scurried back to England. Before long, Thompson became bored and requested permission to travel to the Continent to seek military action, but not for England. Thompson was encouraged by the granting of his travel request and a half-pay pension, and was unexpectedly knighted by King George III (here's your hat, what's your hurry?). In Europe, Sir Benjamin Thompson made contact with Elector Karl Theodor of Bavaria, who was just about to take office in Munich. Sir Benjamin became a special confidante to the elector and ultimately his principal adviser. During his time in the elector's service, Sir Benjamin reformed the Bavarian military, rid the streets of beggars by giving them jobs making military uniforms, and obtained nine hundred acres of prime land for a beautiful park in the center of Munich that became known as English Garden. In his spare time, he continued his scientific pursuits by performing experiments and writing papers for the *Philosophical Transactions*. The elector rewarded his efforts by making him a count in 1791. Sir Benjamin chose the title Count Rumford to honor the site of his first success. A short time later, both Austrian and French armies were on the move near Munich, and the elector fled for his life. Count Rumford stayed and, after a few tense weeks, he and his newly outfitted and trained army convinced the passing armies to keep right on going.

Back in New Hampshire, Rumford's wife died, but their daughter, Sally, had reached her majority. In 1797 she visited her father in Munich and was made Countess Rumford by the elector, possibly the first American-born countess.

In one of his many duties for the elector, Count Rumford supervised a factory where brass cannons were manufactured. Drilling out the barrel was a hot process that required a lot of cooling water, and Rumford noticed that the heating continued as long as the drill was turning, even if no metal was removed. According to the caloric theory, caloric was leaving the brass, but since the heating effect appeared to be limitless, Rumford concluded that there was no such thing as caloric—the mechanical action of the drill was converted to heat. He wrote a paper for the Royal Society titled "An Experimental Enquiry concerning the Source of the Heat which Is Excited by Friction" (Rumford certainly made the best of a boring job). This paper spelled the beginning of the end for caloric theory, whose champion, Lavoisier, had been beheaded in 1794 during the French Revolution.

Not long after, the elector died. Since he had been roundly disliked by the Bavarians, this appeared to be a good time for Count Rumford to move along. He went back to England and helped found the Royal

Institution (not to be confused with the Royal Society, which has similar aims, but is older). The first lecturer of the Royal Institution, Humphry Davy, was firmly on the side of the kinetic theory, saying heat was the "vibration of the corpuscles of bodies." Before long, Count Rumford had personality conflicts at the institution, so he left. He traveled to Paris, where he attended entertainments given by Lavoisier's wealthy widow. Their friendship blossomed into marriage, but their styles turned out to be quite different. She enjoyed the social life, and he liked to tinker with household gadgets. The quarrels started almost immediately, and before long they separated. Rumford took up residence in Lavoisier's old villa in the Paris suburb of Auteuil. With many visits from his spinster daughter, Sally, he puttered around for several years, then died suddenly in 1814. One biographer said Count Rumford "had the most unpleasant personality in the whole of science since Isaac Newton." On the other hand, Franklin Delano Roosevelt said, "Thomas Jefferson, Benjamin Franklin, and Count Rumford are the three greatest minds America has produced."

The caloric theory was finally put to rest completely by Sir James Prescott Joule in 1849. Through careful experimental work, he obtained an accurate determination of the amount of mechanical work that is the equivalent of heat. In recognition, the work (and energy) unit was named after him (1 calorie = 4.18 joules). The term *heat energy* is often used to reinforce the equivalence of mechanical work, kinetic or potential energy, and heat.

Before we delve into classical thermodynamics, where macroscopic properties are used exclusively, let's continue at the submicroscopic level to see what happens to the molecules when heat energy interacts with matter.

Heat Energy Is Added to a Solid and It Remains Solid

Kinetic Energy Increase: At the submicroscopic level, the molecules gain kinetic energy through collisions. When kinetic energy increases, temperature increases.

Heat energy added to a solid that remains solid raises the temperature, and the heat energy required depends on the body's mass and the details of its molecular structure.

$Q = mc\Delta T$

where Q = heat energy added
m = body's mass
c = coefficient of specific heat
ΔT = temperature increase

This is the familiar effect in which a block of frozen food heated in a saucepan increases its temperature. The bigger the block, or the greater the food's specific heat, the more heat energy it takes.

Substance	Specific heat, J/kg °C
Aluminum	900
Copper	387
Lead	128
Wood	1,670
Ice	2,050
Marble	858
Gold	130

Potential Energy Increase: When a solid's kinetic energy increases, the molecules move faster and also range farther from their equilibrium positions, so the entire solid expands slightly. The amount of expansion in one dimension depends on the original length, the change in kinetic energy (and hence, temperature), and a constant depending on the characteristics of the molecules that make up the body.

A temperature increase causes a body to expand in length, and the change in length is directly proportional to the body's change in temperature, its original length, and a constant that depends on the molecules of the body.

$\Delta L = \alpha L_0 \Delta T$

where α = coefficient of thermal expansion
L_0 = original length
ΔT = temperature change

Substance	$\alpha \times 10^{-6}/°C$
Aluminum	22.6
Copper	16.5
Lead	28
Marble	12
Gold	12.2
Wood	5.0
Concrete or brick	12

From a practical standpoint, the length change is usually so small that it is hardly noticeable, except for large temperature changes or long objects. For example, a slab of concrete in a highway or a steel girder in a bridge may exhibit length changes of several centimeters (an inch) from winter to summer. Strips of different materials inserted into long runs of concrete or steel allow expansion and contraction to occur, but occasional temperature extremes beyond the design limits may cause unwanted effects such as buckling or cracking. The length change constitutes a strain, which causes stresses that may exceed the material's capacity. Other examples include railroad tracks and oil or gas pipelines. You might even notice long electric wires sag a little more on a hot day, if you're not too busy sagging as well.

A clever technological use of linear expansion is the *bimetallic strip*. Two metals with different coefficients of expansion are bonded together to form a strip. As the temperature of this strip rises, the metal with the larger expansion coefficient expands more than the other; this makes the whole strip bend toward the lower coefficient metal, the way runners on the outside lanes of a track must run farther. This metal strip may be used as part of an electrical circuit, where the bending breaks contact to turn off the circuit. This can be seen in many homes that use old-style round thermostats. When the coiled-up bimetallic strip inside the thermostat housing cools off sufficiently, it bends the other way, restores contact, and turns on the burner and blower.

Two- and three-dimensional expansion has enabled a clever manufacturing technique. Metal parts are heated before they are assembled. When they cool off, they fit tighter.

Heat Energy Is Added to a Solid and It Becomes Liquid

Instead of adding kinetic energy through collisions, the heat energy breaks bonds between molecules in the solid. Some molecules are thus free to move about, while others are still bonded. This state of matter is *liquid*, and the molecules' mobility allows them to move around sufficiently so they conform to the shape of the container. The amount of heat energy required to break the solid's bonds sufficiently—to form a liquid—depends on the amount of material involved and the molecules' bond strength.[2]

Heat energy added to a solid that becomes liquid breaks bonds within the solid without increasing the temperature; more heat energy is required depending on the body's mass and the bond strength of its molecular structure.

$$Q = mL$$

where Q = heat energy added
m = body's mass
L = latent heat of fusion

The term *latent heat* is used to imply a change in potential energy as opposed to the kinetic energy change that would accompany a temperature change. The temperature remains constant while the bonds are being broken. Think about it this way: if the kinetic energy tried to increase, it would just cause more collisions that would break more bonds.

Substance	Melting point (°C)	Latent heat of fusion kJ/kg
Water	0 (32 °F)	334
Sulfur	115 (239 °F)	54
Oxygen	–219 (–362 °F)	14
Lead	327 (621°F)	24
Gold	1060 (1940 °F)	64.5

Once *all* the solid's molecular bonds are broken, the substance becomes pure liquid. Adding energy then increases the collisional kinetic energy and thus the temperature. The temperature increase continues until collisions begin breaking the liquid bonds, and another phase change occurs.

Do you see the pattern that's developing? Let's summarize the situation by drawing a heating curve for water. If you start with a chunk of ice at –20 °C and keep adding heat energy, here's what would occur.

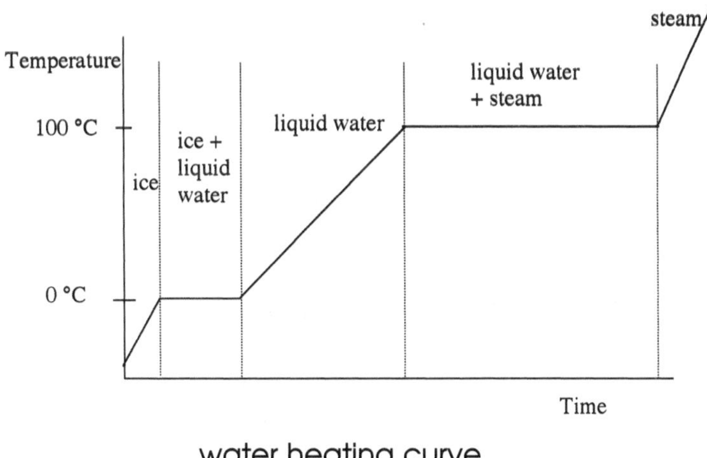

water heating curve

This logic sets up a process called *calorimetry*. The heat energy lost by one substance is gained by another. In Experiment 17, boiling water at 100 °C lost heat energy, while ice at 0 °C gained heat energy. Liquid water's loss of kinetic energy was used to break molecular bonds in ice, changing its phase to liquid water. If we put in the numbers, it turns out that it takes an amount of liquid water that has 80 percent of the mass of ice to melt the ice completely, given that no heat energy is lost to anything else.

THE LAWS OF THERMODYNAMICS

Thermodynamics studies the macroscopic effects of heat energy, internal energy, and work on a system. If this seems vague, that's because it is intended to be general. To make the idea clearer, we will study the laws of thermodynamics, using a heat engine or some part of one.

First Law of Thermodynamics

Head energy added to a system is converted completely into an increase in internal energy of the system or work done by the system.

$$\Delta Q_{added} = \Delta U_{internal} + W_{done\ by\ system}$$

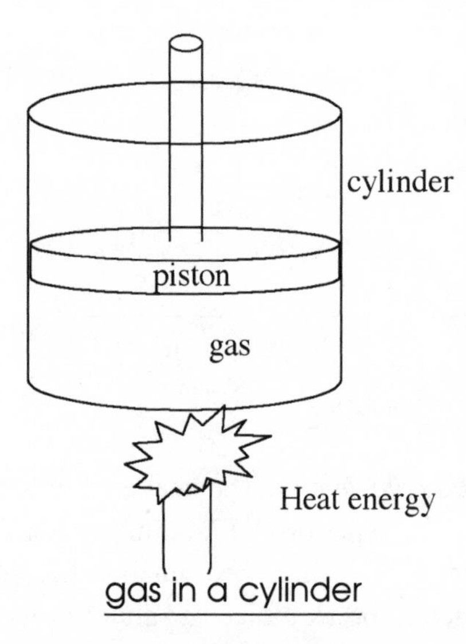

Doesn't this seem reasonable? Put energy into a system and it either stays within the system or comes out in the form of work. It's just another way of stating the *principle of conservation of energy*. To see how this works, let's look at an example. For a system, let's use an ideal gas, contained in a cylinder by a piston and heated by a burner.

The first law of thermodynamics says that the heat energy added to the system increases the gas's internal energy and does work by raising the piston. Earlier (chapter 14), we noted that adding heat energy raised the temperature of a *solid* body according to $Q = mc\Delta T$. That worked just fine for a solid because the molecules occupy fixed average positions, so all the heat energy went into increasing the kinetic energy of the molecules rather than pushing them around and doing any work. (There is a tiny expansion, but it is negligible for these purposes.) With a gas, it's not so simple. The molecules are not fixed in their positions, so what happens depends on the details of the thermodynamic process. Let's look at several cases.

Constant Volume Process

If the piston is prevented from moving, then the gas's overall volume is constant and all the heat energy goes into raising the kinetic energy of the gas. The relationship that describes this process is a little different from the equation that applied to solids:

$\Delta Q = nC_v\Delta T$, where n = moles,
C_v = molar specific heat at constant volume, and
ΔT = temperature change

Since the piston is fixed in place, no work can be done, so the first law becomes:

$$\Delta Q = \Delta U = nC_v\Delta T$$

Constant Pressure Process

If the piston was released but the pressure was kept constant, the heat energy input will be divided between raising the internal energy as it did before and pushing the piston up, accomplishing work. As we know from chapter 5, work is the force exerted times the distance moved. If the piston moves up an amount Δy, the work done is $F\Delta y$, or in terms of standard gas quantities, $pA\Delta y$. Geometrically, $A\Delta y$ is just the change in volume, ΔV, so the work done is $p\Delta V$. The first law becomes:

$$\Delta Q = nC_v\Delta T + p\Delta V$$

Since the system is an ideal gas, $p\Delta V = nR\Delta T$, so the first law yields $C_P = C_V + R$.

$\Delta Q = nC_p\Delta T$, where n = moles,
C_p = molar specific heat at constant pressure, and
ΔT = temperature change

Constant Temperature Process

If the gas inside the cylinder is kept at constant temperature—perhaps by immersing the whole apparatus in an atmosphere that maintains a single temperature—then the ideal gas obeys the relation pV = constant. Since constant temperature means that the gas's internal energy remains the same, the first law of thermodynamics tells us that all the added heat energy goes into work done by the gas.

Zero Heat Energy Input Process (Adiabatic)

If the heat energy input is simply turned off, then the first law of thermodynamics tells us that the sum of the internal energy change and work done must add up to zero. In other words, any work done must be accompanied by a decrease of internal energy. This is similar to the energy conservation we saw earlier where kinetic and potential energies each vary, but their sum remained constant.

A technique that helps us to visualize these processes is a pV diagram:

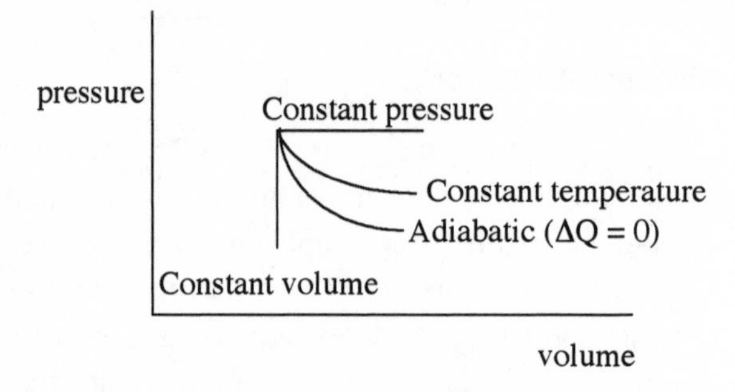

pV diagram

If thermodynamic processes are performed in sequence—bringing a gas through a series of steps that return the gas to its original state—this is called a *cycle*, and the device that accomplishes these processes is called a *heat engine*. On the pV diagram, a cycle forms a closed figure, and the work done is the area enclosed by the figure. As an example, let's look at the *Otto cycle*, an idealized version of the process that occurs in automobile internal combustion engines:

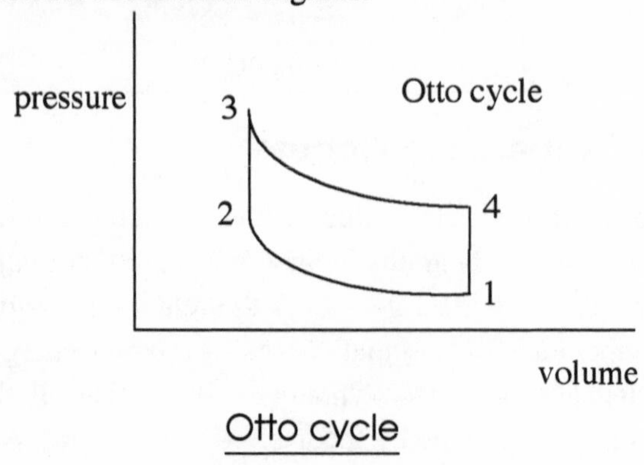

Otto cycle

From 1 to 2, air is compressed with no heat addition; from 2 to 3, the fuel/air mixture is ignited, generating heat energy input at constant volume; from 3 to 4, the air expands with no heat input, pushing the piston (power stroke); from 4 to 1, waste heat is exhausted, returning the air to its original condition (many details are omitted from this simplified treatment). Other cycles also exist, and there is a way of rating them on a common scale known as *efficiency*, which is given the symbol e.

> The efficiency of a heat engine operating in a cycle is the amount of work done by the engine divided by the amount of heat energy input.

Making the idealizations that a heat engine works between two sources called *reservoirs*, which are so large that their temperatures are unaffected by losing or gaining energy, we use a diagram to summarize the operation of heat engines:

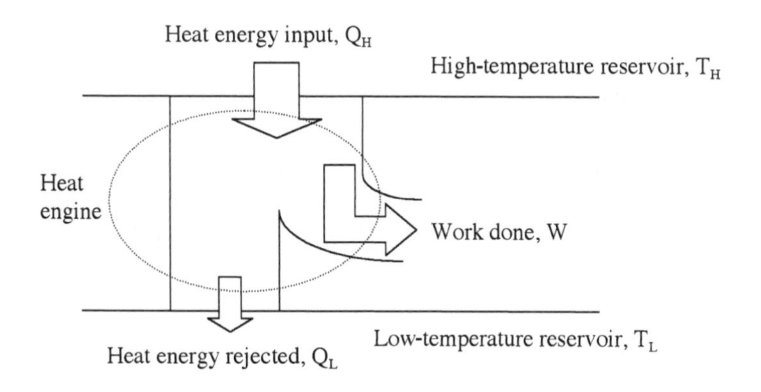

heat engine diagram

In terms of quantities shown in the diagram, the heat engine's efficiency is

$$e = \frac{W}{Q_H}$$

If we assume ideal conditions, an Otto cycle might achieve an efficiency of approximately 50 percent, but power train, friction, and aerodynamic losses bring the overall automobile efficiency to around 10 percent.

There is a way of determining the maximum efficiency of an idealized heat engine. In 1824 Sadi Carnot (1796–1832), a French engineer on leave from the military, wrote a popular book to explain and analyze engines, especially steam engines. The steam engine had been around for a while, but its design and function were more a matter of intuition than careful engineering. Within his book, Carnot idealized the concept of a heat engine and found an expression for its maximum possible efficiency:

$$\text{maximum possible efficiency} = \frac{T_H - T_L}{T_H}$$

This development was particularly remarkable, since the first law of thermodynamics was not yet known, and Carnot based his ideas on the now-defunct caloric theory. Although his ideas had little immediate impact, their significance became recognized later when they were reformulated and integrated into more general thermodynamic theory.

An interesting variation is what happens if a heat engine is run in reverse. Heat energy is taken from a cold reservoir, work is added to the system, and heat energy is exhausted to a higher-temperature reservoir. Does that sound strange or familiar? It is a *refrigerator*. The quantity used to measure a refrigerator's efficiency is called its *coefficient of performance*, COP, and it is the amount of heat energy taken from the cold reservoir divided by the amount of work needed to be done to accomplish this transfer:

$$\text{COP} = \frac{Q_L}{W}$$

The higher the coefficient of performance, the less work is required to extract heat energy from the cold reservoir. Practical refrigerators have COP values that range from about 2 to 8. The Carnot COP is the ratio of the temperature of the low-temperature reservoir and the difference in temperature between the high-temperature and low-temperature reservoirs.

$$\text{Maximum possible COP} = \frac{T_L}{T_H - T_L}$$

Second Law of Thermodynamics

The first form of the second law is the Kelvin-Planck statement:

> It is impossible to take heat energy from a high-temperature reservoir and convert it entirely to work. Some heat energy must be rejected to a low-temperature reservoir.
>
> In terms of heat engine efficiency, e is always less than 1.

Another way to state the second law is due to Rudolf Clausius (1822–1888):

> It is impossible for heat energy to flow spontaneously from a low-temperature reservoir to a high-temperature reservoir without any work being done on the system.
>
> In terms of refrigerator coefficient of performance, the COP can never become infinite.

Yet another way to express the second law is in terms of the often misunderstood quantity *entropy*.

The roots of entropy relate to the idea of conservative forces discussed in chapter 5. A force is conservative if potential energy can be generated, and the energy's value depends only on the starting and ending points rather than on any details of the body's path. Similarly, one thermodynamic quantity is a function only of end-points: the internal energy, which depends only on temperature. It doesn't matter how a thermodynamic system gets from one temperature to another; the change in internal energy is the same. Work, on the other hand, is strongly dependent on the detailed path. If points 1 and 3 in the Otto cycle are linked by two different pathways, 123 and 143, the upper pathway involves much more work than the lower. The first law of thermodynamics may be manipulated mathematically (using calculus) to find a new thermodynamic quantity called *entropy*. Since entropy is path-independent, it is very useful. But it arises from mathematics rather than physical considerations, so its physical interpretation becomes difficult. Some try to link entropy to disorder or unavailability of energy to accomplish useful work.[3] Since the entropy of the universe increases, this leads to the notion that the universe is becoming more disorderly and that energy is becoming less available. Carried to a logical extreme, this signals the *heat death* of the universe, in which the temperature of all bodies becomes uniform, so there is no heat flow. (A student once compared this situation to Robin Hood, who robs from the rich and gives to the poor. In the long run, Robin Hood is out of a job because everyone becomes middle class.) The only published estimates for the time required for the universe's heat death are upward of 10^{14} years in the future. Since the universe is only 10^{10} years old now, we have a bit of time to ponder the question.

Third Law of Thermodynamics

The third law of thermodynamics was formulated by Walther Nernst.

It is impossible to use any finite process to cool a body to absolute zero

Beat poet Allen Ginsberg paraphrased the three laws of thermodynamics as:[4]

First law: "You can't win."
Second law: "You can't break even."
Third law: "You can't quit."[4]

Meanwhile, back to heat energy. There are several ways for heat energy to be transferred from one body to another.

Conduction

If two reservoirs are separated by a connecting medium, energy can flow through the medium by colliding with the molecules. The reservoir with higher temperature has faster-moving molecules, so when they collide with the molecules of the connecting medium, energy is transferred, and moves toward the low-temperature reservoir.

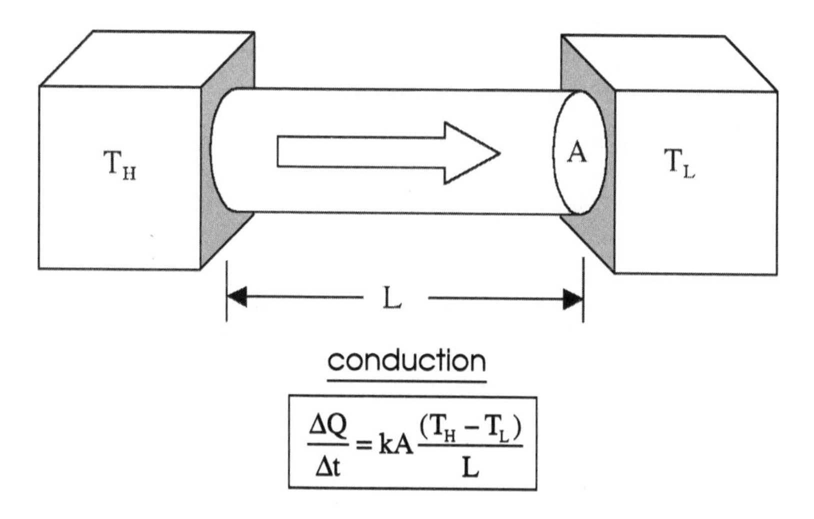

There is a wide variation in the thermal conductivity of various substances, from good conductors to poor conductors, also referred to as *insulators*.

Substance	**Thermal conductivity, k, J/ms°C**
Silver	420
Aluminum	220
Steel	45
Ice	1.9
Glass, brick, concrete	1.0
Soil, liquid water	0.6
Wood	0.1
Fiberglas	0.04
Air	0.024

Here are some examples that you might recognize from personal experience:

- Metal objects seem cold to the touch because they conduct heat energy away from your hands quickly, whereas wood objects seem much warmer because they are insulators.

- Some silver tea or coffee servers have rubber inserts in the handles to keep the heat from flowing into your hand as you pour your favorite hot beverage.
- To keep heat energy from flowing out of buildings, various insulating techniques have been used, such as additional layers of insulation, blown-in insulation, and expanding caulk.
- Insulation is rated according to an R-value, which is proportional to 1/k, so the higher the R-value, the lower the thermal conductivity.
- Insulated glass often consists of two panes with air trapped between, since air is an even better insulator than glass.
- If you live in a cold enough climate to collect frost or snow on roofs, you may notice that melting takes much longer if the attic is well insulated.

CONVECTION

If heat energy is transferred by mass movement of molecules from one location to another, this is called *convection*. A familiar example of convection is the action of boiling water in a pan on the stove. If the water is in a rolling boil, you can see the churning movement, transferring heat energy from one part of the water to another. Clearly, convection is not possible in a solid, since the molecules are fixed. Liquids, gases, and plasmas support convection very nicely, however. Another familiar example is an interior heating system. Warm air is circulated either by fans or by the natural upward movement of warmer, less dense air in an attempt to equalize the temperature of air throughout a room. Larger examples include the convection currents caused by hot spots within the Earth's mantle. The mantle flows slowly and drags plates around, causing earthquakes in the process. On an even bigger scale, there are hot spots in the Sun's atmosphere that lead to variations in temperature at the Sun's photosphere; we know these as sunspots.

Convection is so dependent on the geometry of the particular case that no general physics principles or equations will be listed here. In fact, many practical decisions of heating and cooling systems are made on the basis of overall thermodynamic requirements, with the occasional hot or cool spots adjusted on a case-by-case basis.

RADIATION

This last type of heat energy transfer requires no medium at all. This is actually electromagnetic radiation and will be analyzed in chapter 26. For now, let us state:

> The heat energy transferred by radiation from a body is directly proportional to the body's emissivity, its surface area, the time elapsed, and the body's temperature to the fourth power.
>
> $Q = e\sigma tAT^4$, where e = emissivity, ranges from 0 to 1
> σ = Stefan-Boltzmann constant
> t = elapsed time
> A = body's surface area
> T = body's temperature

Emissivity is a measure of how efficiently a body emits energy.

If thermodynamics got you heated up, you will really get a charge out of the next topic: electricity.

Experiment 18: Static Cling

Equipment needed: balloon, scraps of paper, metal foil

Classic experiments in static electricity involve rubbing amber with fur. Not far behind in generating charge is rubbing a balloon with hair or clothing. What happens next is familiar to anyone who has experience with clothes dryers. The balloon acquires a charge and attracts other bits of matter. But the attraction is not uniform, as you will see.

PROCEDURE

- Inflate the balloon to a level near its maximum size.
- Rub the balloon on your hair or clothes several times.
- Bring the rubbed balloon near the small pieces of shredded paper and note the effect.
- Bring the rubbed balloon near the the small pieces of shredded foil and note the effect.

Chapter 16. CHARGE IT — ELECTRICITY AT REST

"If there is no other use discovered for electricity, this however is something considerable, that it may help to make a vain man humble."

—Benjamin Franklin

"Electricity is really just organized lightning."

—George Carlin

Earlier, we dealt with the physics of familiar things—things that we could see and test. But you may have noticed that as you continue reading this book, more and more abstract ideas have arisen. We're about to cross another threshold between the visible and nonvisible, so hang onto your hat.

Like so many other things in physics, electricity was first noticed by the Greeks, who recorded their thoughts about it. Thales of Miletus (about 600 BCE) found that amber, when rubbed with fur, would attract light objects, such as feathers. This wasn't quite the beginning of a scientific treatment of electricity, however. Many others (not Thales) thought this was just an interesting magical property of amber. Magic or not, little was done with this effect until the scientific revolution, more than two thousand years later. The next step was taken by Dr. William Gilbert in 1600. At the time, Gilbert was personal physician to Queen Elizabeth I. Having only one patient allowed him some time to study natural philosophy. His major work, *De Magnete*, was mostly concerned with magnetism. Dr. Gilbert experimented with amber and named the force that rubbed amber exerts on light materials the *electric* force, after the Greek term for amber, *elektron*. In 1733 Charles Du Fay theorized that electricity consisted of two fluids, which he called vitreous and resinous. Before long, an extremely useful piece of electrical apparatus was invented. The Leyden jar, invented in 1745 by Pieter van Musschenbroek (and Ewald Georg von Kleist independently), was improved by William Watson within a year. The Leyden jar consisted of a glass cylinder, with metal foil coating on the inside and outside. Although they didn't quite know what electricity was, Musschenbroek and von Kleist knew that a Leyden jar would store it for later experiments. This sets the scene for a giant contribution that was made by an unlikely source.

Benjamin Franklin (1706–1790) Mini-Biography[1]

Benjamin Franklin

Benjamin Franklin's father, Josiah Franklin, was married to Anne Child in England in 1677. They had three children there, then emigrated to Boston in 1683 for religious or economic reasons. Anne Child Franklin bore four more children, then died in 1689. Within a year, Josiah Franklin married Abiah Folger, and they had ten children together, the last one born in 1712. Josiah Franklin was a soap boiler and candle maker who had a shop very near their residence on Milk Street. Within the family, Benjamin was the fifteenth child of seventeen and the tenth and last of the males. As a youth, Benjamin was precocious, but the family resources were not sufficient to send him to school for two full years, and even then he failed arithmetic. By age ten, he became apprenticed to his father, but hated every minute of it. To avoid losing Benjamin to the sea, as had happened to Josiah Jr., his father allowed him to become an apprentice to his brother James, who was a printer. In his teens, Benjamin was intellectually curious and read everything he could get his hands on. Almost six feet tall, he had great physical strength and was an especially good swimmer. To extend his range, he floated on lakes, towed by a string of kites.

Shortly after Benjamin joined his older brother, James Franklin launched a weekly newspaper, the *New England Courant*. It was the first colonial newspaper that publicly ridiculed the government. After several scrapes with authorities, some because of letters Benjamin wrote as if he were a middle-aged widow, Silence Dogood, James was imprisoned and denied the right to print the paper. He transferred ownership to Benjamin, so the paper continued. Since the ownership transfer ended his apprenticeship, Benjamin plotted his escape from Boston. Soon, he left and settled in Philadelphia. Within a year, Franklin (age eighteen) had so impressed Pennsylvania governor Sir William Keith that he was sent to London to obtain the necessary equipment to become official printer for the Pennsylvania government. In London, Franklin thrived personally by working at the printing business, but no money was sent to buy the promised equipment. In London, he met Thomas Denham of Philadelphia, who befriended him and promised him a job when he returned to the colonies. Upon his return to Philadelphia, Franklin worked as a bookkeeper for Denham and started the American Philosophical Society. Here, Franklin's intellectual curiosity was served as well as his sense of social responsibility. In 1728 Denham died suddenly, so Franklin returned to his previous trade and started a printing business of his own. Soon, Franklin became Pennsylvania's leading publisher, printing the *Pennsylvania Gazette*, legal and governmental documents, and preachers' sermons. He wrote *Poor Richard's Almanack*, which had a substantial circulation for twenty-five years and made a fortune for Franklin.

Once the thriving printing business reached a point where it could be turned over to others to manage, Franklin did so and pursued his other interests. In 1747 an itinerant English lecturer showed Franklin some electric experiments, which fascinated Franklin. He bought the apparatus when the lecture tour ended and sent to England for more, including Leyden jars for electrical storage. At this time, scientific lectures were seen as entertainments, and lectures were sometimes billed as "electricity parties." Electricity was still a wide-open field, unlike medicine or astronomy, which required specialized knowledge or equipment. This was an ideal area for the self-taught but energetic Franklin, and he dived in with gusto. Rejecting the two-fluid theory of Du Fay, Franklin and his chief associate, Ebenezer Kinnersley, favored the idea of electricity's being a single fluid. Normally, bodies have just the right amount of this fluid, but bodies with too much were called by Franklin positive, and bodies with too little were termed negative. The fluid approach to electricity, treating it like other fluids, was eventually discarded, but the naming of positive and negative was retained.

Franklin's larger contribution to electricity was his recognition that lightning was the same electricity that could be generated by rubbing amber with fur, just on a much larger scale. In keeping with Franklin's general attitude about basing physical theories on reality, he performed his famous kite-flying experiment in 1752. A few writers have since questioned the actual performance of this experiment because the only witness was his (illegitimate) son William, but theirs is a distinct minority view. Franklin flew a silk handkerchief kite in an impending rainstorm and charges flowed down the string and jumped from a key attached to the string to Franklin's knuckle and charged a Leyden jar. The experiment has been repeated many times, often with disastrous consequences—large currents wreak havoc on humans. (No experimentino here.) Franklin also noticed that electricity leaked from pointed bodies more easily than blunt ones, so he invented the lightning rod. The rod attracted electricity to its point and conducted it to the ground, saving buildings and people from the "mischief" of lightning. Superstitions and religious arguments raged for quite a while over Franklin's lightning rods. Some called Franklin a "modern Prometheus" because he stole lightning from the gods.

Besides his scientific and technical accomplishments, Benjamin Franklin played a significant role in the American Revolution, which is well chronicled in history books. He was especially known and revered in France, where he was considered a scientist turned statesman, the reverse of his reputation at home. Not long before completing his nine-year ambassadorship to France, he witnessed one of the balloon flights of Jacques Charles (see chapter 14). Franklin was enthusiastic about the potential of balloons. When a skeptic in the crowd of balloon watchers wondered about their practicalities, Franklin quipped, "What is the use of a new-born baby?"

Benjamin Franklin was the only one of the United States' Founding Fathers to sign all four of the major documents: the Declaration of Independence, the Treaty of Paris, the Treaty of Alliance with France, and the United States Constitution. Even after his death in 1790, Franklin's influence continued. A French mathematician wrote a satire called *Fortunate Richard,* based on *Poor Richard's Almanack,* in which he had Fortunate Richard leave a small amount of money in his will with the stipulation that it not be awarded for five hundred years. The idea was that the interest could accumulate and turn a small sum into a large amount of money. Franklin thanked the mathematician for the good suggestion and left the cities of Philadelphia and Boston £1,000 each to be disbursed in two hundred years. In 1990 the accumulated funds were more than $7 million and they were used for high school student loans and for the establishment of a trade school that became the Benjamin Franklin Institute of Technology.

The next development in electricity also came at the hands of a talented amateur. Charles Augustin Coulomb was a French military engineer specializing in the design of fortifications.[2] Participating in a remarkable range of activities in his career, Coulomb applied advanced mathematical techniques to soil mechanics and structural problems and worked out the theory of very sensitive torsion balances. He also developed a scheme for reorganization of the military engineer corps and conducted an in-depth study of friction that defined the field. At age fifty, he was posted to Paris, where he worked on electricity and magnetism, arriving at the inverse square law for the attractive and repulsive force in 1785. After that, he took on hospital reform, then was placed in charge of much of Paris's public water supply and fountains. Finally, he set up the countrywide system of public instruction. His first child was born when Coulomb was fifty-four, and he married his child's mother when he was sixty-six.

Fascinating as these bits of history may be, let's jump ahead and view Coulomb's law in more modern terms. Charge comes in two kinds, called positive and negative. If two charges are the same kind, the force they exert on each other is repulsive; if they are unlike, the force is attractive.

Two charges exert forces on each other. The magnitude of the force is directly proportional to the product of the two charges and inversely proportional to the distance between them, squared. The direction of each force is along the line joining the charges.

$$F_{electric} = \frac{kq_1q_2}{r^2}, \text{ where } k = 9 \times 10^9 \text{ Nm}^2/\text{C}^2;$$

q_1, q_2 = charges, coulombs = C;
and r = distance between charges, m

You may recall from chapter 3 that the gravitational force equation looks quite similar:

$$F_{gravitational} = \frac{Gm_1m_2}{r^2}, \text{ where } G = \text{gravitational constant} = 6.67 \times 10^{-11} \text{ Nm}^2/\text{kg}^2$$

The mathematical structures of electrical force and gravitational force are parallel: both have a product of two quantities in their numerators and r^2 in their denominators. But the size of the constant that adjusts the units and gives the force its overall magnitude is radically different—by a factor of 10^{20}. This huge difference in strength could make you wonder. If gravity plays such a role in the universe, what about the electric force, which is so much bigger? Several factors come into the picture here. In your personal experience, masses of a few kilograms are common, and if you want more, just pile them up—you could make a planet, or something even bigger. A few coulombs of free charge are not nearly so common and they don't pile well. We'll learn more about this problem shortly.

In order to make charges and electrical forces easier to understand and visualize, the concept of the *electric field* was introduced by Michael Faraday in the latter 1830s. The electric field of any system of charges is found by placing a +1 coulomb test charge in various locations and drawing an arrow to indicate the magnitude and direction of the electric force the test charge would experience. The arrows are then connected. For point charges, the field diagrams look like this:

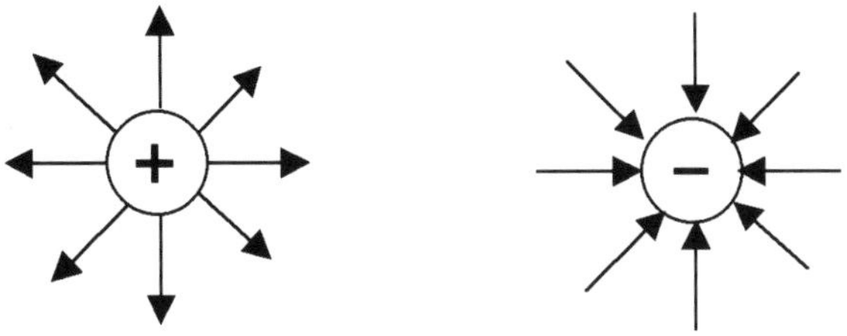

electric fields

For more complicated charge systems, the field diagrams become much more complex. You can see additional field diagrams on the Web sites listed in the endnotes.[3]

Although it is tempting to think that fields are just a diagrammatic convenience, rather than something as substantial as matter, Faraday thought of fields as actually influencing space itself, which is a surprisingly modern attitude.

Shifting to a modern perspective, we must confront the basic issue: charges are not a fluid or fluids; charge carriers are particles. They are parts of the atom, the positive proton and the negative electron. We'll postpone detailed discussion of atoms until chapters 26 and 27, but you can see how confusion would reign more than sixty years prior to the discovery of the electron. In addition, atoms make up the molecules that comprise matter, so matter is electrical at the deepest level. The reason some materials allow charge to flow freely and other materials don't is all tied up in the electrical structure of the molecules. Let's see how this explains the results of Experiment 18.

Rubbing the balloon with hair pulls electrons off the hair molecules, leaving them positively charged and rendering the balloon negatively charged. When the negatively charged balloon comes close to the paper scraps, the molecules of paper have no free electrons. The molecules simply reorient themselves so that their electrons are as far as possible from the electrons on the balloon because of their mutual repulsion. This leaves the positive part of the molecule closer to the electrons of the balloon, and hence makes a net attractive force, forcing the paper to stick to the balloon. Although the molecule is still electrically neutral overall, its negative charge has moved to the other side of the molecule from the positive charge. This is termed an *induced dipole*. In the case of the metal foil scraps, things are a little different. Metal has free electrons available, so they move to the far side of the foil, similar to the paper. However, because metals allow charges to flow, electrons from the balloon jump to the metal foil, making the overall charge of the foil negative. The negatively charged foil is then repelled by the negatively charged balloon, so they move apart.

The forces that join atoms to form molecules (intramolecular) and the forces that bind molecules to form matter (intermolecular) are electrical in nature, so the chemical behavior of matter is electrical at the most fundamental level. For example, ions form the strongest bonds and they are the most highly charged. Next in bond strength is the polar molecule, with a charge less than a full ion; then comes a nonpolar molecule that has only induced polarity, which has the weakest bond strength.[4]

Seemingly small differences in electrical properties of molecules have profound implications for biochemistry. For example, the hemoglobin molecule has spaces that are usually occupied by several nonpolar O_2 molecules as it rides around in the bloodstream of humans and other oxygen-breathing animals. Because oxygen is nonpolar, it is very loosely held by hemoglobin and can be easily dislodged, much to the benefit of oxygen-craving muscles. If a similar-sized molecule of carbon monoxide, CO, manages to find its way into hemoglobin's spaces, it fits the opening well enough, but its small dipole forms a bond that doesn't release easily, so the muscles (or brain) don't get the oxygen they need. This causes carbon monoxide poisoning. The solution is to replace hemoglobin molecules that have their O_2 templates filled with CO. Fortunately, the body carries out this maintenance function regularly.

Another example lies in the olfactory senses in humans and animals. In a lock-and-key sort of arrangement, particular molecules from the outside world "fit" into openings in molecules in the olfactory sensory area. The filled/nonfilled status of these molecules is communicated to the brain, and the brain interprets the presence of particular molecules as certain smells.

Now, let's get back to the problem of piling up charges. Most of the problem lies in the fact that charges of the same sign repel each other, while masses come in only one flavor—attractive. Actually, there is no shortage of charges. A tiny droplet of water contains 1,000 coulombs of negatively charged electrons. The problem is that they are all busy holding onto 1,000 coulombs of positively charged protons. If only we could find a way to separate them, we could apply a force, move the electrons through a distance, and do some physics work. But electrons and protons are subatomic particles, too small to even see, much less to manipulate manually. Chemistry to the rescue. Chemical reactions are all about the transfer of electrons. All we need are chemical reactions that pile negative electrons in one spot and positive ions in another. Fortunately, many such reactions exist and were put to use generating electrical charge separation on a practical basis in the 1780s and 1790s by Luigi Galvani and Allesandro Volta. A device that uses chemical reactions to separate charges is called a *battery* (a term first used by Benjamin Franklin for a slightly different application). Batteries are rated in terms of the amount of electrical potential energy per unit charge they put out. Their unit, the joules/coulomb is called a *volt.*

$$\frac{\text{potential energy}}{\text{charge}} = \frac{\text{joules}}{\text{coulomb}} = \text{volts} = \text{V}$$

In their early days, batteries involved dangerous, corrosive chemicals and were a rarity. Modern batteries still have some dangerous aspects, but they are often sealed to avoid chemical spills, though such spills may still occur. Batteries are used for providing electrical energy to a vast array of devices, from tiny watches to heavy equipment, but they all have clearly marked positive and negative spots, with the positive terminal being the higher potential energy and the negative terminal being the lower.

Once charges have been separated, they can be stored in another electrical device that doesn't rely on chemical reactions: the *capacitor*. The simplest capacitor consists of two parallel conducting plates. If a battery is connected to the capacitor, positive charges sit on one plate and a corresponding amount of negative charges sit on the other.

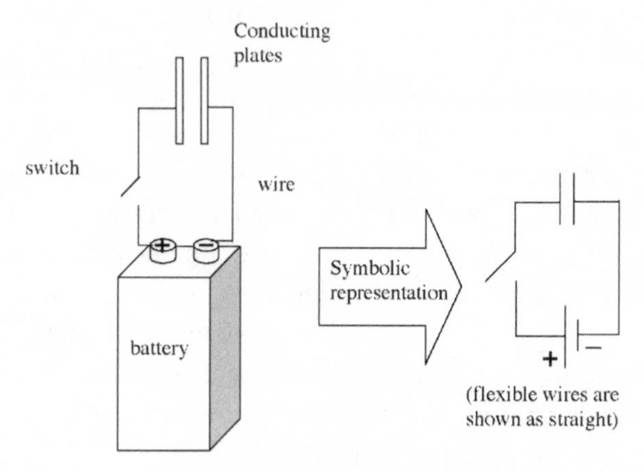

battery and capacitor

Once the switch is closed, the charges move through the conducting wires and sit on the capacitor's plates. The relationship between the charge and voltage is a simple one:

$q = CV$, where q = charge, coulombs C = capacitance, farads, F V = voltage, volts, V

The unit of capacitance, the farad (named after Michael Faraday), is quite large. Many practical capacitors are very small multiples of 1 F, such as mF, μF, nF, and even pF. For parallel plate capacitors, the value of the capacitance is directly proportional to the area of the plates and inversely proportional to the distance between the plates.

$C = \varepsilon_0 \dfrac{A}{d}$, where $\varepsilon_0 = 8.85 \times 10^{-12}$ F/m A = plate area, m^2 d = distance between plates, m

Two or more capacitors may be connected in different ways to make a capacitor network for practical electronics applications. The connections may be series (both capacitors have the same charge) or parallel (both capacitors have the same voltage).

Besides storing charge, capacitors act as energy storage devices, since work must be done to place charges on the plates. The amount of energy stored in a capacitor depends on the capacitance and the voltage that drives the charges into place according to

$$\text{energy} = \frac{1}{2} CV^2$$

The unit of energy is the familiar joules.

Many different capacitor geometries are used, as well as a wide variety of materials. In numerous cases, the volume between the plates is filled with a material called a *dielectric*. Most dielectrics increase the capacitance and provide mechanical separation of the plates, since touching plates would discharge a capacitor. Dielectrics also increase the voltage at which the material between the plates becomes conductive. This is referred to as the *breakdown voltage*. At high enough voltage, all materials become conductors, even air. (The electric field strength that causes air to become conducting is approximately three million V/m.) When air breaks down and becomes a conductor, you know you're witnessing lightning.

Experiment 19: Meter Reader

Equipment needed: 4 resistors (all rated between 700 and 1,000 Ω), a multimeter, a 9 V battery, 4 clip heads

In this experiment, charges will flow from a region of high potential: through wires, through a resistor, through more wires, and then arrive at a region of low potential. The problem is that the wires and resistor will look no different when the charges flow than when they don't. We have left the visual zone, so we need help to find out what is occurring. That's where the meter enters the scene. The purpose of this experiment is to become familiar with the use of meters, resistors, wires, and batteries.

PROCEDURE

1. Determining Rated Resistance from Resistor Color Code

- The colored bands painted on a resistor allow us to determine the manufacturer's rating of the resistor. Examine your resistor for the colored bands, and enter them in table 1 below:

Table 1	**First band color**	**Second band color**	**Third band color**	**Fourth band color**	**First digit**	**Second digit**	**Zeroes**	**Rated resistance Ω**	**Tolerance ±Ω**	**Resistance range, Ω**
Sample	Brown	Black	Red	Gold	1	0	00	1,000	50	950–1,050
R_1										
R_2										
R_3										
R_4										

Table 2. Resistor color code for first three color bands

Color	**Black**	**Brown**	**Red**	**Orange**	**Yellow**	**Green**	**Blue**	**Violet**	**Gray**	**White**
Number	0	1	2	3	4	5	6	7	8	9

- To decipher the color code, the first two colors determine the first two digits in the resistor's resistance and the third color tells how many zeroes to place after the digits. Applying this to the sample, brown, black, and red correspond to 1 0 00 for a total of 1,000 Ω. The fourth color tells the manufacturer's tolerance on the rated resistance. Silver means ± 10 percent; gold means ± 5 percent. For the sample shown, the fourth band is gold, so ± 5 percent must be applied to 1,000 Ω, which yields ± 50 Ω. This makes the sample resistor's rated resistance somewhere between 950 Ω and 1,050 Ω. (Fine print: Other possible colors for the fourth band correspond to more precise resistors, but they are used less frequently. There may be a fifth or sixth band, but we can safely ignore the additional information they convey. Here's a clue about where to start reading: gold and silver are never in the first band, so you never start reading with gold or silver.)

2. Measuring Resistance Directly

- Connect the resistor to the meter with clip leads as shown in the diagram (page 204).
- Make sure all connections are tight.
- Turn the multimeter control to a setting to measure resistance, Ω. Depending on the meter design, there may be different settings for different anticipated values of the resistance, so consult the meter instruction book for details. A battery internal to the meter will pass a small current through the wires and resistor to determine the total resistance. (The resistance of the clip leads is presumed to be negligible.)
- Determine the resistance of the first resistor from the meter reading.
- Enter the value of the directly measured resistor in the appropriate column in the data table (page 205).
- Does the directly measured resistance fall within the range specified by the manufacturer? If not, recheck the colors and reading. If it still doesn't match, discard the resistor as faulty. If it does fall within the specified range, place a check mark in the appropriate column in the table.
- Repeat the process for the other three resistors.

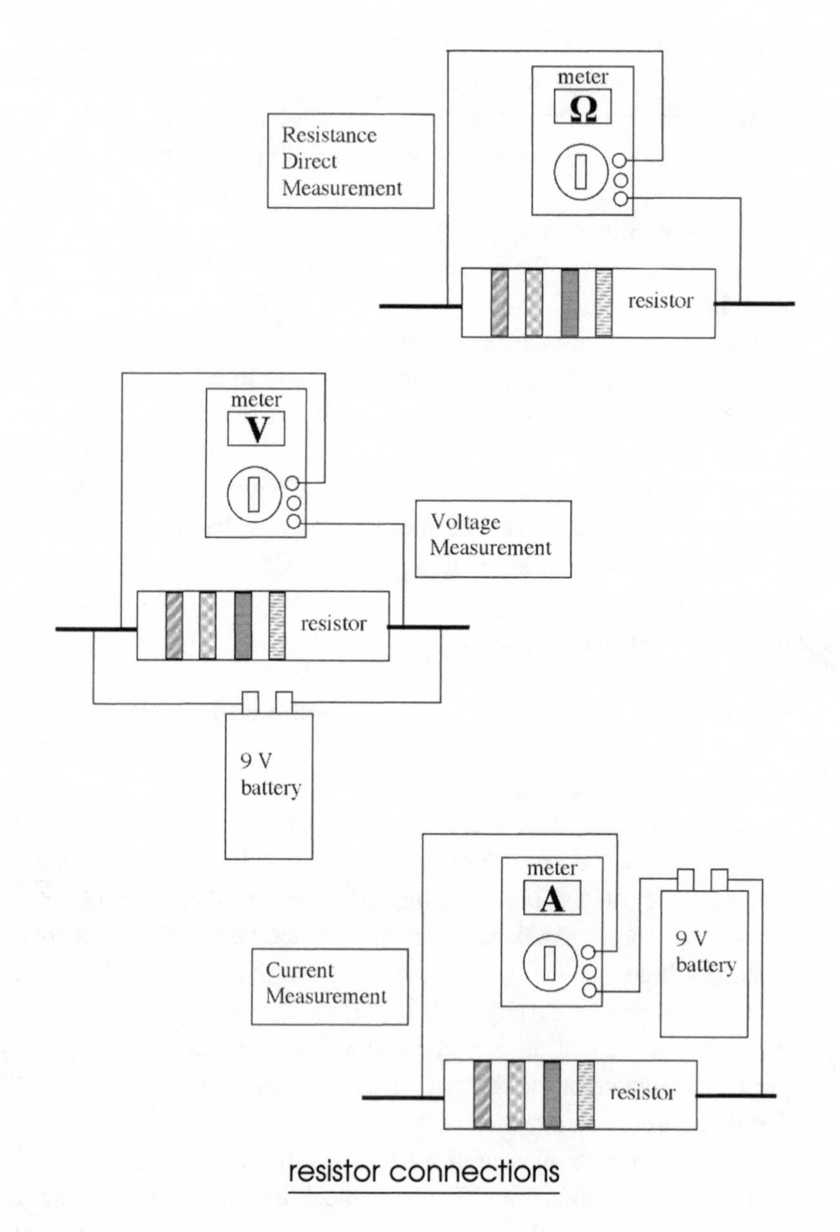

resistor connections

3. Safety Check

Running a current through a resistor deposits energy, which increases the resistor's internal energy and hence its temperature. In commercial electric circuits, heat energy is carried away to its surroundings by conduction (heat sink), convection (cooling fan), or radiation. When a resistor's temperature reaches a critical point, it fails. As you will see in the next chapter, the rate of energy conversion for a resistor is given by

$$P = i^2R = \frac{V^2}{R}$$

where i = current, R = resistance, and V = voltage

Although power ratings aren't shown on a resistor, most common ones are rated at 0.25 watt. Higher-power resistors are available but are not as common. To be safe, let's presume your resistors can dissipate 0.25 watt. If they can handle more, our experiments are even safer. (A typical goal is to run a resistor at roughly 50 percent of its rated power, because that extends its useful lifetime.)

- Calculate the maximum safe voltage for each resistor from $V = \sqrt{PR}$.
- Enter the maximum safe voltages in the data table.
- Calculate the maximum safe current for each resistor from $i = \sqrt{P/R}$.
- Enter the maximum safe currents in the data table.
- Presuming a 9V battery, calculate the expected voltage and current.
- If the anticipated voltage and current is less than the maximum safe values, continue.

Data Table	**Rated resistance, Ω (from table 1)**	**Directly measured resistance, Ω**	**Max safe V, volts = √PR**	**Max safe i, mA √P/R**	**Directly measured voltage, V**	**Directly measured current, i amps**	**Calculated resistance =V/i**	**√**
Sample	1,000 ± 50	985	15.7	15.8	9.17	9.08 mA	1,010	√
R_1								
R_2								
R_3								
R_4								

4. Voltage Measurement

- Connect the resistor, the battery, and the meter as shown in the diagram (page 204) for voltage measurement. With this connection, the voltage across the meter is the same as the voltage across the resistor. This is referred to as a *parallel* connection.

- Make sure the meter is set to read volts and that all connections are tight.
- Record the voltage in the data table.
- Repeat for the other three resistors.

5. Current measurement

- Connect the resistor, the battery, and the meter as shown in the diagram (page 204) for current measurement. Connected this way, the same current flows through the resistor and the meter. This is a *series* connection.
- Make sure the meter is set to read amps and that all connections are tight.
- Record the current (amps) in the data table.
- Repeat for the other three resistors.

6. Calculating Resistance Based on Measured Voltage and Current

- Ohm's law says V = iR. Solving for R, R = V/i. Using your measured values of V and i, calculate R for each resistor and enter the results in the data table.
- Check to see if both the directly measured resistance and the resistance calculated from measured voltage and current lie within the range promised by the manufacturer. If yes, place a check in the last column of the data table. Of the three values of resistance found in this experiment, the directly measured value is probably the most accurate (usually ± 2 percent or less), next is the manufacturer's value (± 5 percent), and the calculation based on measured voltage and current is probably the least accurate because there are uncertainties in both V and i. Further, the presence of the meter in the circuit adds resistance, making the measured current lower than the actual current through the resistor. The resulting value of resistance obtained by calculation tends to be too large.

Chapter 17.
WHO LET THE CHARGES OUT? —ELECTRIC CURRENT

George Simon Ohm

"Most reformers wore rubber boots and stood on glass when God sent a current of common sense through the universe."
—Elbert Hubbard

"We must take the current when it serves. Or lose our ventures."
—William Shakespeare, *Julius Caesar*

In the last chapter, we left some charges nicely separated by a battery into two piles, one positive and one negative, with potential energy stored as a result of the separation. Since charges are too small to see, it is difficult for us to sense what is happening. And the visibility problem isn't going to improve. Suppose the high potential energy storehouse of charges is connected to the low potential energy charges by some material that will allow charges to flow. What will they do? If we were dealing with gravitational potential energy, this would be like a skier at the top of a snow-covered hill. Down the hill the skier would go, gleefully turning gravitational potential energy into kinetic energy. Let's carry this illustration further. Suppose the hill is frictionless. By the time the skier reaches the bottom of the hill, a lot of potential energy has been converted into kinetic energy and the skier is moving quite rapidly. That's great fun, but eventually the skier must stop and climb onto the chairlift and get a ride to the top of the hill for another run.

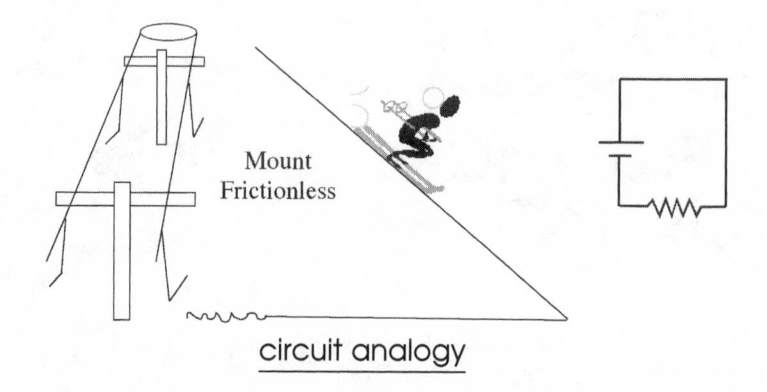

circuit analogy

From an energy standpoint, the skier gains potential energy from the chairlift, converts it to kinetic energy on the way down the hill, then loses it to frictional work before getting back onto the chairlift. This bears a rough similarity to an electric circuit, in which charges gain potential energy from a battery, convert it to kinetic energy by moving in a wire, then lose it to frictional work in a resistor. The skier makes a complete loop and the charges flow in a complete circuit. In a similar way, if there is no complete circuit, no charges would flow, just as a sudden avalanche carving away a section of the hill would cause the ski patrol to close the run. To extend the analogy a bit further, think about a whole series of skiers, one right after another. If you sat in one spot and counted the number of skiers per unit time, this would correspond to the charges per unit time in the electrical case. The charge per unit time that flows is called the *current* and is given the symbol i. In equation form,

$$i = \frac{\Delta q}{\Delta t}$$

The unit of current is coulombs/second = amperes = amps = A. The question that intrigued scientists in the early days of electricity was the relationship between the voltage that drives the circuit and the amount of current that flows. It would seem sensible that greater voltage would drive more current, but there needed to be experimental studies to check the details of the relationship. We have just the person for the task.

Georg Simon Ohm (1789–1854) Mini-Biography[1]

Georg Simon Ohm was the eldest son of master locksmith Johann Wolfgang Ohm and Maria Elisabeth Beck Ohm. Only three of their seven children lived to maturity, and Maria Ohm died in 1799. Georg Simon Ohm was born and raised in the university town of Erlangen, Bavaria. Ohm's father had no formal schooling but educated himself to a remarkable degree, and taught his sons (Georg and Martin, three years younger) to the point that they had a solid grounding in mathematics, physics, chemistry, and philosophy. He even demanded that they learn the locksmith trade. The boys' mathematical abilities were recognized at an early age by Erlangen mathematics professor Karl Christian von Langsdorff, who advised them during their academic careers.

From 1800 to 1805 Ohm attended the gymnasium (high school) associated with the University of Erlangen and became a university student in 1805. He enjoyed college life a lot, but after three semesters of too much dancing, ice-skating, and billiards, Ohm's father pulled the plug. In 1806 seventeen-year-old Ohm was sent to Switzerland to teach math in a private school. After five years of teaching, tutoring, and private study, Ohm's father allowed him to return to Erlangen where he completed his doctorate in less than a year. Upon graduation, the university offered Ohm a lecturer's position, but the salary was so low that Ohm left after two years. He took a job with the Bavarian government, teaching at a poorly regarded school in Bamberg. The school sunk lower and was dissolved in 1816. Determined to demonstrate his abilities, Ohm wrote a geometry textbook. It was not well received. As part of his work for the Bavarian government, he wrote a report detailing some of his educational philosophy. He believed that mathematics should not be taught as an already-finished structure imposed from outside but rather as a free product of one's own mind. He believed this would encourage self-reliance and self-respect.

Shortly thereafter, a Jesuit gymnasium in Cologne was re-formed, and Ohm took a job teaching mathematics and physics there. He continued his private mathematical studies, but teaching duties consumed much of his time. In 1820 he became inspired when he read Ørsted's paper linking electricity and magnetism. The gymnasium had good equipment, so Ohm began some electrical experiments, mostly for his own interest. As the school's standards began to slip, Ohm saw his experiments as critical to his own development, so he pursued them more systematically. Finally, in 1825, he obtained a half-pay leave to go to Berlin and work with his brother Martin.

Martin Ohm's career had followed a much smoother path. After his schooling was complete, he worked as a private tutor in Berlin for a time, made some good contacts at Berlin University, and was eventually appointed to a faculty position. His field was number theory, and the textbooks he wrote were quite successful. One of Martin Ohm's research interests was the "golden ratio," which figures in the popular Dan Brown book *The Da Vinci Code*. Some of his colleagues thought Martin arrogant and socially inept, but the brothers worked well together. Georg Simon Ohm published his major work in 1827, *Die galvanische Kette, mathematisch bearbeitet*. This treatise lays out Ohm's complete theory of electricity, including his famous law stating the simple linear relationship between voltage and current, $V = iR$. Ohm's experimental work was confirmed by several other researchers, but the overall paper was criticized for two major reasons. First, Ohm's ideas were stated in much too mathematical a format to suit his colleagues. The paper began with the necessary mathematical background to understand the rest of the work, but most German physicists weren't convinced that physics needed to be so mathematical. Second, experiments formed the basis of his work rather than the more philosophical approach favored by contemporaries such as Hegel. The negative response of his colleagues, and the prospect of returning to the gymnasium in Cologne, discouraged Ohm greatly. He resigned the Cologne appointment and stayed in Berlin with his brother, taking temporary jobs teaching mathematics for the next several years. Finally, in 1833, he was hired for a position at the Polytechnic Institute in Nuremberg. Although he held the title of professor, his duties were very similar to those of his job in Cologne. As he acquired additional administrative duties, his scientific activities diminished. It looked like he would never get the university faculty position he had always wanted. In 1841 his career got a boost from an unlikely direction: his major paper was translated into English. It was very well received by the active research community in England. Ohm was awarded the prestigious Copley Medal and made a foreign member of the Royal Society. In 1849 Ohm finally was given a faculty position at the University of Munich and was appointed to the chair of physics in 1852 at age sixty-three. Ohm continued working beyond the mandatory retirement age and lectured the day before he died in 1854.

What Ohm found, through his experiments, was that the current that flowed through some materials was directly proportional to the voltage across the material and a constant that depended on the composition of the material and its geometrical arrangement.

> The current that flows through a material depends on the potential difference between the ends of the material and a constant that is determined by the molecular structure of the material and its geometrical arrangement.
>
> $V = iR$, where V = voltage difference, volts
> i = current, amps
> R = resistance, Ohms = Ω

Ohm's law, $V = iR$, demonstrates a relationship that is similar to Hooke's law for springs, $F = kx$. They are both good approximations that apply in some cases rather than revealing something fundamental about the way the universe functions. Materials for which Ohm's law applies are mostly metals and are referred to as Ohmic. Just as there were non-linear springs, there are non-Ohmic materials, for which Ohm's law doesn't apply. Even something as simple as a lightbulb is non-Ohmic in that the relationship between voltage and current changes substantially with temperature. Carbon is another example of a non-Ohmic material that has unusual characteristics. One of the original lightbulb filament materials was carbon-coated string. As the filament lit up, its temperature increased and its resistance decreased, drawing more current and making it heat up more, so its resistance decreased more, then, oops, failure. That's why early filaments had such short lifetimes.

Returning to the skier analogy for a moment, let's think about how the voltage and current are related. Voltage is the energy/charge and current is the number of charges/time. Multiplying these together yields energy/time, which is *power*. Combining this relation with Ohm's law gives several useful relationships for the rate of electrical energy conversion in a resistor:

$$P = iV = i^2R = V^2/R$$

In Experiment 19 you constructed an electrical circuit, checked its power, used a meter to measure its resistance, voltage, and current, and tested Ohm's law. Not all electrical currents are so well behaved that they flow safely through a tiny resistor, thin wires, and a pocket-sized meter. Let's look at two examples that go way beyond our simple experiment.

LIGHTNING

The endnotes lists several Web sites that have more detail about lightning, including locations of strikes and galleries of lightning pictures.[2]

A complete understanding of the wealth of physics in lightning still eludes us, even though lightning strikes Earth about one hundred times every second. While all researchers agree that a cloud of water molecules is involved in the initial stages of lightning formation, the precise mechanism of charge separation is not completely clear. It is possible that vertical air currents drag water molecules upward, where they collide with other water molecules, stripping off electrons. If the cloud is tall enough, air currents deliver water molecules into higher, cooler regions where they form ice crystals. In a complicated chain of events, not fully understood, ice crystals accumulate more water by collisions and form a slushy snow/ice mixture called *graupel*. Since the graupel particles acquire mass through accretion, they become more dense and sink. Collisions between upward-moving water molecules and downward-moving graupel cause both to become charged, with graupel gaining electrons and becoming negative, while the ice crystals lose electrons and become positive. The upper part of the cloud is thus positively charged, and the lower portion is negatively charged, setting the scene for all the activity that follows.

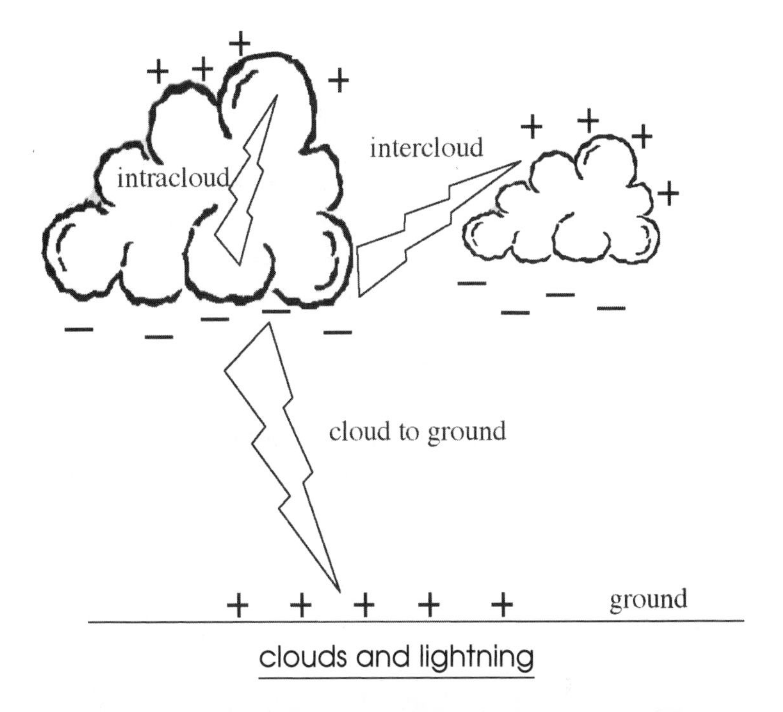

clouds and lightning

1. **Intracloud lightning:** Charges flowing from one part of a cloud to another result in the most common kind of lightning. When enough voltage builds up because of charge separation, the intervening air becomes conductive, allowing current to flow. This is not a smooth process, however. The air becomes conductive in stages, so the current flows in a discontinuous way. This gives the lightning its jagged appearance and is called the stepped leader. Once the full conductive path is complete, the major current flows, which is called the return strike. This kind of lightning, called sheet lightning, often lights up the whole cloud and produces no lightning bolt visible from outside the cloud.
2. **Cloud to ground lightning:** This is the next most common kind of lightning and can be dramatic. Although the ground normally contains a balance of positive and negative charges, the presence of large amounts of negative charge in the cloud base drives negative charges away and attracts positive charges to the region below the cloud. This is the induced charge phenomenon you saw in Experiment 18 with the balloon. Now, instead of the stepped

leader and return strike occurring within the cloud, these take place in the atmosphere. The voltage difference between the cloud and the ground may be as much as 1 billion volts, the current may be 10,000 to 100,000 amps, and several hundred megawatts of power may be involved. The return strike travels in a 5 cm diameter channel and may go as far as 200 km (120 miles) at a speed of 30 million m/s (10 percent of the speed of light). As this few million joules of energy is dumped into the long, narrow cylinder of atmosphere that is the return strike channel, several things happen quite rapidly: there is a flash of light as electrons, stripped by the current, are recaptured by oxygen and nitrogen atoms. Then, the temperature along the return strike channel shoots up to 20,000 K (36,000 °F, hotter than the Sun's exterior visible layer, the photosphere), temporarily converting the gas to a plasma. As the temperature rises, its pressure rises also ($pV = nRT$), and the resulting pressure pulse propagates through the surrounding air as a shock wave that quickly becomes the sound wave that we hear as thunder. Because of the great disparity between light speed and sound speed, we see the light substantially earlier than we hear the thunder. If you estimate this "flash to bang" time to be about thirty seconds, that means the lightning strike occurred about 10 km (6 miles) away ($d = vt$, and $v = 340$ m/s for sound). If the thunder sound continues over a long duration, it might mean there were multiple return strikes or else the sound was reflected, creating an echo.

3. **Intercloud lightning:** The clouds may be separated by a significant distance, and the bolt is visible and appears almost horizontal, or the clouds may be so close that they seem to light up when the charge flows. The sound of thunder produced may be delayed because of the height of the clouds.
4. **Other kinds of lightning:** Less than 5 percent of lightning falls into this category, which contains some very unusual forms.

- **Positive lightning** is a very intense form, often accompanied by an electromagnetic pulse.
- **Heat lightning** may simply be distant lightning where the thunder is masked by ambient noise.

- **Ball lightning** is extremely rare and has not been reliably reproduced in the laboratory. It is reported to be a glowing spheroid about the size of a basketball. Many different colors have been witnessed, and this phenomenon has been observed to persist for seconds or even minutes. Nikola Tesla claimed to be able to produce ball lightning artificially, and others have offered ball lightning as possible explanations for UFOs.
- **Megalightning** occurs above the clouds and takes many fascinating visual forms.
- **Triggered lightning** occurs when some humanmade object assists in the formation of the channel for conducting the charges. This has occurred unintentionally in the space program and by design when sounding rockets trailing wires were launched into thunderclouds. The lightning traveled along the wires, producing an unusually straight bolt.

Since lightning kills eighty to one hundred people in the United States each year and injures many more, there are some safety tips you should follow during an electrical storm. If the "flash to bang" time is less than thirty seconds, you should take cover, preferably in a large, well-constructed building with electrical wiring and plumbing, because those conductors will be more likely to carry a lightning strike than a poor conductor like you. If you are outside and far from a building, get into a car and don't touch anything metal. If a car isn't available, don't seek shelter under a tree. Trees extend high into the atmosphere, so lightning might strike them, then jump through you on its way to the ground. Better to get as low as possible and curl up into a ball to minimize your area and height. The most dangerous storms are weak-appearing ones with few flashes. Dangerous lightning is often produced at the beginning or the end of a storm of any size or at the edges of a severe storm. I've actually experienced an important warning sign of impending lightning when the induced charge buildup caused my hair to stand on end. This was more than just a bad hair day. The person who holds the record for being hit the most times by lightning was a US Park Service ranger, Roy Sullivan. He was struck by lightning seven times in his thirty-four-year career. He lost a toenail, had his hair set on fire twice, and sustained a few minor burns, but luckily survived it all.

ELECTRICAL SAFETY

What happens if a human being becomes part of an electric circuit? Taking a clue from this chapter's title, it's the current, not the voltage. Right. But they are related.

Current	Effect
1 mA	tingle; barely feel the sensation
10–20 mA	"cannot let go"
50 mA	interferes with nervous system: breathing, heart action
100 mA–300 mA	ventricular fibrillation if through heart, especially if 60 Hz ac; can be fatal if defibrillation is not accomplished quickly
6,000 mA	defibrillation current; heart may restart; possible respiratory difficulty and burns

For safety's sake, how can the larger current flows be avoided?

1. **No circuit, no current:** Don't ground yourself. If you're going to work on electric gadgets, stand on a thick rubber mat or wear the thickest rubber-soled shoes you can find. Better a safe geek than a burned dandy.
2. **Don't set up a possible circuit path that includes the heart or lungs.** A good way to handle this is to work with one hand in your pocket, and wear no jewelry that might provide an unintended circuit path. If you use an insulated probe with one hand, current may flow through the probe. (I've had this happen. The surprise factor is great, but a melted screwdriver is preferable to burned fingers.)
3. **Avoid high voltage.** There are high-voltage capacitors in the back of TV sets that must either be discharged or avoided. If you must go near large capacitors, learn how to ground them safely because they can be dangerous.
4. **Keep your resistance high.** A normal person with dry, unbroken skin has a resistance of 100,000 to 1 million Ohms. A household 120 V circuit would pass a current of $i = V/R = 1$ mA or less. This tingle should warn you to quit doing whatever you're doing but

should cause no permanent harm. However, there are activities that lower a person's resistance, such as having wet skin or cuts in the skin that allow current access to the inner parts of the body. Inside a human body, all those rich fluids with dissolved ions make for a low resistance. A story that may be an urban legend tells of someone who was interested in measuring his "internal resistance." He used a multimeter (like the one from Experiment 19) that contained a 9 V battery and inserted the pointed probes under the skin of his thumbs. His internal resistance was about 100 Ω, so the current that flowed was 9 V/100 Ω = 90 mA. The story doesn't have a happy ending. (Electrical safety considerations for home circuitry will be discussed in chapter 21.)

You may wonder what is happening at the submicroscopic level while the charges flow. Let's see.

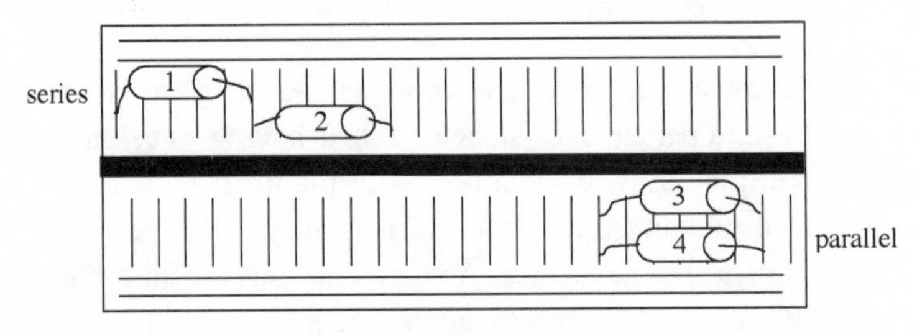

resistors

Experiment 20: Resistance Is Futile

Equipment needed: 4 resistors (all rated between 700 and 1,000 Ω), a multimeter, a 9 V battery, 4 clip leads, breadboard

The half a billion resistors per year that spew out of electronics manufacturers come in several different varieties. The most common (and cheapest) is the carbon film resistor, in which a thin film of carbon is deposited on a ceramic cylinder with metal contacts at the ends. Part of the carbon is then machined away in a spiral pattern, depending on how much resistance is desired. The carbon is coated with an insulating layer, then colored bands are added. Metal film, metal oxide, and wire-wound resistors are also available, but they have even more tightly controlled resistance values and are often able to dissipate more power. Resistance and power dissipation values are manufactured in standard amounts and may not match what is needed for a particular circuit. Rather than having custom resistors made, available resistors are often combined to fulfill a nonstandard requirement. The combinations take two forms: *series* or *parallel.* Resistors in series have the same current flow through each one in turn, and resistors in parallel have the same voltage.

$$R_{\text{equivalent series}} = R_1 + R_2 \text{ and } \frac{1}{R_{\text{equilvalent parallel}}} = \frac{1}{R_3} + \frac{1}{R_4}$$

Operating on the second equation algebraically yields:

$$\text{R equivalent parallel} = \frac{R_3 R_4}{R_3 + R_4}$$

PROCEDURE

Rated Resistance

- Designate the four resistors from Experiment 19 as R_1, R_2, R_3, and R_4.
- Use the formulas above to calculate the rated resistance for the parallel and series resistor combinations.
- Record your results in the table on the following page (the sample values used four resistors, all rated at 1,000 Ω).

Directly Measured Resistance

- Build the circuits on the breadboard as shown on the previous page. The illustration shows how columns of holes are connected on the bottom of the board. Hole sizes were designed to allow one wire per hole to be tightly connected without solder.
- Just as in Experiment 19, set the meter to measure resistance (Ω) and connect it to the ends of each circuit.
- Record your results in the table.

Maximum Safe Circuit Voltage

- For the series combination, both resistors carry the same current. Knowing the maximum safe current for each resistor from Experiment 19, the smaller maximum current should be supplied to keep both resistors safe. Calculate the corresponding voltage, using

$$V_{\text{max safe}} = i_{\text{smaller max safe}} R_{\text{combination}}$$

- Enter your result in the table.
- For the parallel combination, both resistors have the same voltage across them. Knowing the maximum safe voltages for each resistor from Experiment 19, the smaller maximum voltage should be supplied to keep both resistors safe.
- Enter your result in the table.

√ Since the voltage supplied is a 9 V battery, if all maximum voltages are greater than 9 V, place a √ in the table.

Directly Measured Voltage

- As in Experiment 19, set the meter to measure volts and connect it in parallel to the series combination.
- Connect the 9 V battery to the series combination.
- Record the voltage in the table.
- Repeat for the parallel combination.

Directly Measured Current

- As in Experiment 19, set the meter to measure amps and connect it in series to the series combination.
- Connect the 9 V battery to the series combination.
- Record the voltage in the table.
- Repeat for the parallel combination.

Calculated Resistance

- Calculate R = V/i for the series and parallel combinations.
- Does the calculated resistance fall within the tolerance percentage of the directly measured resistance? If yes, place a √ in the last column.

Data Rated	Rated resistance, Ω	Directly measured resistance, Ω	Max safe circuitV, volts*	Safe?	Directly measured voltage, V	Directly measured current, I miliamps	Calculated resistance =V/I	√
Sample series	2,000	1970	31.6	√	9.22	4.66	1,980	√
$R_1 + R_2$ series								
Sample parallel	500	492	15.8	√	9.14	18.23	501	√
$R_3 + R_4$ parallel								

Chapter 18. SILENT STRUGGLES IN THE WIRES—RESISTANCE

"The path of least resistance makes all rivers, and some men, crooked."

—Napoleon Hill

If we could shrink ourselves down to the size of an electron and station ourselves deep within a solid, what would be happening? Depending on the kind of solid, the fundamental unit might be molecules, atoms, or even ions. Regardless, these units would all be vibrating about fixed positions, with more energetic vibrations as the temperature increased and less activity as the temperature decreased. The vibrations would be so rapid that the whole assembly might appear to be fixed in place. If the solid was a crystal, the arrangement would appear quite regular, with an occasional defect spoiling the perfect symmetry of ions, atoms, or molecules. If the solid was noncrystalline, the arrangement would be more helter-skelter, with molecules of different sizes and shapes making up a kind of three-dimensional fieldstone wall. But there's another kind of solid besides crystals and noncrystals, and that is metals. Metals have electrons that are free to roam throughout the material, almost like a gas. So a metal consists of ions (atoms missing electrons) vibrating around fixed positions and electrons in constant random motion. The presence or absence of free electrons makes all the difference in the world from an electrical standpoint.

If our solid were suddenly immersed in a strong electric field because of high electrical potential energy at one end and a low potential energy at the other, what would happen? If there are only a few free electrons, not much occurs—only a small current flows. But if we're stationed inside a metal where there are some free electrons, it's a whole new ball game. The electrons respond to the potential difference by moving toward the positive or high potential side. Yes, that's right. You thought current flowed from the positive toward the negative, but remember, electrons are negative, so they move toward the positive. If only Benjamin Franklin had reversed the naming, but it's too late now. Electrons flow in the opposite direction from the current. Electrons don't have smooth sailing in their quest to reach the high potential. There are all those big, lumbering ions in the way. There are collisions galore, but the electrons continue onward. By analogy, it's like a pinball game. Starting at the top of the slope, the pinball moves generally downward, although collisions may send it in various directions along the way. Even though it may take a while for a given electron to travel across the whole material, the electrons already at the other side complete the circuit quickly. So what happens to the ions that are on the receiving end of the collisions? They gain energy and thus heat up, increasing the overall temperature of the material. Generally speaking, metals are good *conductors*, while nonmetals are poor conductors, also known as *insulators*.

Although this model is a bit oversimplified, you can appreciate some aspects of electrical conduction as being fairly accurate, as in the way temperature affects resistance. As temperature increases, the ions' oscillations grow larger, so they become bigger targets for collisions with electrons, which increases the resistance. Conversely, as the temperature decreases, the resistance decreases. Interestingly, you might think that as absolute zero is approached, the resistance should approach zero also. Resistance does decrease, but there is an unexpected bonus: some materials have zero resistance at temperatures slightly above absolute zero. This phenomenon is called *superconductivity* and is the subject of continuing research. Of particular interest is a material that would superconduct at higher temperatures than near absolute zero. There has been progress in high-temperature superconductivity, and superconductivity at a temperature of 138 K (–135 °C or –211 °F) has been achieved recently.

Between conductors and insulators, there is an interesting class of materials known as *semiconductors*. The simplest semiconductor is a *diode*, which allows current to flow in one direction but not in the other. We'll see more details about diodes in general and specifically light-emitting diodes (LEDs) in Experiment 24.

Armed with this picture of what is happening at the submicroscopic level, let us retreat to the more familiar macroscopic world and deal with the resistance in Ohm's law, V = iR. The resistance depends on the details of collisions between electrons and the matrix of ions in the conductor, which may seem hopelessly complicated by all the possible energy interchanges and paths. But huge numbers and statistics save us again, just like they did in our analysis of gases in chapter 14. Because there are so many collisions, averages give an accurate approximation of reality. The resistance of a conducting cylinder depends on the material's collision characteristics and the geometric properties of the resistor. The longer the cylinder, the more collisions that can occur, so resistance increases with length. The bigger the cross-sectional area, the larger the channel for the electrons to flow through, so resistance decreases with an increasing cross-sectional area. In summary,

$$R = \frac{\rho L}{A}$$, where ρ = resistivity, Ωm
L = length, m
A = cross-sectional area, m^2

Material	Resistivity, Ωm
Silver	1.6×10^{-8}
Copper	1.7×10^{-8}
Gold	2.2×10^{-8}
Aluminum	2.7×10^{-8}
Tungsten	5.4×10^{-8}
Carbon	4×10^{-5}
Silicon	100
Glass	10^{9}–10^{14}
Hard rubber	10^{13}–10^{15}
Teflon	10^{23}

As the table shows, the extremely wide variation in resistivity clearly defines the difference between conductors and insulators. With a factor of

10^{30} difference in resistivity, you can understand how current would prefer to flow through an extremely long wire rather than a thin layer of insulation.

Now that resistors are better understood, let's apply some of our knowledge to a practical case, the automotive electrical system. Here's a portion of a schematic drawing for the automobile electrical system:

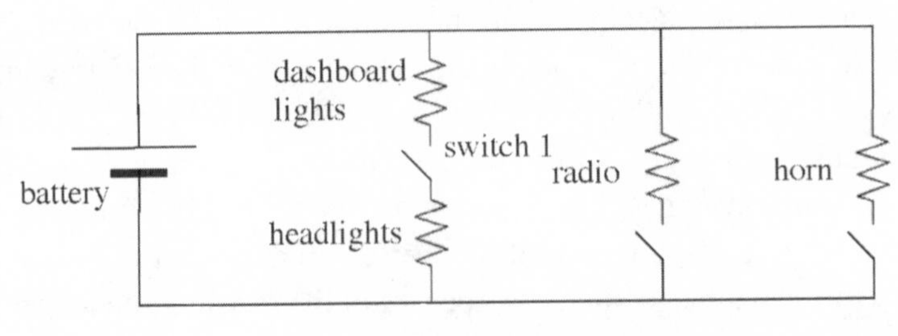

auto circuit schematic

(Electrical schematics always look neat, with straight lines representing zero-resistance wires, batteries that appear to have no internal resistance, simplistic switches, and resistors that look like little sawtooths. In reality, wires are flexible, batteries have internal resistance, switches can be quite complex, and resistors are often cylindrical.)

Once switch 1 is moved to ON, the circuit is complete, and current flows from the battery through the headlights and the dashboard lights. The same current flows through both, so when one set of lights is on, the other is also on. This is called a *series* connection (recall Experiment 20). Any number of resistors connected so the same current runs through them are in series. Since a whole string of resistors in a row would act just like one long resistor, you could replace a series of resistors by a single resistor whose resistance is the sum of all the individual resistors.

$$R_{\text{equivalent series}} = R_1 + R_2 + R_3 + \ldots$$

The radio and the horn, on the other hand, are independent. If one is turned on, it doesn't affect the other. Yet they do have the same voltage across them, namely, the battery voltage. This is a *parallel* connection (recall Experiment 20). If you had a group of resistors connected in parallel, it would give the current much more cross-sectional area through which to flow, but the area is proportional to

$\frac{1}{R}$, so the equivalent resistance would be found by adding reciprocals.

$$\frac{1}{R_{\text{equivalent parallel}}} = \frac{1}{R_1} + \frac{1}{R_2} + \frac{1}{R_3} + \ldots$$

In summary, connecting resistors in series makes the equivalent resistance larger, reducing the power consumption ($P = V^2/R$). Connecting them in parallel makes the equivalent resistance smaller, increasing the power consumption.

Capacitors can be connected in series or parallel, just like resistors. Interestingly, the equivalent capacitance works exactly the opposite from resistors. The equivalent capacitance of capacitors connected in series is given by:

$$\frac{1}{C_{\text{equivalent series}}} = \frac{1}{C_1} + \frac{1}{C_2} + \frac{1}{C_3} + \cdots$$

Capacitors in parallel simply add:

$$C_{\text{equivalent parallel}} = C_1 + C_2 + C_3 + \cdots$$

Similar to resistors, nonstandard values of capacitance may be achieved by combining standard capacitors in series or parallel.

But series and parallel aren't the whole story. It turns out that resistors can be connected in ways that are neither series nor parallel. This is especially true in multiloop circuits with several batteries. In these complex cases, circuits are analyzed by using Kirchhoff's rules.

In a resistor and battery network, start by assuming currents flow in every resistor. Give each current a different name and an assumed direction, then apply:

Kirchhoff's Loop Rule: *Around a loop in a circuit, losses in potential = gains in potential.* Pretend you're carrying a positive charge and walk around a loop, keeping track of whether you gain or lose potential as you go through resistors or batteries.

Kirchhoff's Point Rule: At any branch point in a circuit, *current in = current out*. Application of these rules determines the current through each resistor in a circuit, which forms the basis of circuit analysis.

As the auto circuit shows, real resistors transform electrical potential energy to some other kind of energy—light, heat, sound, and so on. Of course, the car battery's electrical potential energy comes from the heat engine that burns gasoline.

Resistors and capacitors can be combined within a single circuit.

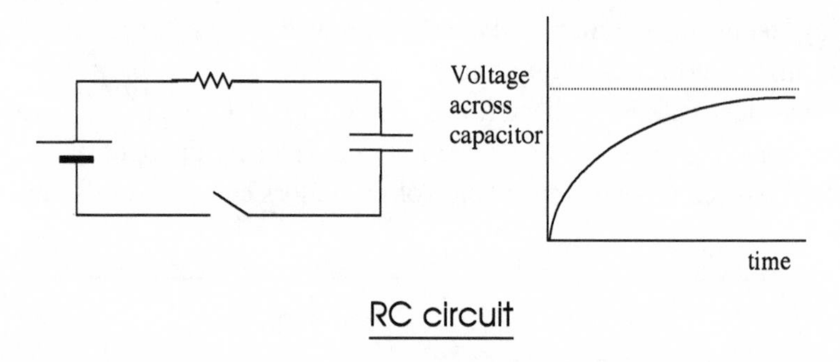

RC circuit

When the switch is closed, current flows through the resistor and then charges the capacitor. The voltage across the capacitor starts out at zero, then increases in exponential fashion, approaching the voltage across the battery. Mathematically, this function is given by:

$$V_{capacitor} = V\,(1 - e^{\frac{-t}{RC}}),$$ where V is the voltage across the battery,
e is the base of the natural logarithms,
t is the elapsed time in seconds,
and RC is the time constant in seconds

RC circuits are used in various timing circuits, such as pacemakers, traffic lights, flashers, and windshield wipers.

If you found electricity attractive, wait until you experience magnetism.

Experiment 21: Do-It-Yourself Magnet

Equipment needed: compass, 9 V battery, wires

A compass is a very handy device. It points north, helping us to find our way. The reason the compass works is because it is bathed in the magnetic field of the Earth. This experiment is designed to fool the compass a little, but not by using a natural magnet. As you will see, we are going to duplicate a classical experiment first performed by Hans Christian Ørsted in 1820.

Procedure

- Place the compass on a flat surface and make sure it moves freely. It should be pointing north.
- Set up the wire so that it runs past the compass, as close as possible, and in the same direction as the compass needle, north and south.
- Bend the ends of the wire so they touch the battery terminals, but don't connect them just yet.
- Watch the compass needle closely, then briefly connect the wires to the battery. There's probably not much resistance in the wires, so a large current will flow, but only briefly.
- You should observe the compass needle swing around to point at the wire.
- Repeat several times to make sure you see the effect.

Chapter 19. APPROACH/ AVOIDANCE —MAGNETISM

Nikola Tesla

"Magnetism is one of the Six Fundamental Forces of the Universe, with the other five being Gravity, Duct Tape, Whining, Remote Control, and the Force That Pulls Dogs toward the Groins of Strangers."

—Dave Barry

Magnetism is tricky. It's the most experienced but least understood force. Gravity we take for granted, electrical force is mainly experienced as annoying static cling, but we can feel the magnetic force directly. Refrigerator magnets, magnetic door latches, and magnetic compasses all demonstrate the reality of the magnetic force. Yet the cause of this force is not apparent and may seem almost like magic to the unsophisticated. People sell magnetic bracelets with supposedly curative powers, magnets that claim to align the gas molecules in your car's fuel line for better mileage, and so-called perpetual-motion machines that rely on magnetic force. Once again, we've run into the gulf between the small and invisible and the large and visible. We'll approach magnetism by first examining the effects of the magnetic field, then its causes.

Effects of the Magnetic Field

If you recall our discussion about electric fields, you'll remember that a positive charge plopped down in an electric field would feel a force. So how about a magnetic field? Would a charge sitting in a magnetic field, minding its own business, feel a force? No, there's no magnetic force unless the charge is moving. Further, if the charge does move and there is a force, the direction of the force isn't in the direction of the magnetic field, nor is it in the direction of the charge's velocity. The magnetic force's direction is perpendicular to the plane formed by the magnetic field and the velocity. As if that isn't complicated enough, that's still not the whole story. There are two different directions perpendicular to a plane. Think about the ground as representing a horizontal plane. Perpendicular to the ground can be either up or down. The direction of the magnetic force is determined by using the "right-hand rule." Point the fingers on your right hand in the same direction as the charge's velocity. Now curl your fingers toward the direction of the magnetic field. Your thumb now points in the direction of the magnetic force on the positive charge.

> The magnetic force on a positive charge moving in a magnetic field is directly proportional to the amount of charge, the charge's velocity, the magnetic field strength, and the sine of the angle between the velocity and the magnetic field. The force's direction is perpendicular to the plane formed by the velocity and the magnetic field, subject to the right-hand rule.
>
> $F_{magnetic} = qvB\sin\theta$, where q = charge, C
> v = velocity, m/s
> and B = magnetic field, tesla = T
>
> The force's direction is given by the right-hand rule.

Now that we have the magnetic field connected to a force, we can use Newton's second law to find the subsequent motion of the charge. But first, let us have a look at the very interesting fellow for whom the magnetic field unit is named.

Nikola Tesla (1856–1943) Mini-Biography[1]

In 1856 Nikola Tesla was born in Croatia. His father, Milutin, was a Serbian Orthodox priest, and his mother, Duka Mandic Tesla, was the daughter of another Serbian Orthodox priest. When Nikola was five years old, he found his older brother's dog dead by the roadside. Dane, his twelve-year-old brother, had been recognized as a child prodigy and was the family's favorite child. He was very upset and blamed Nikola for the dog's death. A short time later, Dane had an accident (a fall, either from a horse or down cellar stairs) and died from the injuries. Nikola thought his parents blamed him for his brother's accident and worked hard to try to make amends. Destined for the ministry, Nikola was a sickly child and became seriously ill several times. As he recuperated from one such illness, he read Mark Twain's *Innocents Abroad.* In his autobiography, he says it lifted his spirits enough to recover. Just after high school graduation, he contracted a serious case of cholera. As he was about to breathe what appeared to be his last breath, he revealed to his father that he hated the clergy and really wanted to be an electrical engineer. His father promised to send him to the best school if he would just recover. He did, and his father made good on his promise.

Tesla attended the Austrian Polytechnic in Graz, and spent many twenty-hour sessions studying electrical engineering, even irritating some professors by advancing beyond their knowledge. A piece of electrical equipment demonstrating the use of a commutator to change the direction of the current inspired Tesla to ask why not have the current flow both ways. Despite ridicule from the instructor, Tesla regarded this idea as a personal challenge and solved it spectacularly within two years. After more studies at the University of Prague (possibly achieving a degree), Tesla, in 1881, became the chief electrician for American Telephone and Telegraph Company in Budapest, Hungary. Here, he suffered another major illness, which made his senses abnormally keen. He says he could hear a watch ticking three rooms away. During his recovery, the solution to the generator with current flowing both ways occurred to him. What he hadn't been able to figure out earlier was that the generator required a rotating magnetic field. The whole design burst upon him complete in all details. What he needed next was a working model, but that would take a while. The telephone station in Budapest was sold, and Tesla went to work for Continental Edison in

Paris, where he became the company's troubleshooter. He thrived on long hours of work, rising at five for a swim in the Seine, a stroll, and breakfast, then arriving at work by eight thirty. He worked till late evening, then often ate at fancy restaurants, picking up the tab for whomever dined with him. He dressed in elegant clothes and cut a fine figure with his tall, slim frame and piercing blue eyes. After Continental Edison suffered an embarrassing failure (a short circuit blew out a wall) during the dedication of their electrical system at the railway station in Strasbourg, Tesla was sent to patch things up. Since it was important to the company, he was promised a bonus if the repairs were completed quickly and well. During the inevitable bureaucratic delays, Tesla found time to work on his project and rented a machine shop near his hotel. There, Tesla built a working model of his new generator. His Strasbourg friends, including the former mayor, weren't enthusiastic about his alternating current generator. When he returned to Paris after completing his assignment, his superiors were similarly uninterested. They pointed out that very wealthy people had invested heavily in direct current systems and wouldn't be interested in competitors. Somehow, Tesla's bonus never materialized.

Tesla's boss suggested he go to America to present his idea to Edison himself. Tesla quit his job, bought a steamship ticket, and gathered his meager possessions for the trip. As he boarded the train in Paris to begin the journey, his luggage was stolen. Thinking quickly, he boarded the train anyway, spent almost all his pocket money to get to the boat, then explained the situation to the boat officials. Tesla told them the time and location where he had bought the ticket and the number on it. He reasoned that either the thief would show up with the ticket, or no one would arrive, and he should be allowed to board. Although they were skeptical, they agreed to wait. When it was time to sail, and no one had showed up with the ticket, they allowed Tesla to be the last passenger aboard. After changing ships and a rough crossing, he arrived in America in 1884 with the equivalent of four cents, the least recorded amount for any immigrant at Ellis Island. Tesla soon met Edison, who was completely uninterested in the alternating current generator but hired Tesla on the spot and put him to work redesigning the company's direct current generators. Tesla said that Edison offered him $50,000 to increase the generators' efficiency. He worked with characteristic vigor—once for eighty-four hours straight—and completed the improvements Edison wanted. When he tried to collect the bonus,

Edison reportedly said, "Tesla, you just don't understand American humor." Tesla quit and formed his own company, Tesla Electric Light & Manufacturing. After satisfying the investors by designing an arc lamp for street lighting and industrial use, they forced him out of his own company when he tried to develop a brushless alternating current motor.

In 1887 and 1888 Tesla had to dig ditches to make a living, but a sympathetic foreman took him to see a Western Union official, who rounded up other financial backers. They organized the Tesla Electric Company, where Tesla was able to work out the mathematical details and build many different dynamos and motors, including a mechanical oscillator that shook neighboring buildings. In 1890 Tesla was invited to give a lecture to the American Institute of Electrical Engineers, titled *A New System of Alternating Current Motors and Transformers.* He became an instant celebrity, and accepted a $1 million plus royalties offer from engineer George Westinghouse for his patents. Half the million went to his investors, but Tesla figured the other half would support him indefinitely. That was not the way it worked out.

From 1890 to 1895 Tesla enjoyed a period of fame and good fortune that he would never see again. He became a US citizen, lived the high life in New York City, worked tirelessly, and befriended several prominent New Yorkers, including his boyhood hero, Samuel Langhorne Clemens, Mark Twain. In 1893 the Tesla/Westinghouse alternating current system lit a hundred thousand lightbulbs at the World's Fair in Chicago. AC had definitely arrived, but DC wasn't done just yet. "The War of the Currents" continued.

Tesla won the battle but suffered a crushing setback. His Houston Street lab was completely destroyed in a building fire. All his plans, equipment, and partially completed projects were ruined and had not been covered by insurance. Not long after, Westinghouse's financial backers questioned his royalty agreement with Tesla. When Westinghouse passed along their concerns, Tesla responded by tearing up the agreement. Although it was eventually estimated to have cost him around $10 million, Tesla was optimistic that he would make much more. Tesla's research interests included wireless power transmission (radio), remote control of mechanical devices (robotics), and single-node vacuum tubes (x-rays). Tesla had a new lab built in Colorado Springs, where he experimented with high-frequency, high-voltage systems, cosmic rays, atmospheric electricity, and electric oscillations of the entire Earth/ionosphere system. Just after he started up his "Magnifying

Transmitter," the power demands knocked out the Colorado Springs Electric Company's generator. The facility was soon closed and dismantled to pay debts. Tesla returned to New York and obtained financial backing from J. P. Morgan to build a laboratory on Long Island, called Wardenclyffe. The Wardenclyffe Tower was designed to be the hub of a World Radio and wireless power distribution system. Tesla's plan was to shoot electrical currents (similar to lightning) into the upper atmosphere. A distant receiving station would conduct and distribute the current. The current would then flow into the ground, over a hundred meters deep, and complete the circuit by flowing back to the distribution station at Wardenclyffe. In his view, the Earth and ionosphere constituted a giant capacitor, with the rest of the atmosphere acting as a dielectric. Delays and cost overruns required more funds, and when Morgan found out that Tesla's ultimate aim was the *free* distribution of wireless power, he withdrew funding and the project collapsed. Tesla's discouragement was severe, and his mental state may have deteriorated. He became more secretive than ever, lived in a succession of hotels (the Waldorf-Astoria, the St. Regis, the Governor Clinton, and the New Yorker), became morbidly afraid of germs (eighteen clean towels per day and dozens of napkins at each meal), and periodically issued grand statements about his current research without any evidence. In retrospect, some think he had obsessive-compulsive disorder. His reputation suffered, and few took him seriously. He died in 1943 and was cremated shortly thereafter. He is still in the public eye, the subject of a new opera that opened in 2004 called *Velvet Fire*. Tesla is depicted as a modern Prometheus, who suffers mightily because he stole lightning from the gods.

Although Tesla made no direct contribution to physics, his invention of the alternating current generator made possible the modern power grid, which enabled a second industrial revolution. Lord Kelvin said, "Tesla has contributed more to electrical science than any man of his time."

Back to the effects of a magnetic field on a moving charge. Let's say we're taking a ride on a positive charge and suddenly encounter a magnetic field. What would happen is we would experience a magnetic force perpendicular to our velocity. But we've seen that happen before. If we rode a yo-yo like the one in Experiment 5, the string force would be perpendicular to the velocity, and the yo-yo would be dragged around in a circle. The same thing would happen to the positive charge in a magnetic field. Its path would be bent into a circle around the magnetic field lines. Actually, if the charge had a velocity component in the same direction as the field, the path would become a spiral. So that's what would happen to our ride: it would go from a straight path to a spiral. A great example of this is provided by the Earth's auroras, the aurora borealis in the north and the aurora australis in the south. Charged particles emitted by the Sun form the solar wind. When the charges in the solar wind reach Earth, they spiral around the field lines of the Earth's magnetic field and head toward the poles. Before they actually reach the poles, the particles crash into the nitrogen and oxygen particles that make up the Earth's atmosphere and knock electrons loose by collisions. When these electrons are recaptured, light is emitted, creating spectacular light shows near the North and South poles. Both Jupiter and Saturn have similar auroras, thanks to comparable conditions. There are several links to aurora photographs in the endnotes.[2]

In 1930 E. O. Lawrence used a magnetic field to bend a beam of charged particles that were then accelerated after each 180° turn. The device was called a *cyclotron*, and it opened a whole new field, called *high-energy physics*.

Sources of the Magnetic Field

Magnets were recognized as far back as six thousand years ago by the Chinese and almost three thousand years ago by Greek philosophers in the form of an iron oxide called magnetite. In 1269 Petrus Peregrinus made magnetite spheres and sprinkled small iron pieces onto them. The iron pieces concentrated around two points on opposite sides of the sphere, which he called *north* and *south poles*. Today, using two refrigerator magnets as our modern version of lodestones, you can demonstrate magnetic attraction and repulsion. This requires large, strong magnets and a little maneuvering. Does this begin to sound familiar—two different entities undergoing attraction and repulsion? Yes, magnetism resembles its cousin, electricity.

In 1600 William Gilbert suggested that both lodestones and the Earth are magnetic. Does that mean that there's a giant refrigerator magnet buried within the Earth? Even though the geographic poles are cold, there's no refrigerator magnet there.

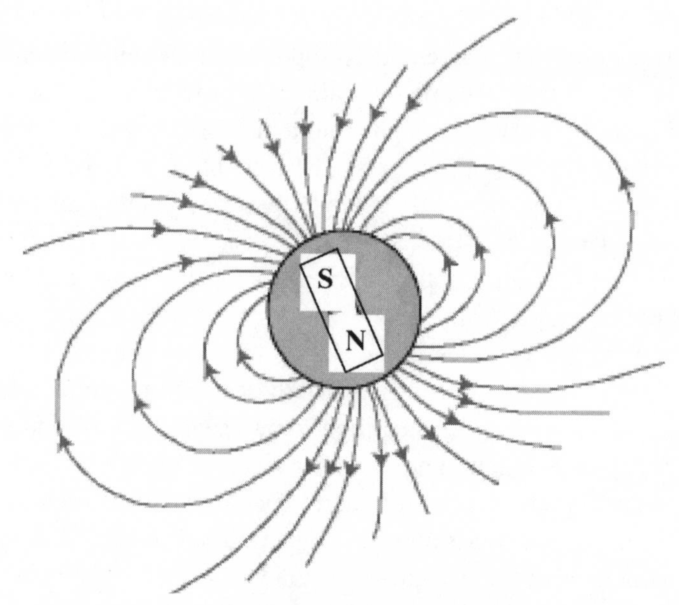

Earth's magnetic field

It turns out that all magnetic fields, even those within permanent magnets, are generated by moving charges. The basis for this was discovered in 1820 by the Danish physicist Hans Christian Oersted (Ørsted in Danish). He was preparing a lecture explaining electricity and magnetism to the public and noticed a compass needle deflect every time an electric current was run through a nearby wire (just as you found in Experiment 21).

> A magnetic field is caused by an electric current. For a straight current, the field strength at any point is directly proportional to the current and inversely proportional to the perpendicular distance to the current. The field circles around the current the same way the fingers of your right hand curl if your thumb points in the direction of the current.
>
> $B_{\text{straight wire}} = \dfrac{\mu_0 \, i}{2\pi \, r}$, where μ_0 is the permeablity of free space
> $= 4\pi \times 10^{-7}$, T m/A
> i = current, A
> r = perpendicular distance to the wire, m
>
> The field direction is given by the right-hand rule.

The permeability of free space is so small that magnetic fields generated by a current flowing through a straight wire are usually tiny. To increase the size of the field, all you have to do is bend the wire into the form of a loop. Then the magnetic field is concentrated inside the loop, and the field at the center of the loop becomes inversely proportional to the radius of the loop.

$B_{loop} = \frac{\mu_0}{2\pi}\frac{i}{R}$, where R is the loop radius, and the field direction is along the axis of the loop. Even better, if you form a *coil* or *solenoid* by stacking up a whole series of loops, the magnetic field becomes larger.

$$B_{solenoid} = \frac{\mu_0}{2\pi}\frac{Ni}{L}$$

where N is the number of turns and L is the solenoid length

If a solenoid is bent to form a shape like a doughnut, called a *toroid*, the field is even stronger.

The current loop is at the heart of *all* magnetic effects. Let's start at the smallest level and build up to the largest.

Iron atoms: Deep inside refrigerator magnets, compass needles, and lodestones is the iron atom. The iron atom is magnetic because of spinning electrons. (Thinking of an electron as an actual spinning ball is a quantum mechanically inaccurate picture but a useful analogy for our purposes here.) In most atoms, electrons spinning counterclockwise are balanced by electrons spinning clockwise, so there is zero net spin, hence no net charges moving in a circle. But because of the way atoms are built, iron (and, to a lesser extent, cobalt and nickel) has a net electron spin and generates a magnetic field. When a large number of atoms align so their fields add to each other, it is called a *domain*. You might wonder what causes them to line up in the first place. Good question. Two things have to happen: the iron's temperature must be high enough for them to have some mobility (called the *Curie temperature*, 770 °C or 1420 °F for iron), and they must be in the presence of an external magnetic field. These conditions are met by meteorites, in lightning strikes, and within the Earth.

The Earth: The best theory of geomagnetism, although not completely worked out, is called the *dynamo theory*. Below the Earth's crust and mantle are the outer and inner cores. The outer core is thought to be liquid iron and nickel that sloshes as the Earth rotates. The sloshing action frees enough electrons that an electric current circulates in the core and produces the Earth's magnetic field. The core's rotation isn't in perfect synchronization with the solid part of the Earth (ever try to rotate a bucket full of water?), so the current and field periodically reverse—north becomes south and vice-versa. This reversal happens sporadically—on the average about every four hundred thousand years. It's been 780,000 years since the last one, so watch your compasses. You needn't worry, though. The Earth's overall rotation won't be affected, but there'll be no more auroras. When the solar wind comes in, there'll be no magnetic field to deflect it while the reversal is in process. So you'd better hope your DNA doesn't get hit and turn the next generation into mutants. The dynamo theory also explains the magnetic properties of other planets in the solar system. Jupiter and Saturn with their large, gassy atmospheres and rapid rotation have the largest magnetic fields in the solar system, and Venus, with a core-mantle-crust system like Earth's, has a very slow rotation (243-day period) and almost no magnetic field. The Sun has plenty of charges in its plasma atmosphere and it rotates faster at the equator than the poles, making a tangled mess of magnetic field lines. This gives rise to sunspots, flares, and all sorts of magnetic mischief. Things get even more interesting when you go a little farther out in the universe and encounter neutron stars. At the end of a large star's lifetime, the atmosphere is blown out and the star's core collapses, which forms a neutron star. However fast the normal star was spinning, the neutron star spins way faster because it has collapsed to a small size. The rapid spin rate and whatever loose charges it picks up from its neighborhood make for large currents and hence a large magnetic field. Any charged particles that try to escape from a neutron star spiral around the field lines and crash into the poles like some aurora on steroids. The only particles that do escape come from the poles. Because of the typical offset between the spin axis and the magnetic poles, the radiation from the poles, referred to in astronomy as *jets*, rotate like a searchlight. These objects have been observed and are called *pulsars*.

By now you may be wondering if magnetic fields are made by currents, and there are so many symmetries between electricity and magnetism, might the situation be reversed and have a current caused by a magnetic field? The answer is yes, under some conditions that we are about to see.

Experiment 22: Battery-less Voltage

Equipment needed: coil, magnet, meter, leads

Usually, voltage is associated with batteries, capacitors, or wall outlets. But in this experiment, we will generate voltage (and current) in a coil with nothing but a magnet.

PROCEDURE

- Set the coil on a flat surface and locate the two ends of the wire that form the coil.
- Make sure the insulation is stripped from the wire ends, possibly using sandpaper.
- Attach the multimeter to the two wire ends and make sure the connection is tight.
- Turn the meter selector switch to the lowest voltage setting.
- Hold the magnet along the axis of the coil and move it quickly toward the coil.
- The meter should indicate a voltage as the magnet moves, then the reading should fall to zero when the magnet stops.
- Turn the magnet 180° and move it toward the coil again.
- The meter reading should reverse direction +/- or -/+.
- If you can accomplish this without disturbing the meter, hold the magnet still and move the coil toward or away from it and note the meter reading when the coil moves.

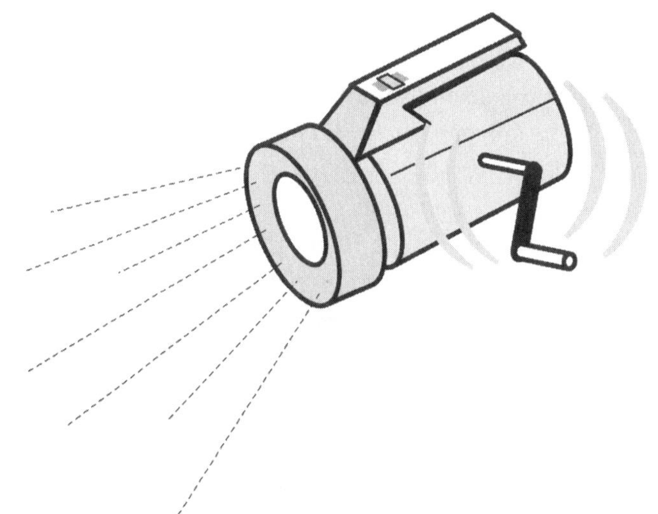

Chapter 20. WE ARE FAMILY— ELECTRICITY FROM MAGNETISM

"I know of no sentence that can induce such immediate and brazen lying as the one that begins, 'Have you read . . . .'"

—Wilson Mizner

Ørsted's 1820 discovery that magnetic fields are generated by electric currents excited (electrified?) scientists worldwide to a flurry of experimental activity. At the Royal Institution in London, Humphry Davy (remember him from chapter 15? He was hired by Count Rumford) discussed the implications of Ørsted's work for experiments conducted by a colleague, William Wollaston. Wollaston's work proceeded slowly, but Davy's assistant, Michael Faraday, carried out some electrical experiments and got significant results quickly. Davy charged plagiarism (Wollaston didn't) and assigned Faraday to work on optics projects. After Davy's death in 1829, Faraday resumed electrical research and established his law of induction in 1831, which was based on experimental evidence.

Extremely similar experiments were carried out in the United States by the American scientist Joseph Henry at the Albany Academy in New York. Although Henry may have carried out his experiments earlier, Faraday published first, so he got the credit.

Faraday (and Henry) generated a magnetic field by sending a current through a coil. To enhance the field, Faraday had the coil wound around an iron ring. Then he placed a second coil onto the ring, hoping the second coil would have a current induced by the magnetic field of the first. The effect didn't quite happen that way. Instead, current flowed in the second coil only when the first current changed. The key to induction turns out to be the time rate of change in the product of the magnetic field times the area of the loop.

Faraday's law: A voltage is induced in a conducting coil of N loops when the coil is immersed in a magnetic field that changes with time. The induced voltage is directly proportional to the number of loops in the coil and the time rate of change of the product of the magnetic field times the loop area. The direction of the induced voltage is such that the current that flows generates a magnetic field that opposes the change that produced it.

$$\text{Induced Voltage} = -N\,\frac{\Delta(BA)}{\Delta t}, \text{ where } N = \text{turns},$$

Δ = change in
B = magnetic field, T
A = loop area, m^2
and t = time, s

The negative sign is called Lenz's law and is a reminder that the induced voltage causes a current to flow in a direction such that its magnetic field opposes the change that caused it.

Lenz's law may be difficult to state but is quite sensible when you think about it. Let's apply it to Experiment 22. If you hold the north pole of the magnet toward the coil, then move it in the coil's direction, that action stuffs more magnetic field lines through the coil. Let's consider each of the two possible directions for the induced voltage and current. If the current direction is such that it would generate a magnetic field in the same direction as the field is already increasing, that would draw the magnet *in*. That, in turn, would create *more* current, and so on. This is an unstable situation referred to as *positive feedback*. On the other hand, if the current produced a magnetic field in the direction that opposed the increase, that would resist the magnet's inward motion, which matches the experimental fact. The experiment also showed that the voltage is generated whether the coil or the magnet was moving; it is the *relative* motion that matters.

If you have a good visual picture of magnetic field lines, you can envision that relative motion between the coil and the magnet has the effect of cutting field lines. I often think of voltage induced in the opposite direction as a toll that must be paid for cutting magnetic field lines.

So far, we've made varying magnetic fields by physically moving natural magnets or the coils themselves. Another way to generate varying magnetic fields would be to use electromagnets and either move them physically or vary the current through them, both of which methods would cause the magnetic field to change. Changing currents happen in many electronic circuits that were not intended to generate any induced voltages at all. Suppose you have two totally unconnected circuits in close proximity. When one circuit is turned on, it creates a changing magnetic field. The other circuit is close enough to that changing field, so a back voltage is induced. This situation is called *mutual inductance* and is described mathematically by the equation:

$$V_{\text{induced in circuit 2}} = M \frac{\Delta i_1}{\Delta t}$$

where M = coefficient of mutual inductance, henrys = H
Δi_1 = change in circuit 1's current, A
Δt = time interval, s

(Note that the unit of inductance is the henry, named for Joseph Henry. Faraday already had a unit named for him, the unit of capacitance, the farad.)

Think about this from circuit 2's standpoint. You're just humming away, minding your own business, and there's a sudden voltage surge because of current changes in some nearby circuit that you didn't know about. How rude. But that's not the worst of it. How about circuit 1 itself? If the current increases, a voltage is induced in the opposite direction, a phenomenon often referred to as a back voltage. Conversely, if the current diminishes, the induced voltage tries to keep the current flowing. This is called *self-inductance* and is given by:

$$V_{induced} = -L \frac{\Delta i}{\Delta t},$$ where L = coefficient of self inductance, henrys = H

Δi = change in circuit 1's current, A

Δt = time interval, s

In some circuits, inductance is undesirable and must be minimized, but in other cases, inductance is necessary. Inductors are commercially available, often in the form of coils. Inductors, resistors, and capacitors are circuit elements available to electrical circuit builders. A circuit consisting of a battery, a switch, an inductor, and a resistor would have its current growth delayed by the inductor's induced voltage.

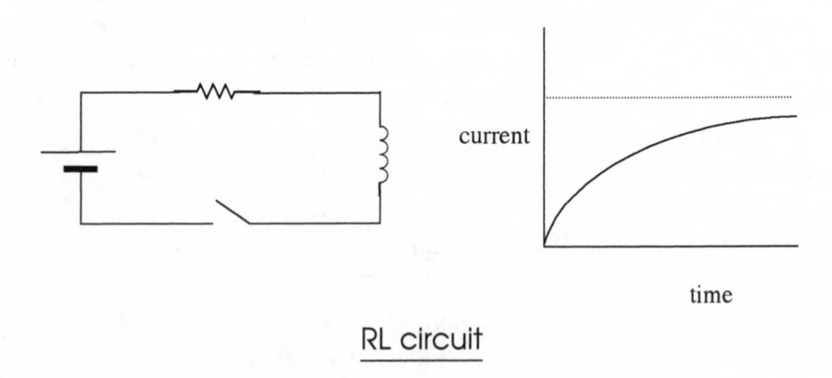

RL circuit

Mathematically, the current is given by:

$$i = i_{max} \left(1 - e^{-\frac{t}{\left(\frac{L}{R}\right)}}\right).$$

You may note the similarity between this expression and the voltage across a capacitor in an RC circuit in chapter 16. Just as the current is delayed from reaching its maximum value when the switch is turned on, induction causes the current to continue awhile even though the switch is turned off. You may have noticed an inductor at work when a radio delays slightly when it is turned on or when it keeps playing for a short time after it is turned off.

A clever gadget that makes use of induction is the *generator*. In its simplest form, a generator is a square loop of wire rotating in a magnetic field. The ends of the loop are kept in contact with fixed wires by carbon brushes, and the fixed wires are connected through a resistor to form a complete circuit.

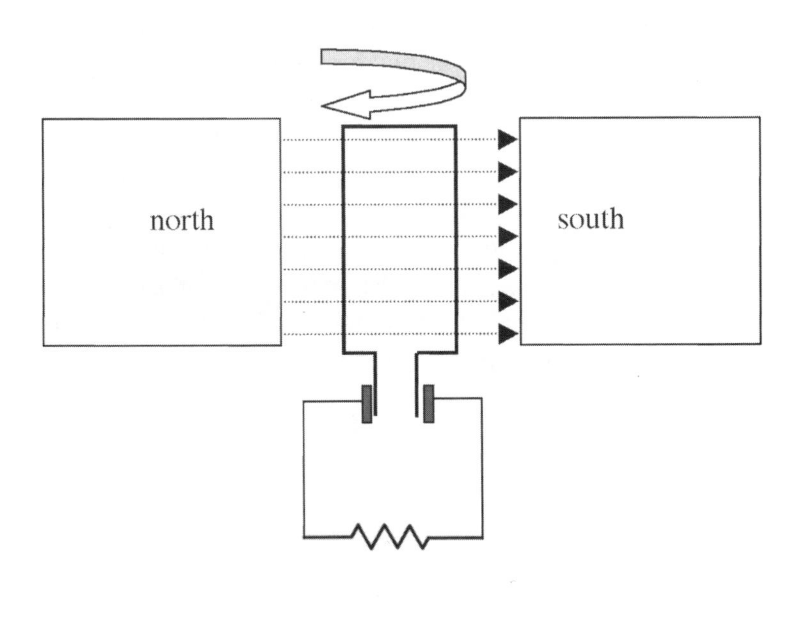

generator

Interestingly enough, instead of turning the coil to generate current, it is also possible to operate this in reverse. You can run a current through a wire and cause a coil to rotate. This is an *electric motor*. There are many technical details that must be solved to make this a practical device, but this is the basic idea.

There are some questions that need to be addressed about a generator:

- What makes the loop spin? As you know, things don't move in circles all by themselves, especially when a force (magnetic, in this case) opposes their motion. The energy needed to turn this coil may come from a variety of sources, such as flowing water, blowing wind, or turbines spun by steam heated by coal, oil, geothermal heating, or nuclear reactions. A generator may be a kind of misnomer, if it implies that the energy starts with it. Energy doesn't originate with the generator; it is simply converted from another form.
- What form does the current take, coming out of the loop? Depending on the details of how the carbon brushes are arranged (slip ring or split ring), the current comes out in one direction or alternating directions. The first form sounds a lot like direct current (DC), but it's not steady as you would expect DC to be. This can be remedied by adding some capacitance to the circuit, which delays the variations. The other form changes the direction constantly and is called alternating current, AC. As for the brushes themselves, they would seem to create a lot of friction and therefore cause losses and maintenance problems. Now you understand the question Tesla asked his college professor. The answer to his question and many others will be found shortly.

Experiment 23: All Shook Up

Equipment needed: shake flashlight

The shake flashlight has entered pop culture.[1] A Google search generated seven hundred thousand hits for shake flashlights, also known as forever flashlights, Faraday flashlights, LED torches, no-battery flashlights, magnetic induction, everlife, eternal, and more. I love them all because there's so much physics in them. The point of this experiment is to get inside one of these gadgets and see how they illustrate the physics we've been discussing. No particular size shake flashlight is required for this experiment, but its case should be transparent, and it should be capable of being disassembled, because we're going to do a bit of reverse engineering.

Procedure

- Shake it (the flashlight, that is) for a while, turn it on, and leave it on for a while to see how long it takes to dim. If possible, do this in a dark room (then turn on the lights).
- Inspect it carefully so you know how it looks assembled, since you will need to reassemble it.
- Disassemble the flashlight and keep track of all the pieces.
- Find the magnet. More expensive models have more powerful magnets. Test the magnet by seeing how strongly it attracts iron-based objects. Don't get it near your credit card or it may scramble your code numbers.
- Find the coil. Can you visualize the field lines from the magnet being cut by the coil as the magnet slides through? This is the heart of the flashlight's operation, since the induced voltage causes the current that creates the light. Does it bother you that the magnet slides both ways, so voltage should be induced in both directions? It should. The next part of the circuit will take care of this problem.

- Refer to the circuit diagram below. Suppose current flows out of the *top* of the coil. It then enters the 4-diode network (called a *rectifier*) at the top. As we discussed earlier, diodes allow current to flow only in one direction, indicated by the direction of the arrow in the symbol. As long as the switch is open, the only place that the current can flow is to the upper plate of the capacitor. Trace the current path on the diagram.

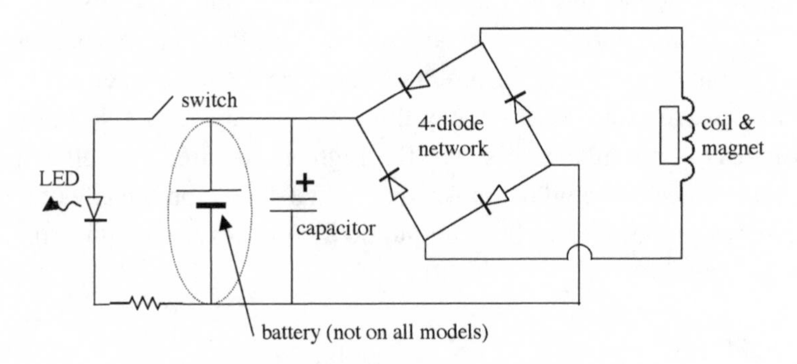

shake flashlight schematic

- If current flows out of the coil *bottom*, it enters the 4-diode network at the bottom and can flow only to the capacitor's positive plate, as before. Trace this path on the diagram. The net result is that either direction of motion of the magnet in the coil charges the capacitor's upper plate positive.
- In most shake flashlights, the circuit components beyond the coil and magnet are located on a circuit board, similar to the breadboard used in Experiment 20. Look on the circuit board and see if you can identify the four diodes. They often have a colored band at the end where the current exits. Try to trace the diode connections.
- Next, locate the capacitor, which is often the largest component on the board. Sometimes you can see the capacitance stamped on the side. If you see it, write it here _______.
- Once the capacitor is charged, the next activity is to close the switch. Current then flows from the positive plate of the capacitor through a special diode, called a *light-emitting diode* (LED). (The operation of a LED will be discussed in the next experiment.)

- Because the LED is physically small (and consumes little power), a lens is used to spread the light, making it more like a normal flashlight beam.
- The last item on the circuit board is a resistor in series with the LED. Its function is to limit current so there is enough to light the LED, but not enough to burn it out. Just for the practice, read the resistor's color code and write it here: ________________. Return to the table from Experiment 19 and determine the value of the resistor. Write it here: ____________.
- (Optional) Some flashlights (often the cheapest ones) have two small "button" batteries, like the ones normally used in a watch or a camera. They are connected in parallel with the capacitor, and make the LED shine even if you don't shake the flashlight. They also keep the brightness almost constant. They are not rechargeable, so they will eventually discharge, leaving the LED powered only by the capacitor. Since the batteries are connected in parallel with the capacitor, you can remove them, and the light will still work. Remove the batteries and see how quickly the light dims (you may want to put them back in before you reassemble the flashlight).

For a detailed analysis of one particular shake flashlight model, see the Web site listed in the endnotes.

Chapter 21.
BIG-TIME ELECTRICITY —AC

"Genius is 1% inspiration and 99% perspiration. Accordingly a genius is often merely a talented person who has done all of his or her homework."

—Thomas A. Edison

"His (Edison's) method was inefficient in the extreme, for an immense ground had to be covered to get anything at all unless blind chance intervened and, at first, I was almost a sorry witness of his doings, knowing that just a little theory and calculation would have saved him 90 percent of the labor."

—Nikola Tesla

As we just saw, the principles behind generators are simple enough that either direct current (DC) or alternating current (AC) generators can be built. So, you may ask, why is the world's power grid almost exclusively based on alternating current? Good question; glad you asked. The answer is twofold: (1) physics and (2) economics, personalities, and politics. The physics we can handle; the rest we'll just sketch and leave the details to others.

The Physics of AC/DC

Once current has been generated, whether it's AC or DC, the resistance of the wire collects its toll. The power lost due to a current in a wire is $P = i^2R$. Since the wire is a cylindrical conductor, its resistance is $R = \rho L/A$. Combining these relations yields: $P = i^2\rho L/A$. Clearly, if you want to minimize the heating losses in the wires, you must limit the current, the resistivity of the wires, and the wire length, and/or maximize the wire area. Although silver has the lowest resistivity (see chapter 18 for a table of resistivities of metals), its cost makes it prohibitive. (Please excuse me. Economics seems to have slipped in here.) Fortunately, copper is nearly as good. The current in a circuit is dictated by resistors used in the circuits connected to the generator, as well as how many circuits there are. The wire's length depends on how far the generating plant is from the customers. And the wire area would be determined by the amount of copper either strung on poles or buried underground. (Oops, economics again.) The biggest payoff would come from limiting the current, since current appears as a squared term, but that seems a forlorn option if you want to supply a large number of customers. Here's where AC diverges from DC.

Do you recall (in chapter 20) Faraday's iron ring with two coils wound on it? Changing current in one coil induced voltage and current in the other coil. Calling the first coil the *primary* and the second coil the *secondary*, let's make a little variation on Faraday's ring. Feed AC into the primary and AC will be induced in the secondary, as you would expect. But if the number of windings on the two coils is not the same, something very interesting occurs. The voltage across the secondary differs from the voltage across the primary. The relationship is a very simple one:

$$V_2 = V_1 \frac{N_2}{N_1}$$

where V = voltage and N = number of turns

This device is called a *transformer*, and there are two kinds, depending on the ratio of the number of turns.

$$N_2 > N_1 \Rightarrow V_2 > V_1 \text{ step-up transformer}$$

$$N_2 < N_1 \Rightarrow V_2 < V_1 \text{ step-down transformer}$$

It might seem like the transformer is somehow manufacturing energy, but it isn't. If there are no losses in the iron core (not quite true, since changing the magnetic orientation of those domains of iron atoms doesn't happen for free), the same power is delivered to the secondary as there is in the primary, so $i_1V_1 = i_2V_2$. Here's where AC's payoff comes in. If you transform the voltage up, the current goes down. When the current decreases, the power losses diminish. So the technique is to generate AC at whatever voltage is convenient, transform it to higher voltage, then send it along transmission lines with small losses. When this high-voltage, low-current AC arrives near the customer's location, it is then transformed downward to a lower voltage and higher current. This minimizes the transmission losses but gives the customers the large currents they need. Since transformers need a changing current to function, no such procedure is possible with DC.

The Personalities, Economics, and Politics of AC/DC

We've already seen a sketch of one of the principal actors in this drama, Nikola Tesla. Next, let's take a brief glance at the others: Thomas Alva Edison and George Westinghouse.

Thomas Alva Edison and George Westinghouse, Mini-Biographies (up to 1887)

Thomas Alva Edison (1847–1931)[1]

Thomas Alva Edison

Thomas Alva Edison was the seventh and last child of Samuel Edison and Nancy Matthews Elliott Edison and was born in Milan, Ohio. In 1854 the Edisons moved to Port Huron, Michigan, where they hoped the lumber business would be better than in Ohio. Shortly after the move, young Alva (as he was called) contracted scarlet fever, which delayed his entry into formal schooling until he was over eight. After three months at school, he came home crying, saying his teacher had referred to him as "addled." His mother withdrew him immediately and took on the task of his education. She encouraged his independent thinking, which flowered under her guidance. Edison worked his way through many classics, including Newton's *Principia,* with the help of a family friend. He acquired a distaste for mathematics because he thought Newton could have appealed to a wider audience if he used less math. In 1859 the Grand Trunk Railroad completed a rail line that included a run from Port Huron to Detroit. Edison became a newsboy on the morning train, selling produce on the way to Detroit and newspapers on the way back. Although he became hard of hearing around this time, Edison was quite successful and even used a spare freight car as a laboratory for his personal experiments and for printing a newspaper of his design, the *Grand Trunk Herald.* By 1863 Edison had gone beyond newspapers and become a telegraph operator. Moving from place to place for several years, working the night shift and studying and tinkering during the day, Edison finally arrived in Boston in 1868. By the end of the year, Edison made a fateful decision. He resigned from Western Union to devote his full effort to "bringing out inventions." His first was a vote recorder for congressional hearings. It pleased no one and was abandoned. His next effort, an improved stock ticker, was only a little better. A system for sending more than one message over a single telegraph wire (called a *duplex*) seemed promising, so Edison borrowed $800 to build the equipment and convinced the Atlantic & Pacific Telegraph Company to let him give it a test over their wires, sending multiple messages from Rochester to New York. It failed miserably, even though the next issue of the National Telegraphic Union's magazine, the *Telegrapher*, reported it as a "complete success." Edison returned to Boston in disgust, borrowed some more money, and went to New York, flat broke and in need of a job. In New York, he contacted his friend Franklin Pope, who was a well-respected telegrapher, author, and editor of the *Telegrapher* (and possible participant in the duplex test). Pope was also chief engineer for the Laws Gold Reporting Company, which ran a service that relayed gold prices via wire from the New York Gold Exchange to several hundred brokers' offices. There was no job for Edison, but Pope arranged for Edison to sleep in the company's basement battery room until he found work. Having no other place to go, Edison accepted. After business hours, he had the run of the place and soon figured out how the machinery worked. Shortly after Edison moved in, the transmitter quit in the middle of a business day, and the office became crowded with messengers from the brokerage houses, wanting current gold prices. Edison was present and almost immediately found and repaired the problem. The following day, Edison was made Pope's assistant and within a month had Pope's job (at $300 per month), when Pope resigned to become an independent consultant. Within three months, Edison improved the operation and applied for some patents. He then found himself an employee of Western Union again when they bought the Gold Reporting Company.

Through several more inventions and improvements on telegraphic and printing equipment, including waxed paper for mimeograph machines, Edison was rewarded with a huge payoff from Western Union: $40,000. With this money, Edison set up his first workshop in Newark, New Jersey. More inventions followed, mostly dealing with telegraphic equipment. Eventually, Edison and his assistants outgrew the early lab spaces and moved to Menlo Park, New Jersey. Within a year of the move, Edison invented the phonograph, and a year later was hard at work on an *incandescent* lamp. The principle was simple enough: run an electric current through a material that would glow brightly enough to provide light. The difficulties included finding a material for the filament that would last, attaching the electrical contacts to the filament securely, finding the right shape for the bulb, and making a vacuum inside the bulb to prevent the hot filament from reacting chemically. Edison's assistants tested many, many materials before settling on carbonized thread in a highly evacuated bulb. By late December 1879, a hastily rigged DC system of electric power generation, distribution, and lighting was set up for public viewing at Menlo Park. It was a huge success, with people braving stormy weather to see "the Wizard of Menlo Park's" latest invention. Converting the initial system to a commercially viable one took a bit of effort, as well as some time. The dynamo had to be set up, bulbs manufactured, and shallow tunnels dug for the wiring to be buried. In September of 1882, Edison's Pearl Street Station in Manhattan was finally ready, and before the end of the year, twenty-four hundred Edison bulbs glowed brightly in offices within New York's financial district. As noted in our mini-biography of Nikola Tesla, he worked for Edison from the summer of 1884 through spring 1885, helping troubleshoot Edison's DC systems. Edison was completely uninterested in Tesla's AC ideas, since he was so thoroughly committed to DC. Besides, Edison told Tesla, AC is "a deadly current, whereas direct current is safe." Besides the usual difficulties of installing new technology, the price of copper became artificially inflated because a speculator was attempting to corner the market. Edison's financial backers were displeased by the slow (in their view) progress. By 1887 Edison was feeling squeezed from several directions.

Now for the Other Participant in the Current War George Westinghouse (1846–1914)[2]

George Westinghouse was born in 1846, in the small village of Central Bridge, New York. His father manufactured farm implements, which exposed young Westinghouse to machinery at an early age. He and his two brothers served in the military during the Civil War. After the war ended, Westinghouse spent a short time at Union College studying engineering. He cut his studies short to become an inventor. In 1865 he obtained his first patent, for a rotary steam engine. The railroad industry caught his attention, and he designed a device for getting derailed cars back on the track, another that prevented derailments at switches, and a fail-safe braking system that used compressed air. Next an automatic system followed, in which electricity was used to signal the passage of trains. Westinghouse patented his inventions after seeing the first few stolen by unscrupulous railroad managers. He started many different companies to produce his inventions and guarded his patent rights fiercely. An innovative employer, Westinghouse paid his workers well, cut the workweek from six days to five and one-half, and was among the first to institute paid vacations and pensions. When an exploratory oil well on his property in Pittsburgh produced a gusher of oil and gas, Westinghouse designed and built a distribution system so the gas could be reduced in pressure and safely piped to many homes.

George Westinghouse

In 1884 Westinghouse hired William H. Stanley Jr., an inventor and patent holder in his own right. Not long after hiring Stanley, Westinghouse read about a transformer (then called a secondary generator) invented by the Europeans Gaulard and Gibbs that was used to step down high-voltage AC to lower voltages suitable for lighting. Westinghouse bought US rights to the transformer, and Stanley proceeded to improve the idea to make it commercially practical. Westinghouse thus entered the electrical business with an AC system. They lit several commercial establishments along the main street in Great Barrington, Massachusetts, where Stanley lived and maintained his lab. The AC generator for this system was initially a European import, but Stanley built an improved one. From the customers' point of view, there was little difference between the Westinghouse system and the rival Edison system, also in use in Great Barrington in 1886. By 1887, after one year in business, Westinghouse had 68 AC systems built or under contract, while Edison had 121. If only Westinghouse had an AC motor, he could have competed with Edison across the range of electrical business, not just lighting.

Current Wars

The AC/DC competition took a strange turn of events in 1888. Edison opened a state-of-the-art lab in West Orange, New Jersey. Westinghouse and Tesla connected, providing Westinghouse with the AC motor he had been needing. And several people were killed in electrical accidents. There followed a scathing letter to the editor of the *New York Evening Post* about the dangers of AC and how the public was in "constant danger from sudden death" because of AC. The letter writer, Harold P. Brown, a seemingly obscure New York engineer, recommended that AC above 300 volts be outlawed in the interest of public safety. Truly, there was some public danger involved, but it was mostly due to the huge number of overhead wires already in place. They had been strung, willy-nilly for arc lighting systems, telegraph, stock tickers (including the Gold Exchange), and other private electrical systems. It seemed like an opportunity was seized to discredit Edison's competition, even though they weren't at fault. Westinghouse understood this and sent a letter to Edison, proposing peace between the companies. Edison's answer was, "My laboratory work consumes the whole of my time."

Westinghouse felt like he had no choice but to fight. He appeared before the New York City Board of Electrical Control and quoted impressive safety statistics that favored his system over Edison's DC. By the end of July, Brown struck back. He held a demonstration at Columbia College in which he subjected animals to various electric shocks, trying to demonstrate the danger of AC. When they realized what he was going to do, many people left the room. Finally, an agent for the ASPCA stood up and forbade Brown to execute any more animals. The hostile audience filed out, with a diatribe against AC bouncing off their departing backs. After several other animal execution demonstrations, Brown began to work on a larger target: electrical execution (electrocution) with alternating current as a "quick, humane" form of capital punishment. Letters, testimony, legal actions, legal fees, and many billable hours flew back and forth. The point was that the DC forces wanted to portray AC as a "killing current" so the public would fear it. In August 1889 the *New York Sun* published an exposé of Harold P. Brown. Someone broke into his Wall Street office and stole forty-five letters that showed he was paid by the Edison Company and the Thomson-Houston Company. Nevertheless, bureaucratic wheels continued to grind onward, proceeding toward the inevitable conclusion. In August 1890 convicted felon William Kemmler was executed in the electric chair at Auburn prison in New York. Driven by a generator voltage of 1,000 to 1,400 volts, electrical current surged through Kemmler's body for seventeen seconds before the prison doctor pronounced him dead. As other attending physicians examined Kemmler, his chest suddenly heaved up and down. Quickly, they reattached the electrodes and ran the current for several minutes. Kemmler was finally dispatched, but it was hardly neat or even humane.

The current war (and a bulb war also) continued in less bizarre fashion, with the battlefield shifted to financial, governmental, and courtroom venues with small victories for each side. The next big skirmish—the Chicago World's Fair of 1893—was a victory for the AC team, but the basis for decision was economic rather than technical. The decisive battle was fought at Niagara. The Cataract Commission was set up to decide on the best method for extracting energy from the raging Niagara River. After keen (and some not-so-keen) analysis, the commission awarded the contract to Westinghouse's AC system, since DC transmission losses were so severe that insufficient power would get through to Buffalo to drive anything. In a sense, AC won the wars, since the power grid is AC-based. But both Edison and Westinghouse were losers, since they lost control of the companies that bore their names. Financial backers and skilled political infighters of both companies were the real beneficiaries of the war, if not the ultimate winners.

Currently, the power grid in the United States features about six thousand generators that produce almost 4×10^{11} watts. A typical generator output of about 30 kV is then transformed up to 138 to 765 kV for long-distance transmission, transmitted, then transformed down in stages, ultimately to 120 volts for home use. About 7 percent of the power generated is lost in heating the wires in the transmission lines and in the transformers at both ends. Since AC cannot be stored like DC (in a capacitor, like the shake flashlight), the game is to transmit AC and accept the small transmission losses. Fortunately, the demand for electrical power is not constant. As you might expect, during the winter, the demand peaks in the early evening hours, when lights are turned on. During the summer, electrical demand peaks around two PM as the heat of the day encourages the use of air conditioning. In all seasons, as midnight approaches, the demand becomes less. But midnight on the eastern seaboard is only nine PM on the West Coast. You can almost picture the peak demand as following sunset around the globe. Making adjustments in generator capacity is difficult, since generators run best at constant speed and so are either online or off-line. This makes the odd situation that the demand for electrical power is variable, while the supply is almost constant. Utility companies either send the power westward through interconnections in the power grid or offer a discount to users at off-peak times.

After all the transformers and wires, when the AC arrives at your home, things get interesting. Wires enter your house and go through a meter. This is an important element, but let's trace things further, then return to the meter. Next, the current passes through a breaker panel, where the circuit may be interrupted for safety reasons. In the early days of electrical circuits, the device that interrupted a circuit was a *fuse*. The incoming electrical current passed through a narrow, thin strip of metal and heated it up. If more than a set amount of current flowed, the metal strip melted, breaking the connection. The fuse would then have to be replaced. If it blew again, the circuit would have to be investigated to find the source of low resistance—which caused the large current draw—called a *short circuit*. Unfortunately, fuses were almost exactly the same size as the penny, so some people replaced burned-out fuses with pennies. The results—overheated wires, melted insulation, possibly fires—often cost much more than pennies to fix. Most modern breaker panels feature the use of *circuit breakers*. A circuit breaker may be a thermal one, which uses a bimetallic strip (see chapter 15 for details) to break the circuit, or electromagnetic, which runs the current through a coil to create a magnet. If too much current flows, the magnet is strong enough to pull an iron switch, breaking the circuit. Either kind of circuit breaker may "pop" if too much current flows and must be reset manually before current can flow again. Continuing to trace the wires, let's track them to a wall receptacle. In the most modern versions, there are three slots in the receptacle that allow three blades of an appliance to be plugged in.

The left slot (slightly larger) is connected to the white wire of the home electrical system, called the neutral wire. The right slot connects to the black (hot) wire, and the round slot is the ground, usually connected to a grounding rod, driven into the ground next to your house. This combination is designed to ensure that accidental grounding will result in electric current flowing through a path of least resistance that doesn't include you. Some receptacles have an additional safety feature, called a *GFCI*. A ground fault circuit interruptor is almost like a circuit breaker for a single receptacle, but it is slightly different. Inside the GFCI, the amount of current flowing from hot to neutral is sensed, along with any current flowing from hot to ground. If too much current "leaks" directly to ground, the circuit is broken, and a button must be reset to restore it. This provides quick, personal protection in potentially unsafe situations.

Now, let's investigate the meter. To record your energy usage, current flows through a meter that measures in units of kilowatthours, kWh. Since physics uses the joule as an energy unit, let's convert kWh to joules. A kilo is the metric prefix for 1,000, a watt is 1 J/s, and an hour is 3,600 seconds, which yields 3.6 million joules = 1 kWh. Since the average cost of 1 kWh is about 10 cents, evidently the electrical pioneers' dream of cheap, readily available electrical energy has definitely arrived.

To complete the discussion about AC and DC, consider those little black power cubes that are plugged into many wall receptacles and surge protectors. Many electronic gadgets run on low-voltage DC, so the role of the power cubes is to convert household AC to lower-voltage AC by using a transformer. A diode rectifier system is then used to convert the low-voltage AC to pulsating DC. If you ever have a spare power cube, it's very instructive to take it apart (not easy to do) and find the transformer and diodes inside.

A small note about the voltage. Usually, you see the voltage listed as anywhere between 110 and 120 volts at a frequency of 60 Hz (50 Hz in Europe). Because the voltage and current in AC are described mathematically by sine waves, you could wonder what the 120 V figure represents. It isn't an average, since the average of a sine wave is zero. This is a *root-mean-square* value. Recall the relationship developed for power lost in a resistor to heating when DC current flowed? It was: $P = i^2R$. But in AC, the current is a sine wave, so the power varies as the square of a sine wave. That means the power is constantly varying, with two peaks every cycle and sixty cycles every second. Rather than deal with the details of all this variation, it would be sensible to work with the average. Further, it would be convenient if the relationships had the same form for AC as for DC. To accomplish these goals requires the use of V_{rms} and i_{rms}

The relationship between the peak values of voltage and current are

$$V_{rms} = V_{peak} / \sqrt{2} \text{ and } I_{rms} = I_{peak} / \sqrt{2}$$

So, the average power in an AC circuit is:

$$i_{rms}{}^2R \text{ and } V_{rms} = I_{rms}R$$

The use of root-mean-square values is so widespread that an electrical quantity that has no subscript is presumed to be a root-mean-square value, just as a chemical symbol with no subscript is presumed to mean one atom, as in H_2O.

In DC circuits, there were three circuit elements, and their effects on the circuit were quite different. A resistor collected energy whenever current passed through it. A capacitor allowed the current to flow initially, but as charge and voltage built up, and the capacitor became full, the current stopped. An inductor slowed the growth of current in the circuit by inducing a back voltage, but eventually the current approached its maximum. In AC circuits, the same circuit elements behave quite differently. Here is an AC circuit that contains a resistor, an inductor, and a capacitor in series with an AC generator. The same current flows through each one, but then things begin to change.

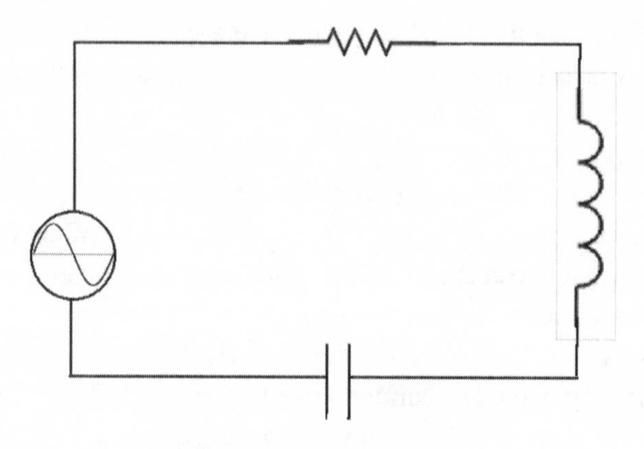

RLC circuit schematic

When an AC current goes through a resistor, the resistor collects its energy regardless of which way the current flows. The current and voltage are both sine waves, in phase with each other. But the inductor has a different story. The induced back voltage doesn't allow the current to peak as quickly, and the current lags 90° behind the voltage. The capacitor does exactly the opposite. Since it takes a while for the capacitor to fill, the voltage lags behind the current by 90°. In AC, the inductance and capacitance shift the phase relationship between the current and voltage in opposite directions. Neither capacitors nor inductors are dissipative elements like resistors, but the phase shifting has an overall effect on the circuit that is analyzed by a technique called *phasors*. The relationship between current and voltage is a little more complicated than a pure resistor network, because it must include the phase-shifting effects of capacitance and inductance. A thorough analysis requires a rotating vector diagram that is beyond our scope here, but we can look at a simpler concept, called *impedance*. For a DC circuit, the voltage and current were related by Ohm's law, $V = iR$. For AC circuits, there is a similar-looking relationship, $V = iZ$. But Z, the impedance, is much more complicated than resistance. Impedance depends not only on the resistors in the circuit but also on the capacitance, the inductance, and the frequency of the AC generator.

V = iZ, where V is the rms voltage in a series RLC circuit
i is the current,
Z is called the impedance, and

$$Z = \sqrt{R^2 + \left(2\pi f L - \frac{1}{2\pi f C}\right)^2}$$

Let's look at simpler cases first, then put them all together.

Case 1: Resistor and Capacitor (No Inductor)

At low generator frequencies, the capacitor fills up and doesn't allow much current to flow. On the other hand, at high generator frequencies, the capacitor doesn't fill up with charge, so the capacitor's contribution to the impedance is very small, and current flows as if the circuit contained only a resistor. A capacitor "likes" high frequency and is often referred to as a *high-pass filter*.

Case 2: Resistor and Inductor (No Capacitor)

At low generator frequencies, the inductor would generate less back voltage to slow down the current, so more current would flow. As the generator frequency increases, more back voltage is induced, so less current is allowed. An inductor "likes" low frequency and is called a *low-pass filter*. A speaker system often has two parallel branches with an inductor in the branch that feeds the low-frequency speaker (woofer) and a capacitor in the branch that feeds the high-frequency speaker (the tweeter).

Case 3: Resistor, Capacitor, and Inductor (All Three)

At low frequencies, the capacitance dominates; at high frequencies, the inductor dominates. Like Goldilocks selecting a bowl of porridge, the frequency that is "just right" (not too high and not too low) allows the maximum current to flow. The minimum impedance occurs when the inductance and capacitance terms cancel each other's effects. That occurs when

$$2\pi f L = \frac{1}{2\pi f C}$$

Solving this equation,

$$f = \frac{1}{2\pi\sqrt{LC}}$$

This is called the *resonant frequency*. When the AC generator has this frequency, the impedance is exactly the same as the resistance. That is, the circuit is purely resistive.

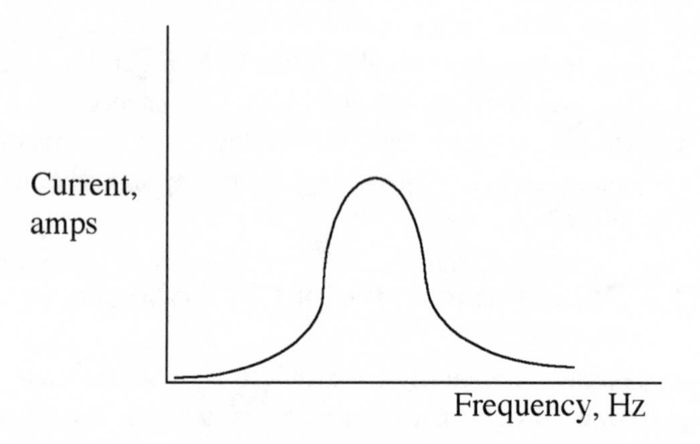

resonance current

The idea of adding inductance to cancel unwanted capacitance is used in telephone systems, since long runs of wires have a certain capacitance built in, so adding a loading coil at regular intervals makes the circuit as close to being purely resistive as possible.

In a series RLC circuit, if the capacitor (or inductor) were adjustable, the circuit could be "tuned" to allow maximum current at a particular frequency. Such a tuning circuit forms the basis of radio and television signals, as we will see in the next chapter.

Experiment 24: LEDs Lead the Way

Equipment needed: circuit board, battery, resistor, leads, LED

We've already mentioned those unusual materials called semiconductors that conduct current in one direction but not the other. In this experiment, we'll see how this maneuver is carried out and note how a particular diode can be made to generate light.

Conductors have extra electrons, and nonconductors have no electrons to spare, so where does that leave semiconductors?[1] As you might expect, they combine features of both. To make a semiconductor, start with a nonconducting material and add impurities to it (called *doping*). If the impurities have excess electrons, the new material is called an *n-type* semiconductor material (n for negative electrons). If the added impurities have vacancies where electrons can fit, they are called *p-type* materials (p for positive spots, also called *holes*). To build a *diode*, all you have to do is bond p-type material to n-type material. At the junction, electrons fall into holes, so there is an area in the middle of a diode where there are no excess electrons or holes. This is referred to as the *depletion zone*. Since the basic materials are nonconductors to start with, electrons or holes some distance away from the depletion zone have little mobility, so they are unaffected. If a battery is attached to this diode, what happens depends on the battery's polarity. If the battery's positive, high-potential end is connected to the p-end of the diode, positive holes will be driven across the depletion zone, toward the negative end, and electrons will go the other way. In short, current will flow, so the diode has become a conductor. This battery setup is called *positive bias*. If the battery is oriented the other way (*negative* bias), holes and electrons will separate further, widening the depletion zone. Now the diode acts like a nonconductor. A *transistor* extends the diode idea by having three layers of semiconducting material bonded together, sort of like a diode sandwich. The particular diode known as the light-emitting diode (LED) is arranged so the electrons that fall into holes (are captured by atoms) emit electromagnetic radiation that are within the visible range of the spectrum. Commercial LEDs are designed to emit the radiation outward for maximum visibility. Because LEDs consume very little power, they last a long time and are used as auto taillights, traffic lights, and flashlights (see Experiment 23).

PROCEDURE

- Build the circuit as shown below, but make sure you connect only the positive battery terminal at first. The positive side of the LED has a longer leg than the negative side. The resistor's function is to limit the current to the LED, so it does not burn out.
- Connect the other battery terminal. The LED should light.
- Reverse the battery terminals. The diode should not conduct, so no light should be emitted by the LED.
- If you have other LEDs, mount them in parallel to see if they light, then experiment with their polarity.

(If you would like to build more circuits, just Google "advanced electronic circuits.")

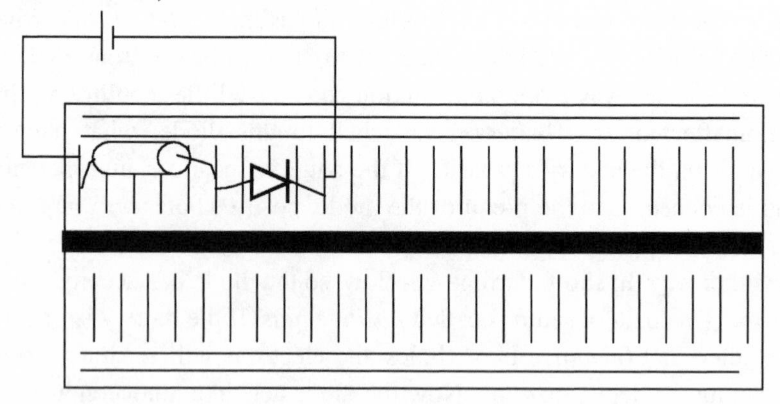

Chapter 22. THE LIGHT DAWNS—ELECTRO-MAGNETIC WAVES

"But soft! What light through yonder window breaks?—It is the east, and Juliet is the sun!"
—William Shakespeare, *Romeo and Juliet*

We've seen charges exhibit some fascinating behavior in the last few chapters, but their bag of tricks is not exhausted. Not by a long shot. What we're going to see next is both familiar and far more subtle than meets the eye: light. Let's start by means of a thought experiment technique. This is called a *gedankenexperiment*, in mixed German/Latin. The term was coined by Hans Christian Ørsted and was used extensively by Albert Einstein. In your mind's eye, imagine two long vertical conductors, some distance apart. Put a positive charge on the left conductor and call it A. Now give the right conductor a charge and call it B. We'll make charge A the active charge and see how charge A's actions affect charge B.

Case 1: Charge A is stationary.

Charge A has an electric field, as we saw in chapter 17. Charge B's response to the electric field set up by charge A is simply to be repelled and move as far away as possible—but that movement is limited by the conductor. Charge A has no magnetic field because only moving charges have magnetic fields. Charge B sits in no magnetic field, so there is no effect.

Case 2: Charge A moves upward with a constant velocity.

Charge A's electric field still causes repulsion, so charge B moves downward. Because charge A is moving, it generates a magnetic field, whose direction is perpendicular to the electric field. Charge B moves in the magnetic field of charge A, so it experiences a magnetic force.

Case 3: Charge A accelerates upward.

Charge A's electric and magnetic fields now both change with time, so charge B is in a region of changing magnetic field. This induces a voltage and therefore a current—if charge B's conductor is part of a loop.

Case 4: Charge A is set into harmonic motion, traveling up and down the conductor like a mass on a spring. The electric and magnetic fields of charge A now both vary in a periodic fashion. The voltage induced at charge B also varies, sending charge B traveling up and down its conductor with the same frequency as charge A. What we just built here is a wireless transmission system. The conductors where charges A and B are located have become transmitting and receiving antennas. Since the traveling disturbance is a wave, and both the electric and magnetic fields are involved, this is called an *electromagnetic wave*. This makes for an interesting picture: oscillating charges in one spot cause charges in another spot to oscillate. Several questions are surely bubbling to the surface of your mind: How fast do these waves travel? How does this relate to visible light that we can see? Waves normally travel in a medium, so where's the medium here? The first two questions will be dealt with here, the last one will require the genius of Einstein in chapter 25.

How Fast Do Electromagnetic Waves Travel?

Our little gedankenexperiment relied on knowledge of electric fields, magnetic fields, and induction, so don't be too harsh in judging ideas from earlier times when these concepts were unknown. Earlier ideas were focused on the most familiar electromagnetic wave, light.

If we start again with Aristotle, his idea about the speed of light was very simple: its speed is infinite. It seemed to him, and to many others, that light arrived instantaneously. Other cultures had many different ideas. The fourteenth-century Indian scholar Sayana (1315–1387) wrote in a comment on a hymn of the Rig Veda: "Thus it is remembered: [O Sun] you who traverse 2202 yojanas in half a nimesa." Translating these ancient distance and time measurements to their modern equivalents yields a value for the speed of the Sun (or is it the speed of light?) to be 3.01×10^8 m/s (187,000 miles/second), amazingly close to the correct value!

After the Scientific Revolution, experimental determination of the speed of light began. In 1629 Dutch physicist Isaac Beeckman suggested the speed of light could be measured by setting off a cannon and watching for the reflection of the muzzle's flash in a mirror positioned about a mile away. The great experimentalist Galileo wrote about a similar experiment in 1638, having possibly performed it earlier. Galileo stationed an assistant on a distant hilltop, and Galileo uncovered his lantern. As soon as the assistant saw Galileo's lantern, he uncovered his. When Galileo could measure no difference between the lantern uncoverings, he concluded that the speed of light was too fast to measure using that apparatus. The failure was almost as bad as dropping cannonballs from the wrong side of the Leaning Tower of Pisa.

Galileo Galilei

(A similar experimental setup was made in 1969 by the *Apollo 11* astronauts, who left a reflector on the Moon's surface. Rather than measuring the speed of light, however, this reflector allows the Earth-Moon distance to be determined to within ± 3 cm.)

The first numerical estimate of the speed of light was made on the basis of measurements made by Danish astronomer Ole Rømer. Rømer was studying eclipses of Jupiter's moon Io and noted that his measurements didn't match predictions based on data from the last eclipse. Going further back, he noted that the eclipse's timing seemed to depend on the season on Earth. Rømer reasoned that the cause of this variation was the light from Io traveling different distances depending on the distance between Jupiter and Earth. The Paris Observatory's director, Giovanni Cassini, ignored Rømer's ideas because he thought light's speed was infinite. Although Rømer made no direct calculation of the speed of light, Christian Huygens used Rømer's data and found light speed to be "more than six hundred thousand times greater than that of Sound." Using the known value for the speed of sound yields a speed for light of 210 million meters per second (131,000 miles/second). Although this is substantially smaller than the modern value of 300 million meters per second (186,000 miles/second), this was the beginning of a great experimental quest. Clever experimenters including Bradley, Fizeau, Cornu, Foucault, Michelson, Rosa, Dorsey, Essen, Gordon-Smith, Froome, and Evenson carried out experiments of increasing accuracy to determine the speed of light very accurately. The modern value of the speed of light is 299,792,458 ± 1 m/s. This constant of nature is given the symbol c.

In 1888 Heinrich Hertz conducted an experiment that demonstrated the existence of radio waves and allowed him to measure their speed. It turned out to be the same as the speed of light. This was the first experimental evidence that visible light was just one of many different electromagnetic waves. Since all electromagnetic waves move at the same known speed (in a vacuum), we can use the standard wave relationship $c = f\lambda$ to determine the range of possible frequencies and wavelengths. The result is a whole spectrum of electromagnetic waves.

The Electromagnetic Spectrum

Long waves	**Radio, TV**	**Micro-wave**	**Infra-red**	**V i s i b l e**	**Ultra-violet**	**X-rays**	**Gamma rays**
Increasing frequency => => =>							Increasing wavelength <= <= <=
ELF, SLF, ULF, VLF Submarine communica-tions	LF, MF, HF, VHF, UHF Navigation, time signals AM RADIO, amateur radio FM, TV wireless	SHF, EHF micro-wave, radar	Far IR, Mid IR, Near IR heat lamps mole-cule absorp-tion	R O Y G B I V	UVA, UVB, UVC	Soft Hard	

We will spend the next two chapters exploring some of the properties of light, many of which will be familiar.

Experimentino 14[1]

A microwave oven is the scene of some very delicious physics. Electromagnetic waves are generated in a special vacuum tube (magnetron) and sent into the oven cavity. The waves are reflected from the oven's walls and form complicated patterns that are similar to standing waves (like the ones we noted for sound waves in chapter 12). The waves bouncing around in the cavity interact with the oven's contents (food, hopefully) by oscillating them at high frequency, usually about 2.45×10^9 Hz. Depending on the molecules' size and electromagnetic properties, they may be unaffected (Styrofoam) or spun wildly (water). Metals, with their free electrons, are affected so strongly they may spark or flame—bad for the oven (that's why metals are banned from microwave ovens). Spinning molecules collide with others and add to the kinetic energy of the food, increasing its temperature (that's why it is called an oven). The complex wave patterns generate a series of hot spots and cold spots, so some microwave ovens have rotating platforms to even out the heating. Knowing that adjacent hot (or cold) spots are located about a half wavelength apart in ovens (without rotating platforms) allows us to measure the wavelength directly and infer the speed of microwaves.

Equipment needed: microwave oven, 3 cups of water, a piece of cardboard that fits into the oven, waxed paper to cover the cardboard, a layer of graham crackers (optional), a layer of marshmallows (optional), chocolate chips (essential), toothpicks, and metric measuring tape

PROCEDURE

- If your oven has a rotating platform, find a way to stop its rotation.
- Place three cups of water on the oven floor. (The cups elevate the experimental material high enough to avoid floor effects, and the water absorbs any excess microwave energy. When you remove the cups after the experiment, be mindful of the hot water.)
- Set the cardboard on the cups. The cardboard provides support for the layers above.
- Place the layer of waxed paper (thanks, Edison) next, just to keep things from getting too sticky.
- (Optional) Place thin layers of graham crackers and marshmallows next. (You'll see their purpose, but it is not essential to the experiment.)
- The last tier is a thin, uniform layer of chocolate chips. (Of course, you may build this apparatus outside the oven and insert it all at once.)
- Turn the power control to full and set the time for twenty seconds. Close the door and microwave away.
- Inspect the chocolate chips. If you *can* see melted spots, remove the apparatus and mark the melted spots with toothpicks. If you *cannot* see melted spots, microwave for ten seconds more.
- Measure the distance (m) between melted spots. This is half a

wavelength. Multiply the wavelength by the microwave's frequency. (Use 2.45×10^9 Hz if it isn't listed on the machine.)

- Compare your answer with the speed of light, 3×10^8 m/s. (Sample: 6.5 cm between hot spots, 2.45×10^9 Hz microwave yielded a speed of light of 3.18×10^8 m/s.)
- Now for the best part. If your answer matches the known value, you can celebrate your success. Rotate the apparatus, return it to the microwave oven, and run it for a little longer. When the chocolate melts, if you have included the marshmallows and graham crackers, you can fold it over and make it into the famous campfire concoction called s'mores.

[Author's note: I had heard of few edible physics experiments before (maybe inducing charges on Rice Krispies might count), so when I pretested this one, it took *many* repetitions before I was satisfied with the result. Maybe I could get the Rice Krispies to jump into the chocolate. . . . I am sure it will require a long development period to perfect this technique.]

Experiment 25: Bent Light

Equipment needed: lens, measuring tape, light source, white card

Oscillating electric and magnetic fields are technically accurate terms for what we normally call light. To get a better look at light, we'll use a magnifying glass. Although it won't allow us to see the fields, it will get us a little closer to perceiving what's going on.

Procedure

1. Lens Focal Length Determination

- Using a strong, distant point source of light (the Sun would be fine, if you could arrange it), hold the magnifying lens so that the incoming light is focused to a point. Youthful scouts sometimes use this technique to start a fire in a pit.
- Measure the distance from the lens to the point where it is focused. This is the *focal length* of the lens. (Most magnifiers have focal lengths between 2 cm and 12 cm, 1–5 inches.)
- Express the focal length in meters and write it here: f = _______ .

2. Real Image

- The source of light is called the *object*. For this experiment, the first object will be a lightbulb (Thanks, Edison) in a socket. Using a permanent marker, write the word PHYSICS on the bulb. This will be significant later in the experiment.
- In a dim room, light the bulb and hold the lens a horizontal distance from the bulb that is greater than twice the focal length determined in procedure 1 above.
- Measure the distance from the bulb to the lens. This is called the *object distance*, d_o. Express the object distance in meters and write it here: d_o= ___________________________.

- On the same horizontal line as the object and lens, hold a white card so that the lens focuses the light from the bulb onto the card. What you see on the card is called the *image*.
- Adjust the card until the image is sharp.
- Measure the distance from the lens to the image. This is called the *image distance*, d_i. Express the image distance in meters, and write it here: d_i = ______________________. The word PHYSICS on the object doesn't look quite the same as the image. If light actually strikes the card, this is called a *real image*. In the space below, draw the image you see:

The relationship between the focal length, the object distance, and the image distance is

$$\frac{1}{d_o} + \frac{1}{d_i} = \frac{1}{f}$$

Using your measured quantities, compare the two sides of this equation to see how closely they match. (Sample: $f = 0.1$ m, $d_o = 0.21$m, $d_i = .195$ m; $1/d_o = 1/d_i = 9.94\ m^{-1}$, $1/f = 10\ m^{-1}$.)

3. Virtual Image

PHYSICS

Using the word PHYSICS as the object, hold the lens one focal length above the object and look down through the lens. You should see an image quite different from the one in procedure 2. This image appears to be behind the paper, where there is no light at all. It is called a *virtual image*. Draw it in the box below:

The chapter will discuss these results as well as other cases.

Experiment 26: Polar Opposites

Equipment needed: light source, 2 polarizing filters, cellophane

The interaction of light and matter takes many forms, and a very unusual one involves *polarization*. Light coming out of this page toward you is generated by an E-field that oscillates perpendicular to the direction of the light, namely, in the plane of the paper. The direction of oscillation is referred to as the *polarization*. Most light has no preferred polarization direction, so it is said to be randomly polarized. However, when light interacts with long thin molecules, the interaction allows transmission of light that is polarized *perpendicular* to the molecules' long axis. Also, the interaction blocks transmission of light polarized *parallel* to the molecules' long axis. Thus, if you run a beam of unpolarized light through a polarized filter, the beam that emerges is polarized in a particular direction. In this experiment, polarized beams will be generated and explored.

PROCEDURE

- Shine light from a strong source (laser pointer?) through a polarizing filter.
- Observe the light that emerges.
- Set up a second filter, oriented the same way as the first and shine the light through both filters. Is there any change in the light?
- Turn the second filter so its polarizing direction is oriented 90° to the first. What change do you see?
- Keeping the two filters crossed, insert *crinkled* cellophane between them. Now what do you see? Record your observation here: ____.

Chapter 23. MIRROR, MIRROR ON THE WALL—REFLECTION AND REFRACTION OF LIGHT

"To hold, as 'twere, the mirror up to nature . . ."
—William Shakespeare, *Hamlet*

Even the guy in the graphic above knows what happens when light hits matter. At the macroscopic level, we see reflection, absorption, transmission, and even refraction. But how does this relate to what happens at the submicroscopic level? We're back to the problem of large and visible versus small and invisible, but it's a bit more complicated. Now, we've got oscillating electric and magnetic fields involved, and they are more abstract than molecules and atoms. The treatment that follows will be simplified, perhaps too much so, but it should serve as a good approximation.[1] Let's recall some of the things we know about waves and molecules. Remember rubbing the top of the glass with your finger to excite the natural frequency of the glass (chapter 11)? Then there was the opera singer who sang a note at the glass's natural frequency and caused it to shatter. Also, pushing the child on a swing needed just the right frequency to be effective. These were all instances where a mechanical system was being forced to oscillate. And the biggest amplitude reinforcement occurred when the frequency of the forcing function matched the natural frequency of the system—in other words, resonance. How about electromagnetic waves forcing the charges in molecules to oscillate? Electromagnetic waves are pretty straightforward to generate: just oscillate a charge and it makes waves. Finding the natural frequencies of molecules, atoms, and ions is much more difficult. For every electrical bond, there can be an oscillation. So intermolecular bonds, intramolecular bonds, bonds within the atom, and bonds within the nucleus all have natural frequencies. To keep our discussion simple, we'll stick with bonds that hold electrons, but see the endnotes for more. To analyze light's interaction with matter, we'll send electromagnetic waves of many different frequencies toward matter, analyze the interaction at the molecular level, then predict what will be seen as a result. Roughly analogous to the mechanical forcing functions that operate on matter, here's the general pattern:

- If the incoming electromagnetic wave's frequency is way smaller or way larger than the matter's natural frequency, the wave is delayed only slightly at each molecule, then travels in the same direction as before.
- If the incoming electromagnetic wave's frequency is nearly the same as the matter's natural frequency, the wave is delayed longer at each molecule, then travels in any direction. Let's apply this logic to several different kinds of matter:

Case 1: This kind of matter has no natural frequencies in the visible range.
Interaction details: No substantial interactions. Each molecule simply delays the light wave slightly, then it continues in its original direction. Depending on the molecular characteristics, the light may be absorbed or transmitted.
Visible at the macroscopic level: If the matter absorbs the incoming white light and scatters no light, we would see a black, opaque object. If the matter transmits the incoming white light, we would see a transparent object.

Case 2: This matter has a natural frequency in the visible range, say, $f = 7 \times 10^{15}$ Hz.
Interaction details: The frequency $f = 7 \times 10^{15}$ Hz is scattered by the molecules in all directions. Waves of all other visible frequencies are either absorbed or transmitted in the same direction that they were traveling when they entered.

Visible at the macroscopic level: Electromagnetic radiation at 7×10^{15} Hz. The eyes report this frequency to the brain, which interprets the sensation as the color blue. If most of the original light goes through, it appears transparent with a blue tint, if most is absorbed, it appears opaque and blue.

Case 3: This matter contains free electrons (a metal) that are capable of vibrating at any frequency.
Interaction details: Electrons oscillate at all frequencies and scatter light in all directions.
Visible at the macroscopic level: All frequencies that go into the interaction come back out.

Case 4: This matter consists of molecules that are long and thin or have a spiral shape.
Interaction details: Molecules absorb waves that oscillate them in the long direction and transmit those oriented along the short direction. Spiral-shape molecules change the plane of polarization of the waves.
Visible at the macroscopic level: Light intensity is dimmed somewhat, but nothing else is visible.

There are many more kinds of molecules, but let's see how these four cases lead to most of the classical optical properties that we observe.

Reflection: When a light wave hits a surface with free electrons, such as a metal, it *reflects*. If the surface is rough, the reflection will be many different directions, but a smooth surface reflects according to the simple law, similar to the way pool balls reflect off side cushions: the angle at which the light strikes the surface (incident) is the same as the angle at which it reflects.

angle of incidence = angle of reflection

Experimentino 15

Measure the distance along a floor from a wall to a point that is a known distance above the floor. Place a mirror halfway to the wall along the floor. Using the laser pointer from Experiment 27, shine the laser light onto the mirror and see if it hits the wall at the same height off the floor. (Safety Point: Make sure the laser beam does not point at anyone's eye.)

Transmission: When a light wave encounters a transparent surface, like glass, most of the wave goes right through and comes out the other side.

Absorption: Depending on the natural frequencies of the molecules, a light wave incident on some materials will be partially or totally absorbed. If the natural frequencies are in the visible range of the spectrum, the object will appear to have some color or it will be a mixture of all colors, sensed as white. If there are no natural frequencies in the visible, the object will appear black.

Refraction: Light travels at an average speed less than c when it moves through a medium other than a vacuum. Oddly enough, light never actually travels at any speed other than c. These statements sound contradictory, so a bit of an explanation is in order. Here's an analogy that may help: The census of 2000 revealed that the average number of children in a US family is 0.9. Yet how many families will say they have exactly 0.9 children? Right. Average doesn't apply to individual cases. Although the real situation is a bit more complicated, try this simple model: Light travels at c from one molecular interaction to the next but then goes nowhere while the interaction is taking place. So light has two speeds: c and zero. The average is less than c, call it v. This leads to the definition of the index of refraction, n, which is always greater than one.

$$n = \frac{c}{v}$$

When light enters a medium with a higher index of refraction, the part of the wave that enters first slows down more and thus bends. It's almost like having the front wheels of a car go onto a soft shoulder and turn the car. The angles and indices of refraction obey Snell's law:

$$n_1 \sin\theta_1 = n_2 \sin\theta_2$$

A particularly interesting case occurs when light goes toward a medium of a lower index. The angle in the second medium could be 90° or more. This leads to the phenomenon called *total internal reflection*, in which no light is refracted into the second medium—it is all reflected. This forms the basis of *fiber optics*.

Experimentino 16

Using a piece of fiber-optic cable and the laser pointer from Experiment 27, shine the laser light into the end of the cable and bend it to see if the laser light stays within the cable.

Dispersion: Some media have natural frequencies past the visible part of the spectrum, in the UV. This leads to a greater refraction of one end of the visible spectrum than the other, namely, the violet end, since it is delayed longer, making n_2 even greater. This is demonstrated by a *prism*. White light goes into the prism, and the light that comes out is separated into a rainbow of colors, but violet is bent more than red.

Polarization: Normally, light consists of fields oscillating in all planes, which is referred to as *unpolarized*. When unpolarized light passes through a polarizing filter, as described in case 4, only polarizations in the one plane pass through. As you saw in Experiment 26, when two polarizing filters are crossed, no light is transmitted. The other part of the experiment used cellophane, whose molecule is spiral-shaped, and rotated the plane of polarization of the light. This allowed some light to pass through the second filter. Many molecules of biological significance have structures that rotate the plane of polarization of light and are therefore used for chemical tests.

Let's see how these properties of light are used in nature and in technological applications.

Optical Gadgets: Mirrors and Lenses

Real objects constantly give off light in all directions. A good way to visualize this is to think about the water waves that would result from tossing a stone into a body of water. The water surface would be a corrugated series of circles with crests and troughs of the waves emanating from the spot where the stone landed. The same is true of light, but light waves proceed in all three dimensions, so they form a spherical set of corrugations in the electric and magnetic fields, with the source at the center of the sphere. The light that proceeds along any line from the source is called a light *ray*. In order to make sense of all the light pathways, physics has evolved a technique called the *ray diagram*. Out of all the many rays that are emitted from any point, only particular ones will be traced.

Let's start with a *plane mirror*. In ray diagrams, light starts from an *object* (we'll use an upright arrow as a standard object) and goes toward the optical gadget, where it is reflected or refracted, depending on the gadget. In the case of the plane mirror, let's trace the progress of two rays from the bottom of the object. One hits the mirror horizontally and is reflected straight back. The other hits the mirror on the axis, at the center. It is reflected so that its angle of incidence matched its angle of reflection. If you place yourself in the position of the arrow, what would you see? To answer that, you first need to know the way the human eye/brain system works: the brain assumes light travels in straight lines. Now, we can project the rays back to where they seem to meet, behind the mirror.

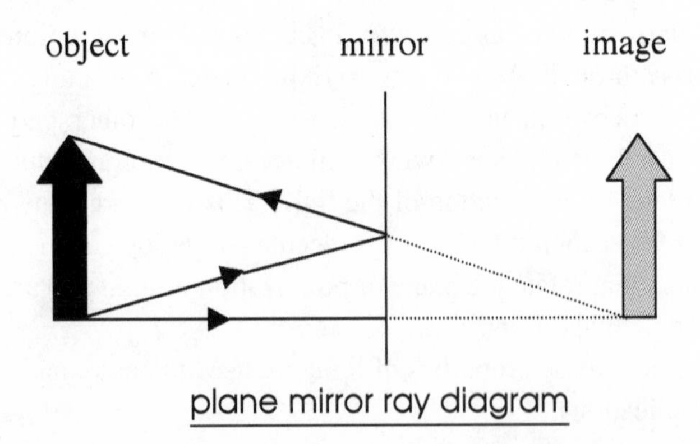

plane mirror ray diagram

The spot from which the rays seem to be originating defines the *image*. In this case, the image is right side up (*erect*), but no actual light is present at its location, so it is called *virtual*. If you substitute yourself for the arrow in the diagram, you can now be convinced of something you already knew. If you look in a plane mirror, you see an image of yourself, and it appears to be behind the mirror and right side up.

Next, let's curve the mirror by making it part of a hemisphere. If it curves toward the object, it's called *convex.* Rays from the object still reflect from the mirror according to the law of reflection, but determining the angles becomes more difficult. This is because the perpendicular to the mirror isn't always horizontal, as it was for the plane mirror. A horizontal ray (called *paraxial*—parallel to the axis) diverges after striking the mirror, rather than simply reflecting straight back. The ray that strikes the center of the mirror axis reflects just as before. The image formed is erect and virtual, as before, but it is much smaller than the object. For practical reasons, convex mirrors are used in corners of convenience stores to spot shoplifters and as "buttons" on right-hand-side truck mirrors to widen the field of view.

A *concave* mirror curves away from the object—think of a cave as something you'd go into. It produces a variety of images depending on the distance from the object to the mirror, called the *object distance.* For the case shown in the diagram, a real, inverted, smaller image is formed, and its distance from the mirror is called the *image distance.* If the object is moved just inside the spot where all the horizontal, paraxial rays are focused, called the *focal length,* f, the image becomes virtual, erect, and magnified. That's why you need to get close to a concave makeup mirror to check for appropriate cosmetic application.

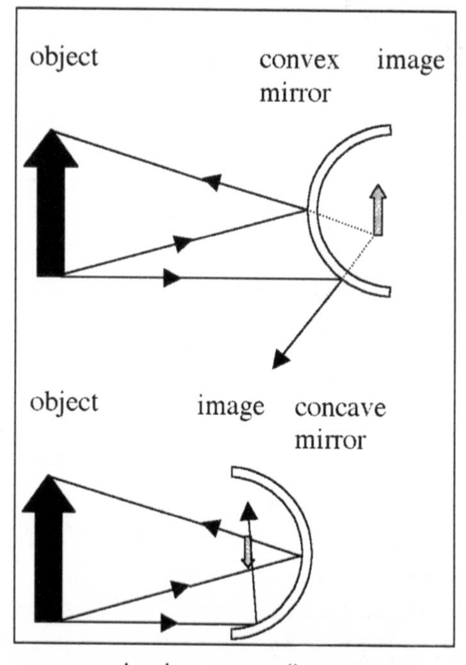

curved mirror ray diagrams

Experimentino 17

Shiny silver spoons, especially soupspoons, provide a wonderful apparatus to check out the optics of concave and convex mirrors. Hold the spoon in front of you and keep track of your image as you move the spoon closer and farther away, using both the convex and concave sides of the spoon. Be mindful that you might be asked to help with the polishing if you get caught by your Aunt Adele. I'm not sure the physics experimentino excuse would work, but it's worth a try.

The object distance, the image distance, and the focal length are related by the equation:

$$\frac{1}{d_o} + \frac{1}{d_i} = \frac{1}{f}$$

where d_o = object distance, d_i = image distance, and f = focal length

Lenses have a lot in common with mirrors in that both thin lenses and mirrors are described by the above equation. You might think them different because light undergoes refraction at the lens rather than reflection at the mirror. Besides, refraction happens twice—once as the light enters the lens and once as it leaves. But the ray diagrams and equation are similar as long as the lens is thin, the refraction is presumed to occur at the middle of the lens, and paraxial rays are brought to a focus at the focal point.

For a *converging lens*, light from a distant object is refracted by the lens, then transmitted through a focal point on the other side. Tracing a paraxial ray and a ray through the center (unrefracted) yields a real image that is inverted and smaller than the object, just as you found in Experiment 25. Just as with mirrors, moving the object closer changes the image greatly. Using a lens as a magnifier with the object placed inside the focal length, the image is virtual, erect, and larger than the object, as you also saw in Experiment 25. A *diverging* lens, as the name implies, refracts entering rays as if they emanate from a focal point on the same side of the lens as the object. As the figure shows, the image produced is virtual, erect, and smaller than the object.

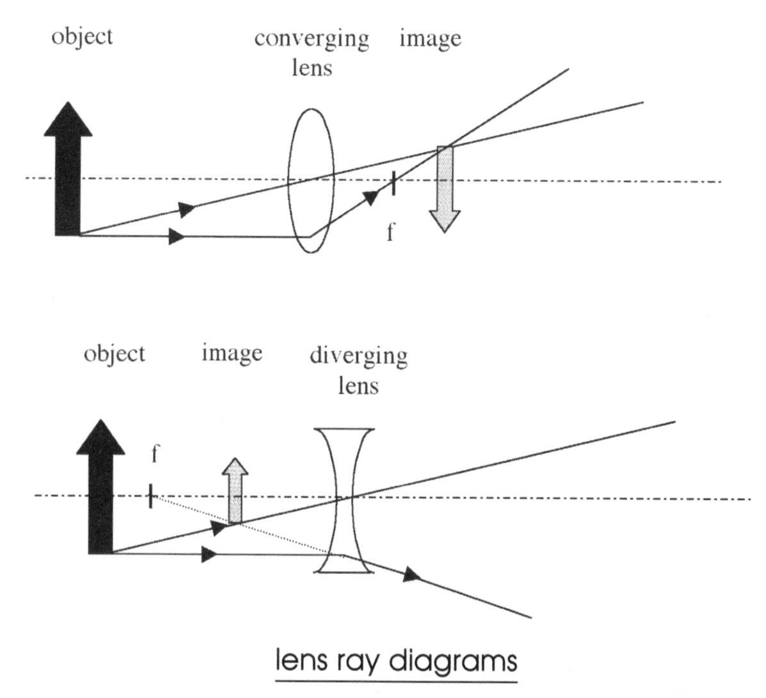

lens ray diagrams

Optical gadgets such as cameras, microscopes, and telescopes often have multiple lenses or mirrors. In principle, they are simple to analyze. Starting with the lens closest to the object, draw ray diagrams for each lens in turn, using the image of the first lens as the object of the next lens, and so on. There are sign conventions, magnification relations, thick lenses, and other details to fully understand this topic, but they are beyond the scope of this book. Consult the "Additional Resources" section for more.

The interaction between light and matter is an extremely common occurrence. Let's analyze a few examples.

Atmospheric Optics

Interesting interactions happen between light and molecules in the atmosphere. Here are a few:

Blue sky: Nitrogen and oxygen molecules have natural frequencies in the UV part of the spectrum but none in the visible. As a result, the blue end of the spectrum is scattered slightly. If white light shines through a lot of air, it scatters enough blue that it appears that color. Small volumes of air, or insufficient sunshine on cloudy days, produce no blue in the sky.

Yellow-white Sun at noon: At midday, the Sun's light shines through a minimum amount of air, so not much blue is scattered away. Almost the whole spectrum is visible.

Red Sun at dawn and dusk: When the Sun shines through the maximum amount of air, much of the blue is scattered, leaving the red end of the spectrum. Also, various air pollutants may be present to give sunlight an overall tint.

Rainbow: When the Sun at your back shines on spherical water droplets in the air in front of you, the white light undergoes refraction and dispersion within the drop. As long as there are enough drops, a particular color will kick around in the drop, then finally emerge at one particular angle, which happens to be aimed right at you. Think of yourself as being at the apex of a cone, and colored light comes at you from the cone base. Since red is bent least by media with natural frequencies in the UV, red is at the top of the bow. (Many other fascinating variations on the rainbow theme exist, such as a secondary bow, a tertiary bow, and even a lunar rainbow.) Good pictures and explanations can be found in the endnotes.[2]

Experimentino 18

Standing with the Sun at your back, shoot a fine spray from a hose into the air, directed away from the Sun. Observe what happens to the Sun's light. Draw what you see (especially noting the order of colors) in the box below:

Mirages and looming: Because the index of refraction changes slightly with temperature, hot air above hot sand makes light bend gradually over long distances. Light from a distant palm tree bends in this progressive way, and the eye/brain system makes its simplistic assumption about light traveling in straight lines. This leads to the interpretation that the tree has an inverted image as well as its normal erect one. The brain interprets that as a sure sign of reflection—and water. Not only does this mirage attract thirsty people on the desert, but you can see it over long stretches of hot asphalt pavement. It looks wet, even though there is no rain. Looming is just the opposite. Cool air over cool water refracts light downward, so the eye/brain straightens it out, forming a ghost ship in the sky. For pictures and more details, see the endnotes.[3]

Technological and Human Optical Systems

The camera and the human eye share several similarities in the way they deal with light. This shouldn't be too surprising, because gadgets often show up in naturally occurring items. The camera has a shutter that opens for an adjustable time. When the shutter is open, a cylinder of light is admitted to the camera. The length of the cylinder is governed by the exposure time, and its diameter is set by the size of the opening (*aperture*). The aperture is also adjustable through a system called *f-stops*. The f-stop number is the ratio of the camera lens's focal length divided by its aperture diameter—thus, the smaller the f-stop, the larger the opening and the bigger the diameter of the light cylinder admitted. Once the light gets through the opening, it encounters a converging lens of fixed focal length. The lens focuses the light and forms an image of the object on a plane at the back of the camera. The plane is called the film plane, although digital cameras have CCDs rather than film at that location. Since the camera uses a lens, it obeys

$$\frac{1}{d_o} + \frac{1}{d_i} = \frac{1}{f}$$

The focal length of the lens is fixed, so the image distance is adjusted by making the lens closer or farther from the film plane. To achieve the right exposure requires adjustment of the f-stop and the exposure time in order to let the right amount of light into the camera. Additional factors include film speed, differences in color response, depth of field, subject or camera motion, distortion of the image near the edges of the lens, and more.

The human eye also has a shutter—the eyelid. Once light gets past the lid, it first encounters the *cornea*, a fixed lens that begins the focusing process and protects the rest of the mechanism. Next, light goes through the iris/pupil system that opens wide to let more light in or closes down to admit only a little. In humans, this process is called *accommodation* and its speed diminishes with age. (As you age, it becomes increasingly difficult to walk into a dark movie theater and see anything at first.) Next, light goes through the lens, where fine-grain focusing takes place. Finally, the image is formed at the *retina*. Since the major optical element is a lens, the system is still governed by

$$\frac{1}{d_o} + \frac{1}{d_i} = \frac{1}{f}$$

In the eye, however, the image distance is not adjustable. What happens instead is that there are ciliary muscles that squeeze the lens to change its focal length in order to produce a clear image at the retina. As people age, the lens gets thicker, similar to the way an onion grows, one layer at a time. A thicker lens is harder to squeeze, so eventually it becomes harder to see objects that are close. Vision correction is accomplished through lenses, either in the form of glasses or contact lenses. The goal is often to correct the near point so that an object held 25 cm (10 inches) from the eye can be seen clearly. Reading glasses are rated in units of diopters, which is 1/f, where f is expressed in meters. As an example, a +2 diopters lens would correct a near point of 50 cm to the standard 25 cm. A more difficult correction is occasioned when the eye's lens has different radii of curvature in the horizontal and vertical planes. This is called astigmatism and requires lenses that are also made with different radii of curvature. Correcting this vision problem with contact lenses posed a challenge until the bottom of the contact lens was made thicker, so that torque would rotate the lens on the cornea until it was placed appropriately (recall our discussion in chapter 8).

The image forms at the retina. Unlike a camera, with its film or CCD sensors, the retina has cells of two fundamentally different kinds: *rod cells* and *cone cells*, named for their shapes. There's only one type of rod cell, and it senses no color but responds to low light levels. Cone cells come in three different varieties, require more light, and their response peaks at different wavelengths. Overall, there are about twenty times as many rod cells as cone cells, but the cones are concentrated near the spot of maximum vision, the *fovea centralis*. This creates a problem if you're trying to view a dim star, such as the North Star, Polaris, by staring right at the location you'd expect to find it. The cone cells at the fovea centralis need more light, so astronomers use a technique called averted vision. If you look just to the side of the star, more rod cells will be excited and you will sense its presence more easily.

Variations among people's cone cell sensitivities lead to a certain amount of subjectivity in defining colors. Partly because of the three kinds of cone cells, it turns out that light of three colors combines to produce what most eyes perceive as white. These three colors are called the *additive primary colors* and are red, green, and blue. Theatrical productions often use all three colors of lights so they can make not only white but also subtle variations to suit the mood of the presentation. Similarly, TV monitors use RGB (red, green, blue) pixels that can be combined in various ways to produce many colors. Working with pigments, however, is a whole different story. A pigment that appears red means that when white light hits it, green and blue are absorbed and only the red is reemitted. So, if you mix red, green, and blue pigments, every color is absorbed and the result will be black. Mixing pigments is a subtractive process, so there are *subtractive primary colors*, cyan (absorbs red), magenta (absorbs green), and yellow (absorbs blue). For example, if you mix cyan and magenta pigments, red and green are subtracted from white, leaving blue. It's difficult to mix real pigments to achieve these ideal results, but a more common example is in the inks that are used in inkjet printers. They are cyan, magenta, and yellow. For clear examples of these ideas, you should go to the Web sites listed in the endnotes.[4]

Now that we've discussed absorption, let's see if we can absorb some new and different properties of light.

Experiment 27: Spreading the Beam

Equipment needed: diffraction grating, laser pointer, measuring tape

A diffraction grating spreads light into a pattern of dark and light spots, depending on the light's wavelength. The spreading is similar to the action of a prism, but the mechanism is totally different, and it will be discussed shortly. Diffraction gratings consist of a set of closely spaced lines ruled in a medium that either transmits or reflects the incident light.

For this experiment, you need a transmission diffraction grating with at least 200 lines/millimeter (5,080 lines/inch). They are available commercially (at science or museum stores, or try the Web) and usually cost $1 or less. Gratings are rated in terms of the number of lines in some unit of length. For example, a diffraction grating used in the sample for this experiment had 200 lines per millimeter. The spacing between lines is 1/200 mm, or 5×10^{-6} m. The point of this experiment is to check the (approximate) equation that describes the diffraction grating.

$$\frac{\lambda}{d} \approx \frac{x}{L}$$

where λ is the wavelength of light, d is the spacing between lines in the grating, x is the distance between the central bright spot and the first bright spot on either side, and L is the distance from the grating to the pattern

PROCEDURE

- In a dim room, measure a distance of 1 meter from a blank wall. This is the distance L.
- With the diffraction grating positioned at the 1 meter mark, shine the laser through the diffraction grating and observe the light pattern on the wall. Draw it in the box below.

(LASER SAFETY: NEVER SHINE A LASER INTO SOMEONE'S EYES OR THOSE OF A PET.)

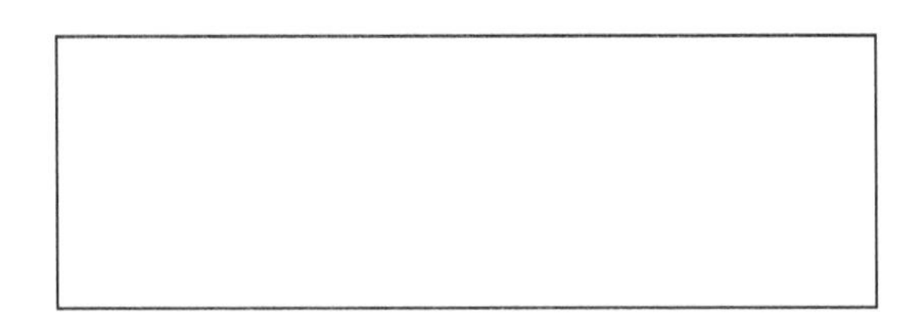

- Measure the distance, along the wall, between the central bright spot and the bright spot adjacent to it. This is the distance x. (Sample: x = 0.13 m.)
- Find the spacing between lines in the diffraction grating, d, using information that is included with the grating. (Sample: Given 200 lines/mm, the spacing, d, is 5×10^{-6} m.)
- Calculate the wavelength of the laser from the equation

$$\lambda \approx \frac{dx}{L}$$

(Sample: $\lambda = 6.5 \times 10^{-7}$ m.)

- Compare your result with the wavelength listed for the laser. Red lasers usually have wavelengths of 630 to 660 nm, green lasers are about 530 nm, and blue ones about 470 nm. (Sample: Laser wavelength given as 650 nm = 6.5×10^{-7} m. Close match.)

Experiment 28: Compact Rainbow

Equipment needed: CD (or DVD), laser pointer, shake flashlight, and room light, fluorescent light, or incandescent light

That marvel of technology the CD not only stores data, music, pictures, and more, but it also happens to be a reflection diffraction grating with about 625 lines/mm. The purpose of this experiment is to explore the light patterns formed when various light sources are shone on a CD.

Procedure

- Turn the CD so the underside faces upward. This is the side with the data burned into spiral grooves.
- Allow natural light (even from a cloudy day), incandescent light, or fluorescent light to shine on the CD. Turn it so you can see how the pattern shifts because of changes in the angle.
- In a dim room, shine the light from the shake flashlight on the CD. Vary the angle of the flashlight and the angle from which you observe and note the differences in patterns.
- Again in a dim room, place the CD on the floor next to a wall. Shine the laser pointer onto the CD and see if you can get a reflected pattern similar to Experiment 27. Change the angle to see what happens to the pattern.

(LASER SAFETY: NEVER SHINE A LASER INTO SOMEONE'S EYES OR THOSE OF A PET.)

Chapter 24. LIGHT'S STRANGEST TRICKS—DIFFRACTION AND INTERFERENCE

"She swore, in faith, 'twas strange, 'twas passing strange, . . ."
—William Shakespeare, *Othello*

Your patience is commendable. We briefly mentioned some strange properties of waves in chapter 11 and we haven't gotten back to them—until now. These properties were diffraction and interference. To some degree, we explored interference to find out about standing waves on strings and acoustically noisy or quiet spots in rooms, but we never mentioned diffraction. Actually, you might have seen diffraction with water waves. If a straight wave crashes into a seawall, most of the wave is reflected back in the direction it came. But if the wall has a small opening in it, the portion of the wave that gets through that opening spreads out, as if the opening was the source of a circular wave. Diffraction also happens at edges. If someone talks to you from around the corner, you still receive the sound waves, partially because they have diffracted by the sharp edge of the corner.

Diffraction posed a big problem for Isaac Newton. "Light," according to Newton, "is composed of tiny particles, or corpuscles, emitted by luminous bodies." The only way he could include diffraction was by somehow "associating" a wave with the particles of light. But a key experiment reinforced the idea of light's wavelike properties. This experiment was originally performed by Sir Thomas Young in 1801. Sir Thomas Young was considered by some to be "the last person to know everything." He made original contributions in medicine, physics, languages, and hieroglyphics as he worked on the Rosetta Stone. *Young's double-slit experiment* involved two small slits cut in an opaque material. Light incident on the material was diffracted at the slits, then the waves interfered after they spread out. The interference produced bright spots along the line of constructive interference and dark spots along the line of destructive interference. A screen shows a series of bright and dark spots that are not explained by any theory other than wave theory. The equation that describes this experiment is

$$\sin \theta_{bright} = \frac{m\lambda}{d}$$

where θ_{bright} is the angle between the central bright spot and another bright spot, m is an integer that indicates the number of the bright spot from the center, λ is the light wavelength, and d = the distance between slits. (Note that if light were a beam of particles, it would just go straight through the holes, and there wouldn't be any multiple bright spots.)

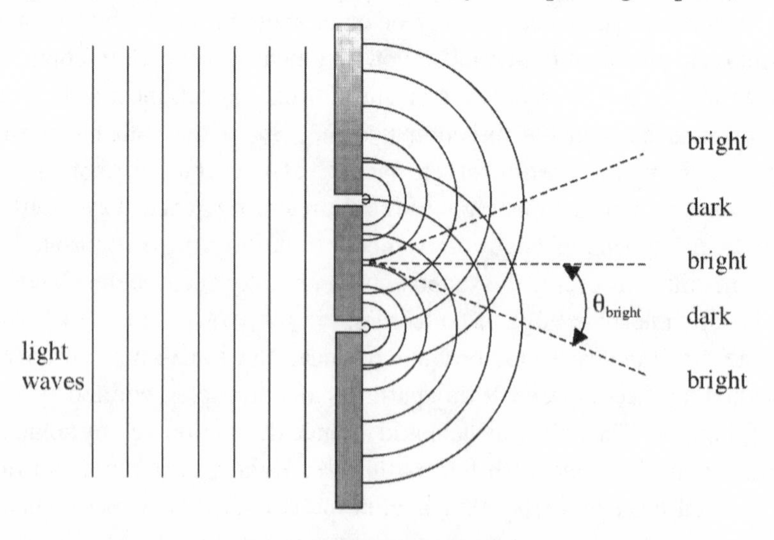

double-slit experiment

As you saw in Experiment 27, shining the laser through a diffraction grating of multiple slits produced exactly this pattern of bright red spots. If the incoming light had been white, the bright spots would have been rainbows, since the longer wavelength (red) would have been bent more. View a white light through the diffraction grating from Experiment 27 (if you can find it) and see if the red light bends more. Experiment 28 used a CD as a reflection diffraction grating and yielded similar results, in that the color depended on the angle of view and the angle of illumination. A new color TV technique under development uses a diffraction grating that is deflected by voltage. So the colors you see will be due to interference effects rather than colors from RGB pixels, as it is now. The dependence of color on the angle of view is often the clue you need to determine whether the color you see is due to a pigment or an optical effect such as interference. For example, consider the beautiful colors of fish, birds, butterflies, and the spectacular peacock, like the graphic at the beginning of this chapter. If you change your viewing angle, the colors seem to change. This shows that the colors are due to interference rather than pigments. The row upon row of tiny filaments that make up feathers or fins act as diffraction gratings and show different colors at different angles, similar to the CD in Experiment 28.[1]

Another source of interesting color is the *thin film*. You may have noticed the phenomenon I call the "rainbow in an oily puddle." In a wet parking lot, if a car leaks a little gas, oil, or transmission or brake fluid, the liquid doesn't mix with water, so the less-dense oil spreads out and forms a thin film on the water surface. White light shining on this oil/water combination goes through some interesting optical gymnastics. Some light is reflected at the air/oil surface and some is transmitted into the thin film of oil. The transmitted portion then encounters another medium change when it gets to the oil/water interface. There, this wave is partly reflected and partly transmitted to the pavement below. The reflected ray goes back up through the oil and is then transmitted into the air. The net result is that the singly reflected ray and the doubly reflected ray interfere with each other. If the oil film thickness is a whole number of wavelengths, the interference is constructive, if the thickness ends in a half wavelength, the interference is destructive. Because the length of light's path in the oil depends on the angle and the index of refraction of the oil, the result will be bright colors and dark spots at different angles. Besides accidental oil spills, thin films are used in technological applications for such things as antireflective radar coatings on airplanes (stealth technology) and antireflective coatings on the lenses of more expensive cameras. The endnotes list a Web reference for more details.[2]

Experimentino 19

Besides their fun component, soap bubbles have enormous physics in them. Here, we're mostly interested in the optical properties of their thin skin. To perform this experimentino, you'll need a soap bubble solution, a shallow dish, and a frame. The bubble solution can either be obtained from commercial sources, or you can make it yourself from water, a liquid detergent, and some glycerin. You'll have to experiment with the proportions because variations of detergent are based on their brand. Sometimes letting the solution stand for a while makes the bubbles more stable. Use shallow dishes and a frame as small as a bent paper clip or as large as a coat hanger, depending on the size of your dish. Make sure the frame forms a complete loop, so you can immerse the loop in the bubble solution completely. Now, immerse the frame and pull it out. Observe the colors in the film that are the result of double reflection and interference. Change your angle of view to notice the way the colors change. Now rotate the loop so its plane becomes vertical. The film becomes a wedge as gravity pulls it down, and the varying thickness will generate a nonuniform color pattern as you look down. Before it pops, blow gently and a bubble will form. If the air from your breath is warm enough, the bubble will be less dense than the ambient air and will float upward like a hot air balloon. (Physics professors are good at this because of the hot air.) As the bubble floats, watch the colors change as your viewing angle changes. There are more references in the endnotes.[3]

Historically, we've just arrived in the late 1800s, when physics was in good shape. Thanks to Newton, the motion of bodies was pretty well understood. Maxwell had reduced electricity and magnetism to four equations. Great strides had been made in understanding heat, sound, and optics. To illustrate physics' feeling of accomplishment, here's an 1894 quote from Albert A. Michelson (who, we'll see, should have known better): "The more important fundamental laws and facts of physical science have all been discovered, and these are now so firmly established that the possibility of their ever being supplanted in consequence of new discoveries is exceedingly remote. . . . Our future discoveries must be looked for in the sixth place of decimals."

The picture was definitely rosy, but as rose gardeners know, there are always a few thorns:

1. Light's status as a particle (Newton) or a wave (Young) was still unclear.
2. When bodies are heated, they give off light. But what is the mechanism for this?
3. Is matter continuous or is it divisible into smaller units called atoms?
4. In what medium does light travel and in what reference frame is its speed c?
5. How does the photoelectric effect operate where light hits a metal and releases electrons?

These thorny, minor difficulties turned out to require fundamental changes in the way physicists viewed the universe. Making these revolutionary leaps in thought required the genius of several outstanding people, and a major catalyst was Albert Einstein.

Chapter 25.
EINSTEIN'S PRODIGIOUS EFFORTS— SPECIAL AND GENERAL RELATIVITY

Albert Einstein

"When you sit with a nice girl for two hours, it seems like two minutes. When you sit on a hot stove for two minutes, it seems like two hours. That's relativity."

—Albert Einstein

The revolutionary insights made by Albert Einstein didn't come in a vacuum. In retrospect, a key experiment was made by Albert A. Michelson. As the first head of the physics department, he had to say something profound for the dedication of the Rockefeller-funded Ryerson Physics Laboratory at the University of Chicago in 1894. Albert A. Michelson (pronounced Michael-son) chose to emphasize accurate measurements rather than revolutionary theory. But it was, ironically, his accurate measurements that led to the revolutionary theory of relativity. Even as an undergraduate at the Naval Academy, Michelson was involved in experiments to measure the speed of light, and this proved a lifelong fascination.

Michelson invented a very sensitive device called an interferometer. This device split a beam of light into two beams and sent them off in perpendicular directions. They were reflected and then recombined. By observing the interference patterns of the recombined beam, differences in path length could be determined to a high degree of accuracy. In 1887, working with chemist Edward Morley at Case Institute of Technology in Cleveland, Michelson set out to measure the speed of light accurately enough to answer one of physics' unsolved problems of the time. Since light is a wave, and waves are a disturbance in a medium, light must have a medium. Light's medium was given a name, lumeniferous Æther (let's call it ether, but don't confuse it with the chemical used as an anesthetic). The properties of this lumeniferous ether had to be pretty strange. Since light goes anywhere, it was reasoned that ether had to exist everywhere. Since light travels extremely fast, ether had to be very "stiff" (think about how string tension has to be large to support a fast wave on a string). Further, astronomical bodies had to go through this ether without any drag, or they would slow down in some observable way. It seemed sensible that the speed of light derived from Maxwell's equations must apply to the absolute reference frame in which the ether would be at rest. So Michelson set out to find the speed of some moving body relative to the absolute frame, in which the ether would be at rest. The fastest object available to him in 1887 was the Earth, so he set out to measure the speed of the Earth relative to the absolute reference frame of the ether. The result of his experimental determination was stunning: the Earth was not moving with respect to the ether. Repeating the experiment six months later, when the Earth was moving the opposite direction, yielded the same null result. Physicists were perplexed. It took an original thinker to work out of this enigma, and that was Albert Einstein.

Albert Einstein Mini-Biography[1]

Hermann Einstein (1847–1902), at age twenty-nine, married Pauline Koch (1858–1920) when she was only eighteen. Their first child, Albert, was born in 1879. Grandmother Koch pronounced baby Albert "fat, much too fat." Before 1881 ended, Einstein had a younger sister, Maria, called Maja. Einstein was a quiet child and began his education at a Catholic school, where he referred to the teachers as "sergeants." After three years, Einstein was transferred to the public school, where he called the teachers "lieutenants." The authoritarian school system was not to his liking, so he achieved mediocre grades. Einstein's youth was filled with family visits, music—his violin and his sister's piano—and sailing on a nearby lake. To fulfill Bavarian law, teenage children had to have some religious education, so Einstein's parents, non-practicing Jews, got a distant relative to teach him about Judaism. Einstein soon gave that up because he couldn't reconcile religious dogma with scientific evidence.

The family's electrical business failed, so the family moved to Italy, leaving Einstein to finish gymnasium (high school) in Munich. Before long, Einstein got a doctor's note, quit school, and showed up at the family's new home in Italy. Young Einstein had a plan. He intended to take the entrance exam at the Swiss Federal Polytechnic (ETH) and start there without finishing high school. That school would prepare him to teach physics or math, so he wouldn't have to worry about poor business conditions. The plan didn't work. He did extremely well in the math and physics part of the exam, but very poorly in languages and biology. It was recommended that Einstein finish gymnasium in Switzerland, at Aarau, then he would be admited. The educational system in Switzerland was quite different from that of Bavaria, and Einstein thrived. He graduated first in his class, and made lifelong friends of the Winteler family, with whom he stayed. Eighteen-year-old Marie Winteler especially hated to see him go.

At the Swiss Federal Polytechnic, Einstein made more friends. One friend, Marcel Grossmann, faithfully attended all the lectures on every subject, even though it was not mandatory. Einstein avoided humanities because he found them uninteresting. Physics began well, but Herr Weber didn't discuss anything past 1850, so Einstein began skipping physics also, spending more time on his own ideas.

Another friend was the only woman in his class, Mileva Maric. Einstein's friendship with Mileva blossomed to the point where she left school and went home (to Hungary) to deliver his child, Lieserl. When final exam time came, Mileva was absent and Einstein was underprepared. Fortunately for him, Marcel Grossmann had taken good notes, which Einstein borrowed. He passed, but barely. Mileva took the exam a year later and failed. Getting a degree was a good start, but the next step was even harder: getting a job. Einstein's professors wrote no let-

Albert Einstein and friends

ters of recommendation for him, possibly because they hadn't seen enough of him. After a year of freelance tutoring and substitute teaching in Bern, Einstein had made some more friends, but little money. The friends gathered at Einstein's apartment to discuss matters of great importance, especially physics and philosophy. They called themselves the Olympian Akademy, fully aware of the irony, since only one of them had a job. Then, there was the matter of Mileva Maric. In desperation, Einstein asked advice of Marcel Grossmann, who had gotten a job teaching at ETH. Grossmann's father had a contact at the Swiss Patent Office and got an interview for Einstein. Einstein landed the job of technical expert, third class. With a job, he was finally in a position to marry Mileva Maric. However, neither family approved. Finally, on Einstein's father's deathbed, he gave his permission. They were married in three months.

The years 1903 and 1904 were very productive for Einstein. With Einstein's new domestic situation, the Olympian Akademy met less, and his job at the patent office required forty-eight hours a week. However, his center drawer at work contained important papers—his physics ideas. Although he wasn't even on the lowest rung of the academic ladder, Einstein reviewed papers for the *Annalen der Physik.* The year 1904 saw the birth of his and Mileva's first son, Hans Albert, a healthy and happy baby who brought great joy to his parents. In the absence of meetings with his friends, Einstein committed his ideas to paper and submitted five of them in 1905 to *Annalen der Physik.* These five papers contained original thought on several different areas of physics: a new way to determine Avogadro's number; an explanation of Brownian motion of particles jiggling around under a microscope; an explanation of the photoelectric effect; the special theory of relativity; and the famous $E = mc^2$. (Most of his ideas will be discussed in more detail in this chapter and the next.) The publication of these papers within a single year was called by some of his biographers the "Annus Mirabili"—the year of miracles. Eventually, it paid off in several ways: he was awarded his doctorate for one paper, a Nobel Prize (sixteen years later) for another, and the recognition of his colleagues for his stunning effort. Things began to change rapidly for Einstein in 1909. First, he resigned his patent office job and became an adjunct professor at the University of Zurich. His second son, Eduard (Tete), was born in 1910, his sister Maja married Paul Winteler (yes, of the family from Aarau) in Italy, and Einstein took a teaching position in Prague in 1911. After a year of enduring bugs and brown drinking water in Prague, the Einsteins moved back to Zurich to the ETH, where he became a colleague of Marcel Grossmann. This was especially convenient because Einstein was working to generalize his special relativity ideas to include accelerated reference frames, and he found that the mathematics was beyond him. Grossmann, now a math professor, helped Einstein work through the necessary mathematics. Within two years, Einstein was offered an academic position in Berlin at the Prussian Academy of Sciences, and he accepted. Mileva and the children survived only three months in Berlin. She thought Berlin was "worse than Prague." Einstein had separation papers drawn up, and Mileva and the boys left for Zurich. After almost two years of an extremely diligent effort, Einstein published his general theory of relativity. Almost immediately, he modified it because of criticisms by astronomer Willem de Sitter. On the political front, things were deteriorating in Germany. It was just prior to the start of World War I, and Einstein's response to the war talk was to sign pacifist petitions. Just as the war started, Einstein became seriously ill with stomach problems. Over a lengthy period, he was nursed back to health by his cousin, Elsa Loewenthal. Finally, the war ended, just about the time Einstein's general theory of relativity received experimental confirmation. By 1920 his marriage to Mileva ended in divorce and he married Elsa. Fame, the Nobel Prize, and travel followed, with Einstein and his violin in demand around the world. Women especially loved to hear him play.

The rise of the Nazis in Germany created more and more trouble for the pacifistic and Jewish Einstein. He eventually fled to the United States in 1933 and took a position at Princeton's Institute for Advanced Studies. After Elsa died in 1936, he was joined by his sister, Maja. For almost the next twenty years, Einstein pursued a unified field theory with no tangible results and served as a physics elder statesman until his death in 1955. Einstein said many quotable things, but this is one of my favorites:

> "The world needs heroes and it's better they be harmless men like me than villains like Hitler."

In his 1905 special relativity paper, Einstein disposed of the lumeniferous ether in the beginning, which is translated here:

> the introduction of a light-ether will prove to be superfluous since, according to the view to be developed here, neither will a space in absolute rest endowed with special properties be introduced nor will a velocity vector be associated with a point of empty space in which electromagnetic processes take place.

Rather than attempt to justify his ideas on the basis of the experimental evidence of the Michelson-Morley experiment (which he doesn't mention), Einstein proposed two simple, innocuous-looking postulates, based on measurements made by observers moving at constant velocity (these are called *inertial reference frames*):

1. The laws of physics are the same to any observer in an inertial reference frame. (Think about tossing a ball into the air while

motionless and while sitting in a moving vehicle. Clearly, there would be no difference in the motion of the ball.)

2. The speed of light is the same for any observer in an inertial reference frame whether the light is emitted by a body at rest in the frame or moving with a constant velocity. (This is more difficult to envision, but the null result of the Michelson-Morley experiment supports this idea perfectly.)

In the paper, Einstein goes on to show the implications of these postulates, rederiving equations already found by Lorentz and FitzGerald. The point of the paper is that fundamental conceptions of space and time are not applicable on an absolute basis but are relative to the observer. As an example of this modification, consider what would happen to the simple concept of two events happening at the same time, referred to as *simultaneity*.

Equip two observers, Rose and Thor, with clocks. Put Rose in the middle of a railroad flatcar in motion to the right with speed v. Put Thor on the ground. Fix a lighting system so that when the front or back of the flatcar passes it, a light is emitted. When the car is in just the right position, Thor will get signals from the front and back of the car at the same time, indicating Thor is even with the middle of the car. The signals are simultaneous from Thor's point of view. On the other hand, Rose is in motion while the light is going toward her, so she is closer to the front signal than the rear. If light from the front signal moved at c + v, and light from the rear signal moved at c – v, they would arrive at the same time for Rose also. But, if light travels at c—independent of the motion of the inertial observer—light from the front of the car will reach Rose sooner than light from the rear because it doesn't have as far to travel. From Rose's perspective, the events happened at different times—first the front light went on, then the back light went on. This example is called the *relativity of simultaneity*, and illustrates the fundamental effects of Einstein's innocuous-seeming principle. If two observers cannot agree about whether or not two events are simultaneous, you know something extraordinary must be happening at a very elemental level.

Einstein's special relativity requires something quite revolutionary: Newtonian ideas of absolute length, time, and mass must be abandoned and replaced by quantities that are determined relative to the observer.

Let's start with length. The length (of anything) depends on the observer according to the equation

$$1 = 1_{proper}\sqrt{\left(1 - \frac{v^2}{c^2}\right)}$$

where 1_{proper} is the length measured by an observer at rest with respect to the length

This is referred to as *length contraction.*

As an example of this, suppose our observer Rose zips past in a spaceship moving at half the speed of light. Rose's spaceship is towing a meter stick behind it, which is handy for making measurements. Our other observer, Thor, sits on the ground, also equipped with a meter stick. As Rose zips past, Thor measures the length of the moving meter stick and finds it to be 0.866 m. In completely symmetrical fashion, Rose measures the length of Thor's meter stick and finds it to be 0.866 m.

The modifications are not restricted to length. Both time and mass are affected too.

$$\Delta t = \frac{\Delta t_{proper}}{\sqrt{\left(1 - \frac{v^2}{c^2}\right)}}$$

where Δt_{proper} is measured by an observer for whom the events defining the beginning and end of the time interval occur at the same location

This is called *time dilation.*

$$m = \frac{m_{proper}}{\sqrt{\left(1 - \frac{v^2}{c^2}\right)}}$$

where m_{proper} is measured by an observer for whom the object is at rest

This is referred to as *mass expansion*, and the proper mass is also called *rest mass*. In other words, a moving clock would seem to take too long between ticks and a moving mass would seem too massive—from the perspective of a stationary observer.

Once the shock wears off and people start thinking constructively about relativity, a number of questions arise. You could think of this as a conversation between Rose and Thor.

- Whose measurement is right? The unsatisfying answer is: They both are. Built into the question is the assumption that there is one absolute answer. Built into relativity is just the opposite notion: measurements are relative. Rose's measurements are right for Rose, and Thor's measurements are right for Thor.
- Why do these examples use spaceships? Don't other moving vehicles like cars look shorter in the direction of motion? Cars do look shorter, but not enough for us to notice. The reason lies in the factor

$$\sqrt{\left(1-\frac{v^2}{c^2}\right)}$$

  A moving vehicle has to be traveling at 10 percent of the speed of light to even make a 0.5 percent apparent difference in its length, as measured by the other observer.
- What about twins? If Rose and Thor were twins, wouldn't each sense the other's clock as running slowly? If Rose was sent on a long voyage at high speed, would both twins age at the same rate? As it turns out, this situation, called the *twin paradox*, is a bit more complex than it sounds. Implied in the question is the notion that the twins will eventually be reunited in the same reference frame. This will require the traveling twin to be turned around; the consequent acceleration would remove it from the realm of special relativity, which demands constant velocity inertial observers. Nevertheless, the short answer is that the traveling twin senses proper time between the events of zipping by Earth and zipping by some distant planet, whereas the stay-at-home twin would experience a longer time. Thus, the traveler would, in fact, age less.

In 1971 an experiment was performed that involved transporting very precise atomic clocks around the world on commercial jets. Jet speeds are considerably less than the speed of light, so the time-dilation effect was expected to be extremely small. But the clocks were accurate to within nanoseconds, so when the traveling clocks were compared with their counterparts left at rest, the traveling clocks had recorded slightly less time, consistent with the relativistic prediction.

- How does relativity imply that the speed of light is some universal speed limit? Think about it this way. If Rose zips by at a speed close to that of light, why can't we give her a little shove and accelerate her right past c? Let's see how big a force would be required. We know F = ma, so we'll just give her a little acceleration. . . . Oops. Rose's mass has gotten huge. (Sorry, Rose, no offense intended.) As Rose's speed approaches that of light, her mass becomes

$$m = \frac{m_{proper}}{\sqrt{\left(1 - \frac{v^2}{c^2}\right)}}$$

so the denominator of this fraction becomes very small, making m very large. It's going to take a lot of force to accelerate such a huge mass.

- OK. Suppose Rose is going at the speed of light . . . hold it right there. Weird things happen at the speed of light. If a body is traveling *at* the speed of light, its length in the direction of motion diminishes to zero and its mass becomes infinite. If you can deal with infinite mass, you must have infinite force. If you've got infinite force, why are we sitting here, discussing physics? You could do something really big.
- What kind of evidence is there for this theory? As you know, if there wasn't sufficient evidence, we wouldn't be dealing with relativity at all. Besides the clocks in the jets experiment described above, there's the matter of the muons. Muons are unstable particles produced in nuclear reactions. They occur naturally in the Sun and arrive at the Earth's atmosphere as part of the solar wind, or can be made and studied in nuclear reactors. Since they move at almost the speed of light, muons provide a handy way to check special relativity. Given that we know their lifetime as measured in the lab, solar wind muons headed for the Earth can be measured in the strat-

osphere and at ground level to see if their lifetime when they are moving matches the relativistic prediction. It does.

Another bit of evidence comes from experimental physicists who deal with high-speed particles on a regular basis. They are interested in high-energy collisions, not relativity. But they cannot use rest masses, because the results are inaccurate. Relativistic mass is required to make accurate calculations.

- What good is relativity? In technological terms, there has been no gadget that is based on relativity—yet. From the standpoint of physics, special relativity deepens our understanding of the universe.

If special relativity was a mathematically simple product of Einstein's youthful exuberance, then general relativity was the mathematically complex result of Einstein's mature professional efforts. It may almost be the difference between inspiration and perspiration. In principle, Einstein's aim for general relativity was easy to state: extend the ideas from special relativity to the more general case of an observer not limited to an inertial reference frame. That's the way it started out, but things got complicated quickly, and Einstein needed some mathematical tools he didn't have: differential geometry and tensor calculus. Fortunately, his friend Marcel Grossmann helped out, and they wrote a paper together (see endnotes). Once Einstein got the math straightened out (we won't do that here), general relativity turned out to be a gravitational theory. Rather than gravitational forces between masses as Newton had conceived, general relativity considered the motion of bodies to be due to curvature in space-time, caused by the presence of mass. (Space and time are related to each other through special relativity and are called space-time.) The principle of relativity says that there is no locally measurable difference between the effects of gravity and the effects of an accelerated reference frame. In other words, inside a closed rocket ship, there would be no discernible difference between the ship being attracted by a massive body or the firing of the engines.

The effects predicted by general relativity are small and therefore hard to measure, but a 1919 expedition by Sir Arthur Eddington to measure a

Very shortly after its formulation, Einstein's general theory of relativity was applied to the whole universe by the Dutch astronomer Willem de Sitter. He found that Einstein's equations had no stable solution: the universe had to be expanding or contracting. After a dialogue between Einstein and de Sitter—it was too civilized to be called a debate—Einstein inserted a new term into his equations. This new term allowed the universe to be static (which he believed it to be), and the universe's curvature determined by the mass contained. He called this term the *cosmological constant* but disliked it because it spoiled the beauty of the equations. De Sitter was none too happy with it either, but it appeared to be necessary. Several years later, Edwin Hubble found evidence for the universe's expansion that allowed a delighted Einstein to remove the offending cosmological constant. But recent experiments show an accelerating expansion for the universe that may require the return of the cosmological constant. Einstein wouldn't be pleased—or would he?

Experiment 29: LASER, not Phasor

Equipment needed: laser pointer, CD, metric tape measure

The acronym LASER stands for light amplification by the stimulated emission of radiation. As we have seen, resonance is an important concept in physics. The laser takes advantage of resonance and several other properties of light and matter to produce a light source that has unusual characteristics. As we will see, electrons have excited energy states above their normal ground state. When the electron makes a transition from an excited state to a lower one, an electromagnetic wave (photon) of frequency f is given off according to $E = hf$. If another electron in the higher energy state happens to be nearby, the presence of the first photon stimulates the emission of another photon, in phase with the first. These two photons now have the same frequency and are in phase, which means they interfere constructively. This addition increases the amplitude, which causes amplification. As long as there are atoms that have an energy level in which the electron lingers for a while, called a *metastable* state, the laser action will work.

Use the result from Experiment 27 to find the energy gap that produces the light emitted by the laser. $\Delta E = hf$ and $f\lambda = c$, so $E = hc/\lambda = 2 \times 10^{-25}$ Jm/λ.

Laser color	λ (meters)	ΔE (joules)
(Sample) Red	(Sample) 6.5×10^{-7} m	(Sample) 3.1×10^{-19} J

(Beware of fakes. Some that are billed as shake flashlights actually contain nonmagnetic slugs and have no diode/capacitor network. The LED is lit by button batteries, and no amount of shaking will generate or store any charge.)

CHAPTER 26. MATTER'S INNARDS—ATOMS AND QUANTUM MECHANICS

Niels Bohr

> "A physicist is just an atom's way of looking at itself."
>
> —Niels Bohr

The echoes from Albert A. Michelson's words "The more important fundamental laws and facts of physical science have all been discovered . . ." had barely subsided when new discoveries began to flow. And flow they did, fast and furious for the next thirty-plus years. If the developments in physics from 1895 to 1930 were summarized into a text message or a tweet, here's what it might say:

2 qik
2 b ez

To provide a picture of the quick pace of physics during this period, we will explore them in chronological order.

In 1895 Wilhelm Conrad Röntgen experimented with Geissler tubes (invented by German glassblower Heinrich Geissler in 1857), in which high voltage was applied to the metal terminals of a glass tube. This tube contained very dilute gases. Besides the colored lights inside the tube, Röntgen noticed that a coated screen about a meter away gave off a faint glow. He experimented further and found that whatever was being emitted by the tube went right through his hand and formed an image of it on the screen. Röntgen's wife, Bertha, appears in a famous photographic plate made on the day of his discovery. It shows the bone structure of her hand and a ring she was wearing. Not knowing the complete nature of the rays he had found, Röntgen adopted the mathematical symbol for an unknown, and called them *x-rays*. Almost immediately, they were used (and misused) for medical diagnosis and even entertainment. Tesla and Edison both experimented with x-rays, but Edison discontinued his research in 1903 when one of his glassblowers who tested x-ray tubes on his hands died from cancer. Röntgen was very modest and resisted the move to rename the rays in his honor. He preferred the term x-rays.

In 1896, Antoine Henri Becquerel, in France, learned about the discovery of x-rays and decided to explore a possible connection between x-rays and naturally phosphorescent materials. Several minerals, especially a collection of uranium salts he had inherited from his father, were exposed to sunlight, then tested to see if they would fog a photographic film covered by opaque paper. The result was positive, but Becquerel was perplexed to find that the images were nowhere near as sharp as those from x-rays. Further, it turned out that the rays were given off even though the minerals were not exposed to the Sun. Initially, they were called Becquerel rays, and he interpreted them as some long-lasting beam within the tube, referred to as *cathode rays*. Thomson's experiments showed that cathode rays were small, negatively charged particles. Thomson called them corpuscles and boldly proposed that they were present in all matter at the deepest level. Many of his colleagues had a hard time with this concept, even suspecting him of "pulling their legs." Soon, the particles were called *electrons*, a term coined by Irish physicist G. Johnstone Stoney to describe a basic unit of charge. Thomson carried the electron idea further, reasoning that it must be present in the fundamental unit of matter, the atom. To make a stable arrangement, Thomson proposed that negatively charged electrons were contained in a sea of positive charge, making the net charge of the atom neutral. This model of the atom, proposed in 1904, was promptly called the *plum pudding model* because it resembled the English holiday treat, with the electrons sprinkled like raisins in a positive pudding. (We'll return to the atom shortly, but first we'll look at more developments that occurred.)

In 1900 thermodynamics, electromagnetism, and mechanics still needed to be integrated in order to give physics a unified view of the universe, and all three branches wrestled with the thorny problem mentioned earlier: light emitted by heated bodies. Oddly enough, the analysis took a fascinating turn based on the work of a classical, conservative thermodynamics specialist, Max Karl Ernst Ludwig Planck (1858–1947). Classical electromagnetic theory had been unable to explain what was called blackbody radiation. Classically, a blackbody is one that absorbs all radiation falling on it, then reradiates it. Think of a situation like that of a cavity, where all light goes in through a small opening and sets up standing waves inside. If all the waves were equally probable, there would be an infinite number of short-wavelength waves that would fit into the cavity. This is hardly realistic, because it leads to infinite energy. On the other hand, since the distribution of intensities had already been measured, the mathematical function that described the distribution was determined by Wilhelm Wien. Max Planck agreed with Wien's result, but found mathematics with no underlying physical mechanisms very unsatisfying and set out to find basic physical properties that would lead to the known mathematical function. Planck's particular interest was to make sure that the fundamental derivation was consistent with the entropy requirements of the second law of thermodynamics. To achieve the mathematical result, Planck had to assume that the energies of the oscillators that make up the body come in small, discrete units. (In retrospect, this is called *energy quantization.*) Tiny energy units (called *quanta* by Einstein; the singular is *quantum*) are integer multiples of a constant multiplied by the frequency of oscillation. Planck said this was "a purely formal assumption and I really did not give it much thought except that no matter what the cost, I must bring about a positive result." The radiation formula was quickly accepted, and the underlying energy quantization necessary to bring it about seemed to escape everyone's notice—including Planck.

In 1905 the stage was set for an important advancement. There were possible solutions to several of the thorns in physics' rosy picture. To solve the problems would require someone with a keen generalist's view of physics, someone up to date in the latest ideas, someone with no long-standing positions to defend, someone not reticent about anything—someone who had nothing to lose. That someone was Albert Einstein. His 1905 papers followed an interesting pattern: he took someone else's work seriously, applied it to some different situation, then looked at the implications. For example, if atoms really exist, then their constant random motion would cause a large body to jiggle visibly in a microscope. Or, if split and recombined light exhibits no interference shift, then its speed must be constant for every observer in an inertial reference frame. This requires different observers to adjust their distance and time measurements before making comparisons. Or, if energy really does come in small quantum packets, then light hitting a metal will give all or none of its quantum of energy to the electrons in the metal. If the light frequency is high enough, electrons will get enough energy to be freed. If the light frequency is too low, no electrons will escape, regardless of how strong the light's intensity. It was this explanation of the photoelectric effect that model didn't last very long. J. J. Thomson's cathode ray experiments revealed the electron's existence, but there was no experimental support for the positive pudding. A former student of Thomson's, New Zealander Ernest Rutherford, solved the problem. Rutherford had become an expert on radioactive particles, having separated Becquerel rays into three parts: the alpha (a positively charged helium nucleus), the beta (an electron), and the gamma (a high-frequency electromagnetic wave). At the University of Manchester, Rutherford assigned the task of bombarding atoms of gold foil with alpha particles to his assistant, Hans Geiger, and a talented undergraduate, Ernest Marsden. If Thomson's plum pudding model was correct, most of the particles would shoot right through the thin positive charge of the atom and almost none would be reflected backward. That's not what happened. Rutherford said about their results: "It was almost as incredible as if you fired a fifteen-inch shell at a piece of tissue paper and it came back to hit you." Rutherford concluded that the positive charge, and almost all of the mass, of the atom was contained in a small volume in the center. He called this the *nucleus*. For stability, the electrons had to move, so Rutherford presumed they orbited the nucleus the way that planets orbit the Sun. In an atom, it is electrical force rather than gravity that keeps the electrons orbiting. This became known as the *Rutherford solar system model* of the atom.

In 1912, Max von Laue, at the University of Zurich, directed an experiment that solved two problems at once. He shone x-rays on crystals and found dark and light spots on the recording film, which corresponded to a diffraction/interference pattern. This showed that x-rays are electromagnetic radiation, that crystals are a regular arrangement of atoms, and that the spacing of the atoms in a crystal corresponds to the wavelength of x-rays.

In 1913 Danish physicist Niels Bohr worked with Rutherford at Manchester and recognized the need to find a mechanism for the radiation of light from excited atoms. The simplest atom, hydrogen, emits specific colors when excited: red, aqua, blue, and blue-violet. Bohr made a bold hypothesis: The hydrogen atom has certain allowed states, for which its electron's angular momentum is quantized according to:

$$mvr = \frac{nh}{2\pi}$$

While the electron is in an allowed state, it doesn't radiate, even though it is an accelerated charge. When it receives energy from any source, the electron can move to higher energy levels, but it lingers there only temporarily. The electron then moves downward in energy, landing only in allowed states, until it reaches the lowest level, called the *ground state*. The energy lost by the electron is given off in the form of electromagnetic radiation according to the Planck relation: $\Delta E = hf$. Experimental evidence supports this hypothesis for hydrogen, and it works somewhat for other atoms with one electron in the outer shell but is quite inaccurate for atoms with larger numbers of electrons.

Bohr became Einstein's friend and carried on a lifelong debate with him about many of the deep issues raised by applying quantum concepts to atoms and smaller particles, especially statistical aspects of the Copenhagen interpretation championed by Bohr. Here's a summary of one of their exchanges.[1]

> Einstein: "I can't conceive of God playing dice."
> Bohr: "Einstein, stop telling God what to do with his dice."

During the 1920s, Schrödinger, Heisenberg, Born, Jordan, and others generalized the quantum theory and arrived at a mathematical relationship that turned out to be a wave equation. Solutions to this equation, called *wave functions*, have been interpreted as being related to the probability of finding an electron at a particular location. Chemistry has adopted the high probability surfaces of wave functions as representing regions where the electron is highly likely to be located. They are called *orbitals*. Implied by the wave equation is a concept called the *Heisenberg Uncertainty Principle*. This principle says that the position and momentum of any particle cannot be determined precisely but only to within certain limits:

$$\Delta x \Delta p_x \geq h/4\pi$$

By this time, physicists had determined that atoms consist of a nucleus containing positive and neutral particles called protons and neutrons. The nucleus contains the majority of the mass and occupies a diam-

the nucleus in patterns that cannot be known. They have definite energies but not definite orbits. In their normal or excited states, they radiate no energy, but when they make a transition from a higher energy state to a lower one, the energy lost is given off in the form of electromagnetic radiation.

Now that we have atoms sorted out, have we learned everything there is to know? Not by a long shot. There is a great deal to learn beyond atoms.

In 1923 Arthur Holly Compton experimented by shining x-rays onto matter. If the x-rays are treated like they are billiard balls, predictions about the results of their collision with matter match experimental results. In other words, x-rays act like particles in addition to acting like waves. The quantum of electromagnetic radiation was called a *photon* by physical chemist Gilbert N. Lewis in 1926, and the name stuck.

In 1924 Louis Victor Pierre Raymond, seventh duc de Broglie, in one of the shortest doctoral theses on record, theorized that particles can exhibit wavelike behavior. According to de Broglie, the wavelength of a particle is

$$\lambda_{\text{de Broglie}} = \frac{h}{mv}$$

If we apply this idea to the electron in the hydrogen atom, the condition for a standing wave would be met if the electron, viewed as a wave, interfered with itself constructively. This would happen if whole waves exactly fit into the circumference of the electron orbit. Mathematically, $n\lambda_{\text{de Broglie}} = 2\pi r$. Rearranging this equation algebraically yields

$$mvr = \frac{nh}{2\pi}$$

the same relation assumed by Bohr. Thus, the electron in a hydrogen atom can be thought of either as a particle, orbiting with quantized angular momentum, or as a wave, constructively interfering with itself. This is curious. From a theoretical standpoint, a particle can be viewed as having

elephant and unsighted observers

interesting irony is that George Paget Thomson, the son of J. J. Thomson, received the Nobel Prize for demonstrating that electrons act like waves, and his father got the Nobel Prize for showing that electrons act like particles.

So what is light/matter exactly—waves, particles, or what? This question sounds similar to the question of which observer is right about relativistic measurements. Try this analogy. Unsighted observers must use senses other than sight to ascertain an unknown object.

We are the unsighted observers and we have a problem. However, when we try to beat light/matter into our molds, it doesn't quite fit. Light/matter is not a particle and it's not a wave. It behaves like a particle sometimes and it behaves like a wave under other circumstances, but it's more sophisticated than our simple models. Our knowledge is incomplete, and even our ways of thinking are too simplistic. We've got lots of room to progress.

You might think this a convenient place to end, but it isn't. There's more.

In 1928 Paul Adrien Maurice Dirac combined quantum mechanics and relativity, forming the Dirac equation. This equation has too much industrial-strength mathematics to exhibit here, but a curious thing happens when its solutions are investigated. It turns out that, for one possible solution, it predicts a particle that is like an electron but that carries a positive charge. By analogy, it is comparable to recognizing that $x = \sqrt{4}$ has two roots, +2 and –2. This possible particle was called the *positron*, or antielectron. Dirac himself was not so confident of the existence of this particle. Why did he lack confidence in this possibility? "Pure cowardice," he later explained. Eventually, all particles were found to have antiparticles, leading to the possibility of antimatter or even antiplanets. It turns out that antiparticles are scarce and short-lived.

Ernest O. Lawrence

In 1932 Carl D. Anderson, born in 1905, was studying cosmic rays (energetic particles, mostly protons and alpha particles, from a variety of sources, some outside the galaxy) in Colorado, where there is less atmosphere to absorb them. Unexpectedly, he found tracks in his detectors that were made by an interesting pair: an electron and a particle of the same mass but opposite charge—the positron, predicted by Dirac's theory. Later, he found another unusual particle, originally thought to be a pion, which had been predicted by another theory. It turned out to be a completely new particle, not accounted for by any theory. It was called the *muon*. Announcement of this discovery was greeted by theoretician I. I. Rabi with the question "Who ordered that?"

Throughout the 1930s, the idea of watching collisions and keeping track of the products was a very fruitful way for physics to probe the deep structure of matter. One difficulty was that cosmic rays were very unpredictable. You never knew when they would arrive, what direction they would be coming from, and how energetic they might be. So Ernest Orlando Lawrence designed a device that made particle collisions more controllable. He called it the *cyclotron*.

Bigger and bigger cyclotrons were built in the 1930s for making controlled collisions. But toward 1940, cyclotron development was put on a back burner because of large, uncontrolled collisions in another realm—

Experiment 30: Half-Life Is Better Than No Life

Equipment needed: 50 coins, box with a lid

The purpose of this experiment is to simulate radioactive decay in a safe manner and demonstrate the concept of a half-life.

PROCEDURE

- Place all fifty coins in the box heads up (not as easy as it sounds).
- Put the lid on the box and give it one good shake.
- Open the box and take out the coins that have tails up.
- Count the number of coins with heads showing and note the number in the table.
- Replace the lid, give another shake, open, count, record as above.
- Repeat until the last coin has been removed.

Number of shakes	Number of heads-up remaining
0	50
1	
2	
3	
4	
5	
6	
7	
8	
9	

Chapter 27. ATOMS' INNARDS—NUCLEAR PHYSICS

"It has become appallingly obvious that our technology has exceeded our humanity."
—Albert Einstein

"I am become death, shatterer of worlds."
—J. Robert Oppenheimer (citing from the Bhagavad Gita, after witnessing the world's first nuclear explosion)

Since we're not capable of shrinking ourselves down to the size of an atom, let's go the other way. We'll enlarge the atom to the size of a football field. The electrons would be similar to the tiniest of gnats, buzzing around in wide arcs at the edges of the field. The nucleus would be like a tightly packed collection of BBs in the very center of the field. But wait. If the nucleus were so jam-packed with positively charged protons, there would be a huge electric repulsive force trying to push them apart. There must be a stronger force present. Called the *strong nuclear force*, it is a very unusual kind of force. When two *nucleons* (the generic name for either of the two constituents of the nucleus, the proton and the

the strong nuclear force drops to zero. In other words, if nucleons are close enough to "touch," they stick together. If they don't "touch," they exert no force at all. By analogy, this is like two pieces of Velcro or the old hook-and-eye fastening of a screen door. The following graph compares the strength of the strong nuclear force with the electromagnetic and gravitational forces. The weak nuclear force (to be discussed later) is included for comparison.

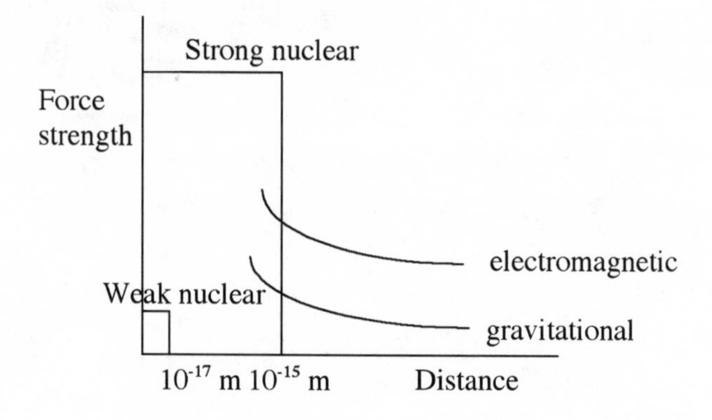

force strength comparison

Experimentino 20

Equipment required: about 100 styrofoam packing "peanuts," a box of toothpicks

The purpose of this experimentino is to build some model nuclei. Use a toothpick to attach one "peanut" to another. Bring in a third "peanut," and use toothpicks to attach it to the first two every place they touch. Add a fourth "peanut," again attaching it with toothpicks at every place it touches the others. Continue adding "peanuts," attaching them with toothpicks at every point of contact. If you are able to build model nuclei as large as iron (56 nucleons), you have done well. I advise you to quit before you arrive at the largest naturally occurring stable isotope, uranium, with 235 nucleons. The model gets wobbly way before that. (You can build nuclei out of magnet marbles, miniature marshmallows, or even tennis balls, as I once did. But it was heavy.)

Don't let these static models fool you. Nucleons move within the nucleus, and there are higher energy levels that excited nucleons can occupy. The chief difference between nucleon energy states and electron energy states is their magnitude. Nucleon energies are about a factor of a million higher. So when a nucleon loses energy coming down from an excited state, the electromagnetic radiation emitted has a million times the frequency of visible light. That's where gamma rays come from.

As nuclei become larger, protons on one side of the nucleus are beyond the range of attraction of protons on the other side, so the entire nucleus becomes less tightly bonded. The result is that larger nuclei are less stable. Since they carry no repulsive electrical force, more neutrons help keep larger nuclei together, but there is a limit. Near the lower end of the periodic table, stable nuclei have roughly as many neutrons as protons, but larger nuclei require more neutrons to act as "glue" to keep them stable. For example, calcium has twenty neutrons and twenty protons, while cesium has fifty-five protons and seventy-eight neutrons. Different forms of the same element that have more or less neutrons in the nucleus are called *isotopes*.

The nuclei of unstable isotopes decay, as Becquerel found in 1896, and the process is called radioactivity. Nuclear decay follows an exponential pattern related to the *half-life* of the nucleus.

$$N = N_0 e^{\frac{-693t}{t_{half}}}$$

where N is the number of nuclei left undecayed, N_0 is the original number of nuclei, t is the elapsed time, and t_{half} is the half-life of the nucleus, in the same unit as elapsed time

This equation works out to be very interesting. For example, if a nucleus of polonium 210 has a half-life of 138 days, 50 percent of a sample would still remain after 138 days, but only 25 percent would still be undecayed after 276 days. This was the point of Experiment 30: every "shake" was but if they are ingested, they are extremely harmful. Beta particles are electrons and may be shielded by a thin metallic layer. Gamma rays require thick lead shields because of their penetrating power. What remains after emission is referred to as *daughter nuclei*. In alpha and beta emission, daughter nuclei correspond to different elements because the atomic number has changed. In gamma emission, the nucleus's energy state is changed, but the element remains the same.

Since all these changes occur way below our ability to see what is happening, some other form of sensor is needed to detect occurrences at this level. The simplest detector is a film that is darkened by radioactive emissions. Such a film is often used in the form of badges worn by people who work near areas where radiation hazards may be present. Phosphorescent screens also respond to radioactive emissions, as do Geiger counters, which are often connected to speakers that crackle whenever a radioactive particle passes through the counter. (Having demonstrated Geiger counters for years in an academic setting, I never fail to be amazed by the high background radiation they always detect.)

We've gotten way ahead of ourselves historically by discussing the nucleus from a modern viewpoint. Let's return to the 1930s when nuclei were still being analyzed.

Once the neutron's existence was demonstrated (by Chadwick in 1932), it became useful as an experimental tool.

> Since the neutron carries no charge, there is no strong electrical repulsion to prevent its entry into the nucleus. In fact, the forces of attraction which hold nuclei together may actually pull the neutron into the nucleus. When a neutron enters a nucleus, the effects are about as catastrophic as if the moon struck the earth. The nucleus is violently shaken up by the blow, especially if the collision results in the capture of the neutron. A large increase in energy occurs and must be dissipated, and this may happen in a variety of ways, all of them interesting.
>
> I. I. Rabi, January 1934

In Italy, a group headed by Enrico Fermi systematically bombarded all elements with neutrons. The most interesting results came from uranium, leading to the possible identification of element 93. A critical response came from a respected German chemist, Ida Noddack: "It is conceivable, when heavy nuclei are bombarded by neutrons, they could break into several large fragments." This was very provocative speculation for 1934.

In Germany, a group consisting of Lise Meitner, Otto Hahn, and Fritz Strassman conducted similar neutron bombardment experiments, and identified sixteen different radioactive substances and their half-lives by 1938. Meitner, the leader of the group, had to flee to Sweden because of the anti-Jewish policies of Hitler's Nazi government. An experiment performed after her departure produced a strange result: the element barium was found in the sample after uranium was bombarded with neutrons. Meitner received this cabled message from Hahn: "Perhaps you can suggest some fantastic explanation. We understand that it (uranium) really can't break up into barium . . . so try and think of some other possibility. Barium isotopes with much higher atomic weights than 137? If you can think of anything that might be publishable, then the three of us would be together in this work after all. We don't believe this is foolishness or that contaminations are playing tricks on us." Lise Meitner figured it out. The uranium nucleus had broken apart, yielding two smaller nuclei, more neutrons, and some kinetic energy. Meitner called this process *fission*, but others referred to it as splitting the atom. (If this constitutes "splitting" the atom, then J. J. Thomson's removal of the atom's electron might be regarded as "chipping" the atom.) Hungarian physicist Leo Szilard read this report with keen interest. He had worked tirelessly to help Jewish scientists escape Germany and was especially attuned to developments in the Fatherland. By 1939 Szilard, a former student of Einstein in Berlin, had fled to the United States, as had Einstein. Szilard foresaw the possibility that neutrons released by fission could cause more fissions, and so on. This chain reaction might occur rapidly enough to cause an explosion—thus a weapon. Since the research was taking place in Germany, the possibility of Hitler's gaining access to such a weapon appalled Szilard. He met Einstein on Long Island in July 1939 to report these developments and convince him to act. According to Einstein, "I really only acted as a mail box. They brought me a finished letter and I simply signed ment of converting a just-discovered scientific idea into a weapon before a rival group could accomplish the same feat. Many established physicists in the United States and England were already hard at work on a shorter-term military priority: radar. It looked like recent émigrés would have to be pressed into service. Foremost among them was Enrico Fermi.

Enrico Fermi Mini-Biography[1]

In 1898 Alberto Fermi (1857–1927), an Italian railways supervisor, married Ida de Gattis (1871–1924), a graduate electrical engineer and schoolteacher. They had three children from 1899 to 1901, Nella, Giulio, and Enrico. Young Enrico and his brother Giulio excelled in school with little effort, allowing them plenty of time for childhood pranks. At age fifteen, Giulio died unexpectedly. Fermi threw himself into study, assisted by books supplied by an engineer friend of his father's, Adolfo Amidei. His college entrance essay amazed the examiner, who thought it could be a doctoral thesis. By the time he was twenty-five, Fermi had a PhD and a job teaching physics at the University of Rome. He headed a group of physicists (who nicknamed him "the Pope"), and they systematically bombarded all the nuclei in the periodic table with the newly discovered neutron. In one of their experiments, they tested uranium but didn't detect any fission products because the uranium was wrapped in a thin foil. In 1930 Wolfgang Pauli theorized the existence of a light neutral particle that was emitted during beta decay. Fermi developed a model of beta decay in 1934 that required a new force, called the *weak nuclear force*. Included in this theory was the small neutral particle suggested by Pauli, but Fermi renamed it the *neutrino*—"the little neutral one." (Now you see why small experiments are called experimentinos.) Fermi was a very unusual physicist in that he was equally talented as a theorist and as an experimentalist. A rental car broke down on one of his trips to the famous summer school at the University of Michigan. He fixed it himself, and the admiring garage owner offered him a job on the spot.

In 1938 Fermi was awarded the Nobel Prize for his work on neutron bombardment. The political situation in Italy had deteriorated, with Mussolini instituting anti-Jewish policies comparable to Hitler's. Fermi's wife, Laura Capon Fermi, came from a prominent Jewish family, and it was clear there was trouble ahead. Fermi and his family (daughter Nella and son Giulio) journeyed to Sweden to accept the Nobel Prize. Rather than return to Italy, they sailed for New York, where Fermi had accepted a position at Columbia. Practically the moment he arrived, Leo Szilard briefed him, and Fermi repeated the experiments of Hahn and Strassman. Then he declared himself ready to join the Manhattan Project.

The Manhattan Project was an enormous undertaking and a challenge at several levels.[2] The physics wasn't completely understood, so a lot of effort was required early on. Next, there was a technology/management challenge. The overall project had two leaders: J. Robert Oppenheimer, noted physicist, and General Leslie R. Groves. The entire project was divided into parts, and teams were assigned to different locations around the country. The other challenge was an ethical one. Any weapon development must be accompanied by considerations about its ultimate use. But this project was clearly aimed at winning a competition with an opponent who planned world domination. As difficult as the project was, it met with a stunning success. Solving myriads of technical problems along the way, Fermi's group produced a *nuclear reactor* (he called it a pile) on the site of what were formerly squash courts at the University of Chicago. Chicago Pile-1 produced enough neutrons to sustain the nuclear chain reaction (it went *critical*) on December 2, 1942. Over the next two years, the other groups also made progress, and a test explosion was scheduled for the summer of 1945. Then, over a four-month period in 1945, the ethical challenge "went critical."

collision

- May 28: Leo Szilard told presidential adviser James Byrnes that scientists wanted the bomb demonstrated before it was used in warfare.
- June 11: The Franck Report was issued by scientists suggesting either complete secrecy about the bomb or rapid development of nuclear arms to make the United States invulnerable.
- July 16: First successful bomb was exploded—code name Trinity.
- July: Military decision was made to drop bombs on Japanese military targets.
- August 6: Hiroshima bombed: one hundred thousand immediate deaths.
- August 9: Nagasaki bombed: seventy-five thousand immediate deaths.
- August 15: Japan surrendered.

[Full disclosure: I hold strong ethical concerns about killing human beings. During the early 1960s, I worked for a short time on intercontinental ballistic missile defense systems. While the work was geared toward defense, I could see how it might be turned to offensive use, and it bothered me. Many Manhattan Project participants experienced far worse issues of conscience.]

Some Manhattan Project scientists left the project as early as VE Day, while others stayed to continue nuclear weapons development. Edward Teller was instrumental in developing a bomb that was nicknamed "The Super." Instead of fission of large nuclei, this bomb used hydrogen *fusion* as an energy source. The fusion bomb, called the *H-bomb*, uses a fission bomb as a trigger to overcome the electrostatic repulsion of hydrogen nuclei (protons). Considerably more energy is released by the fusion process—equivalent to millions of tons of TNT. Some scientists opposed the development of the fusion bomb: "It is clear that such a weapon cannot be justified on any ethical ground. . . . The fact that no limits exist to the destructiveness of this weapon makes its very existence and the knowledge of its construction a danger to humanity as a whole. It is necessarily an evil thing considered in any light." This quote is from a 1949 General Advisory Committee report by I. I. Rabi and Enrico Fermi. The rest is history.

From a physics standpoint, fusion is a very interesting process. Overcoming electrostatic repulsion is the biggest hurdle faced by controlled fusion, although clever schemes involving cold fusion, magnetic bottles, and laser fusion are being pursued. Even the most optimistic technologists don't forecast any breakthroughs soon. Another interesting feature of fusion is its connection with the early development of the universe and the cores of stars.

Let us return to Enrico Fermi. He joined the University of Chicago faculty in 1946 and became, first and foremost, a teacher. His style was one of clarity and simplicity, never using more complicated mathematics than was necessary to solve a problem. He fit into American life very well. Not only did he become a naturalized American citizen in 1944, but his English was excellent, and his square dancing was first-rate. Fermi was more comfortable being called Enrico than anything else and was so secure that he never expressed jealousy about another physicist, although he became annoyed at the press's fawning over Einstein. But, then, so did Einstein. One of his favorite activities was to pose a question for discussion at lunch, and listen carefully to everyone's opinions before he ventured any of his own. He was known for making rough estimates of various quantities. But when he did serious work, he liked to use oversized drafting sheets of paper two by three feet, far from the back-of-the-envelope calculations you might expect.

In late 1954, Fermi was diagnosed with rapidly metastasizing stomach cancer. His friend Eugene Wigner (1963 Nobel Laureate in physics) said, "Ten days before Fermi had died he told me, 'I hope it won't take long.' He had reconciled himself perfectly to his fate."

Many physicists left the Manhattan Project to return to academic life. One was Ernest O. Lawrence, who resumed his work with cyclotrons with more surplus equipment, more contacts, and more vigor than ever.

Richard P. Feynman

Chapter 28. DOWN TO THE NITTY-GRITTY —THE STANDARD MODEL OF THE UNIVERSE'S SMALLEST CONSTITUENTS

> "Things on a very small scale (like electrons) behave like nothing that you have any direct experience about. They do not behave like waves, they do not behave like particles, they do not behave like clouds, or billiard balls, or weights on springs, or like anything that you have ever seen.
>
> —Richard Feynman

After World War II, Ernest O. Lawrence got his cyclotrons fired up with a vengeance. The point of a cyclotron is to accelerate a charged particle in a strong electric field, then bend its path with a magnetic field so it can be accelerated again. Once the particle's speed is high enough, it is deflected and crashes into a target. The collision takes place record the events for later analysis. Besides the technical difficulty of maintaining a high vacuum and generating a strong magnetic field, the faster the particle speed, the larger the cyclotron radius. Thus, each new machine needed to be bigger to achieve higher energies. Eventually, Lawrence's team struck pay dirt. Cloud chamber pictures revealed tracks of particles that had never been seen before: the charged pions (π^+, π^-) and kaons (K^+, K^-), neutral pions and kaons, the lambda particle, the sigma particle, and antiparticles (recall Dirac's prediction and Anderson's discovery) of already known particles. Although these new particles were unstable and decayed into more familiar ones after a brief existence, this was a clear indication that matter had more surprises in store.

The particle race heated up. More cyclotrons came into operation and their designs improved. In a device called a *synchrotron*, the accelerating field was synchronized to ensure a constant particle beam radius. The cloud chamber was replaced with a bubble chamber so that particle tracks were made visible by the formation of bubbles in superheated liquid hydrogen. It was almost like analyzing photographs of an exploding haystack and searching for briefly existing needles. One of my fellow graduate students, for his thesis project, analyzed 240,000 bubble chamber photographs, one picture at a time!

The result of all this effort was a virtual explosion of particles: well over a hundred were found. Nobel Laureate Enrico Fermi remarked to his student Leon Lederman (eventually a Nobel Laureate in his own right), "Young man, if I could remember the names of all these particles, I would have been a botanist."

In 1964 Murray Gell-Mann and George Zweig, working independently, proposed a new scheme. All the heavy particles could be represented as composites of three smaller particles and their corresponding antiparticles. Gell-Mann dubbed these new fundamental particles *quarks*, after the line from James Joyce's novel *Finnegan's Wake:* "Three quarks for Muster Mark." These (first) three quarks were called up (u), down (d), and strange (s) and carried fractional electric charges of +2/3, –1/3, and –1/3, respectively, with opposite charges for the corresponding antiquarks.

According to this model, protons and neutrons are built from three quarks: *uud* and *udd*, respectively. The large group of newly found mesons could be made from quark-antiquark pairs. For example, the negative pion would consist of a down quark and an anti-up quark. The quark idea was proposed tentatively, and though it solved the problem of organizing the vast collection of particles in a mathematical sense, the reality of quarks was suspect, since none had been observed—yet.

An MIT/Stanford team, working at the Stanford Linear Accelerator, investigated the nucleus by shooting electrons at hydrogen and deuterium (the heavier isotope of hydrogen whose nucleus contains one proton and one neutron). They measured the angle and energy of the electrons after the collision. At lower electron energies, the scattering was consistent with protons and neutrons being "soft" structures that would deflect electrons only slightly. But when they used electron beams of record high energy, they found that some electrons lost most of their initial energy and were scattered at large angles. With remarkable similarity to the alpha-scattering work of Rutherford in the initial identification of the nucleus, Richard P. Feynman and James Bjorken interpreted the electron-scattering data as indicative of an inner structure to protons and electrons—namely, the earlier-theorized quarks. Now, the quark hypothesis had to be taken seriously.

There is a great drive in physics to simplify unnecessary complications by combining theories. Near the end of the nineteenth century, James Clerk Maxwell's recognition that electricity and magnetism were simply two facets of the same phenomenon allowed the two to be united. The combination was called electromagnetism. In the 1950s, Richard P. Feynman, Julian Schwinger, and Sin'Itro Tomonaga combined electromagnetic theory with quantum mechanics to form *quantum electrodynamics*, referred to as *QED*. In this theory, electrons interact with each other by exchanging photons. The photons cannot be observed because the electrons emit and absorb them within a region governed by the Heisenberg Uncertainty Principle. Because they are not capable of being

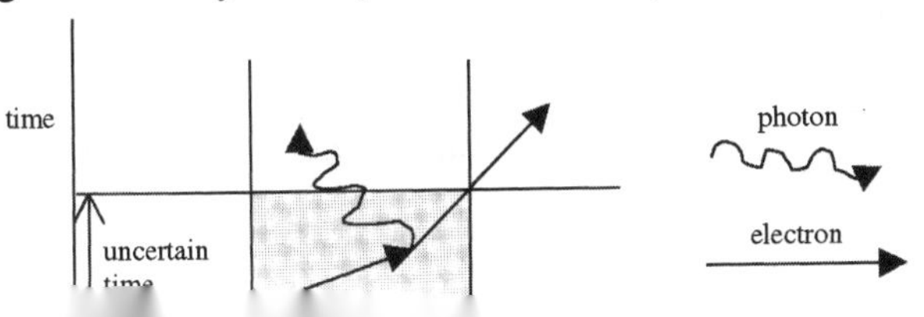

observed, they are called *virtual* photons. As a visual aid, Feynman diagrams depict electrical interactions in ways that provide a guide for making difficult calculations.

Feynman's teaching style is captured in "The Feynman Lectures," while his personal side comes through in his bongo playing and his books—*Surely You're Joking, Mr. Feynman!* and *What Do You Care What Other People Think?*[1]

While quarks were being hunted experimentally in the late 1960s, another theoretical unification scheme was proposed, this one involving two of the four fundamental interactions. Steven Weinberg, Sheldon Glashow, and Abdus Salam, working separately, formulated a theory that unified the electromagnetic and weak interactions into one called the *electroweak* interaction. Besides explaining already observed events in a more general context, this new theory added to the particle list by predicting the existence of several new particles: a neutral, weakly interacting particle (now called the Z^0), the W^+ and W^-, and a massive particle called the Higgs (more about this later).

In 1973 yet another theoretical development occurred. A quantum field theory of the strong interaction was formulated by Murray Gell-Mann and Harald Fritszch. This theory was called *quantum chromodynamics* and was similar to QED in that basic particles—*quarks*—interacted by exchanging *virtual* (in the uncertainty zone) massless particles called *gluons*. Since no one had seen a gluon, that made one more predicted particle that needed supporting evidence.

By the mid-1970s, theoretical and experimental developments were digested and summarized into a single theory called the *Standard Model.* (Mathematical reasoning beyond the scope of this book underlies many of these ideas, so be aware: there is heavy-duty math lurking in the background.)

Fundamental to the Standard Model is the notion that *the basic building blocks of the universe are fields, not particles.* As we discussed earlier, fields were introduced to help visualize forces, but they now play a central role.

According to the Standard Model:

- the universe's fundamental building blocks are *fields*,
- tiny packets of energy (quarks or leptons) result when quantum principles are applied to fields, and
- interactions between quarks or leptons are carried by the exchange of other energy packets (bosons) that cannot be observed because of uncertainty considerations.

So the classical action-at-a-distance picture of a force between particles has been replaced by an interaction consisting of the exchange of virtual energy packets (formerly waves) between quantized bundles of field energy (formerly particles). Now, *there's* a radical departure from earlier thinking.

The Standard Model includes two interactions, the strong and the electroweak.

The Strong Interaction: The particles that result from applying quantum principles to one set of fields are called quarks. There are six quarks (and associated antiquarks) grouped into three families. They have been named (whimsically):

Family 1: up and down
Family 2: charm and strange
Family 3: top and bottom

Quarks interact with each other by the strong interaction, which involves the exchange of virtual particles called gluons.

The Electroweak Interaction: The particles that result from quantizing another set of fields are called leptons. There are six leptons (and corresponding antileptons) grouped into three families. They are called:

Family 1: the electron and electron neutrino

Leptons interact by exchanging virtual particles called the photon, two Ws, and one Z.

In summary, here are the fundamental particles and interaction carriers:

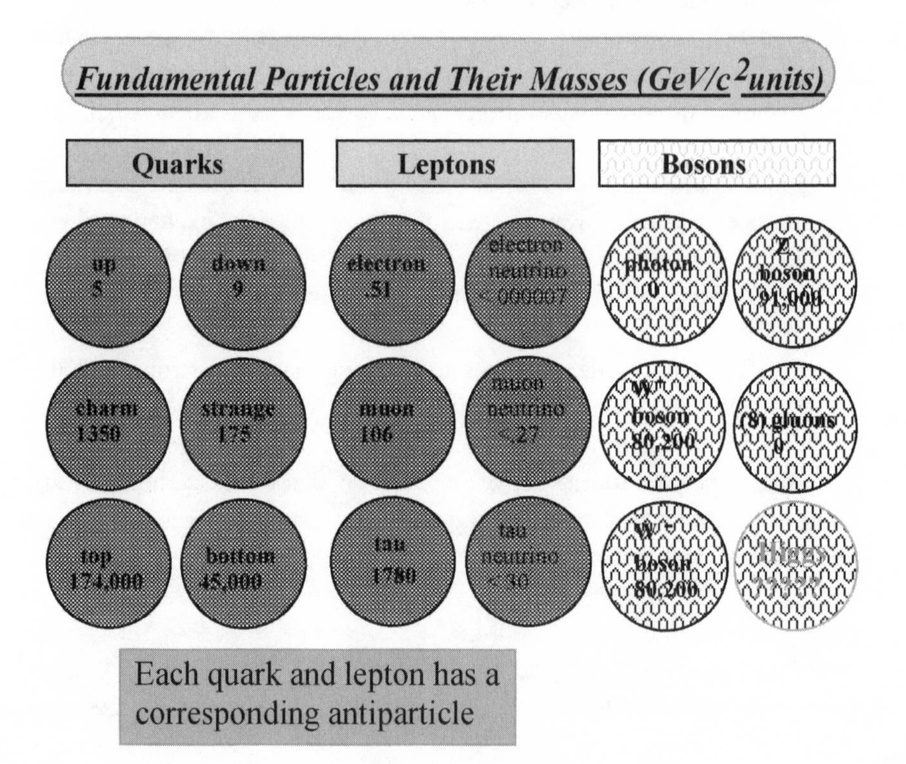

fundamental particles

According to the Standard Model, here's the way an atom works. Protons and neutrons are bound in the nucleus by exchange of virtual gluons between the quarks that make up the protons and neutrons. Electrons are bound to the protons in the nucleus by virtual photon exchange. Note that three families of quarks are matched exactly by three families of leptons. Of course, no one knows why there are exactly three of each. The first families of quarks and leptons are stable and constitute all the matter around us. The other two families are unstable and after a short time decay into their more stable cousins. If you're wondering about the possibility of more families of quarks and leptons, two experiments have confirmed that the number of these families is three. One experiment is based on 1998 accelerator results for the number of neutrinos in a particular particle (lambda zero) decay, and the other rests on astronomical data (more about this in the next chapter).

All the particles listed have nonzero mass except for the gluon and the photon. The zero mass of the photon accounts for the long range of the electromagnetic interaction because its carrier can move at the speed of light. The weak interaction has much shorter range because its carrier particles have substantial mass and cannot move as fast as photons. All quarks and leptons obey a set of statistical rules set up by Fermi and Dirac, and are collectively called *fermions*. Interaction carriers are governed by another set of rules formulated by Bose and Einstein, and they are called *bosons*.

The Standard Model was first named in 1974. At that time, seven particles predicted by the theory had not yet been detected. During the subsequent twenty-year experimental search, all but the Higgs particle were found in experiments with more and more energetic collisions.

Besides observations of the particles themselves, many particle properties predicted by the Standard Model have been tested experimentally and have shown remarkable agreement between prediction and experiment. One example is the Lamb shift. In 1947 Willis Lamb measured the difference in frequency of radiation absorbed or emitted when hydrogen makes a transition between two closely spaced energy levels. Much later, using the Standard Model, the frequency of light given off by this transition was predicted to be 1057.860 ± 0 .009 MHz. Lamb originally measured the frequency to be 1057.845 ± 0 .009 MHz. The two values differ by only about one part in one hundred thousand. When the listed uncertainties are taken into account, prediction and experiment overlap substantially. This remarkable agreement is true of many other prediction/experiment sequences involving the Standard Model, which provides this theory with extremely strong support.

On a purely mathematical basis, Scottish physicist Peter Higgs and molasses sticks to them as they march. Another analogy is that of a cocktail party where the guests are uniformly distributed throughout a room. When a Very Important Person walks in, her nearest neighbors cluster around her, increasing her effective mass. The more important the person, the more her mass is increased by this clustering.

According to the theory, different particles have different couplings with the Higgs field, granting large masses to the Ws and Z, and zero mass to the photon and gluon. But how can we tell if the Higgs field is real or just a mathematical convenience? A sufficiently large jolt, such as an extremely energetic particle collision, within the cosmic molasses known as the Higgs field can set the field quivering. This field vibration could be detected. There should be a Higgs particle that "carries" the Higgs field in the same sense that a photon "carries" the electromagnetic field. In the simplest theory, there is only one Higgs particle that "carries" the Higgs interaction. More complicated theories have multiple Higgs particles, but there's still one particle that is lightest. It is possible that this lightest Higgs particle lies within the range of current accelerators.

For several years, the European laboratory for particle physics in Geneva, Switzerland (CERN), searched for the Higgs particle in an accelerator known as the LEP (Large Electron Positron collider). In 2001 CERN shut down its accelerator to build a more energetic one using the same tunnel. The new accelerator, called the LHC (Large Hadron Collider), has enough energy (8,000 GeV/beam) to be a much more effective probe. The Superconducting Super Collider (SSC), approved by President George H. W. Bush in 1987, had the Higgs search as its principal objective and would have had plenty of energy (20,000 GeV/beam) for the task. It was cancelled, however, by the US Congress in 1993.

Anticipated possible results of the Higgs search:

- If the Higgs particle is found, and its mass is within range of current accelerator capability, the Standard Model can be extended to include its effects.
- If the Higgs particle is found and has a mass beyond the anticipated range, the Standard Model breaks down, because it makes the impossible forecast that certain events are more than 100 percent probable. This will necessitate a substantial overhaul or replacement of the Standard Model.

- If multiple Higgs particles are found, new theories beyond the Standard Model will be needed.
- If *no* Higgs particle is found, this necessitates replacement of the Standard Model.

As it stands, there is no experimental support for any theory beyond the Standard Model, but there are a few theories waiting in the wings.

Grand Unification Theory (GUT) and *Theory of Everything (TOE)*: These names are misleading because they promise more than they can deliver. In fact, they are umbrella terms that refer to an integration of the known interactions into a single, comprehensive theory. GUTs unite the electroweak and strong interactions. The more ambitious TOEs include not only strong and electroweak interactions but the gravitational interaction as well.

M-theory: Theorist Edward Witten says, "M stands for 'Magical,' 'Mystery' or 'Membrane,' according to taste." Several earlier theories have been shown to be subsets of this overall theory—so-called strings, superstrings, and brane theories. Instead of treating quarks and leptons as point (one-dimensional) particles, this theory proposes that they have two dimensions (lines or strings) or even more dimensions (membranes, or "branes"). These related theories unify all forces, including gravity, and contain no embarrassing infinities that require renormalization like the Standard Model does. Since they all require more than four dimensions (ten, eleven, and twenty-six are the current major choices), the extra dimensions may range from being curled up too small to measure, within range of current measurement techniques, to being too big, approaching infinity. According to one theory, all the dimensions of the universe started out at similar size but then separated and changed size as the expansion proceeded and the temperature decreased. One difficulty with choosing from among the many theories in this category is that our experience cannot be applied to dimensions beyond the four in which we live.

Supersymmetry (SUSY): If fermions and bosons are interchanged, be a selectron, a proton's would be a sproton, and so forth. An "–ino" suffix would be added to name bosons' superpartners—a photon's superpartner would be a photino, a W particle's superpartner would be a Wino, and so forth.)

Technicolor: This theory considers quarks and leptons to be composites of smaller particles. Since this theory predicts new particles, this idea might be testable.

Twistor theory: Both the Standard Model and general relativity are reformulated using complex-number representations of space-time in this theory. (A complex number is defined as a + ib, where i = the square root of –1 and a and b are real numbers.) The significance of complex numbers in the real world is unclear—however, they cannot be applied to count or measure any real entities.

As usual, to avoid being cast into the scrap heap of discarded theories, any scientific hypothesis must make predictions that are supported by experimental evidence. Some of these new theories are too abstract to make any predictions that can be tested, some are too difficult to allow calculations to be made, and others involve quantities too far removed from our familiar surroundings for us to apply any constraints based on our experience. To provide experimental evidence for some of the extremely massive particles predicted would require an accelerator bigger than the entire solar system. Ah, but that's the point of the next chapter.

Chapter 29. BUT WAIT, THERE'S EVEN MORE—THE UNIVERSE'S BIGGEST CONSTITUENTS

Edwin Hubble

"There is more stupidity than hydrogen in the universe, and it has a long shelf life."

—Frank Zappa

"The Universe is made mostly of dark matter and dark energy, and we don't know what either of them is."

—Saul Permutter

More energy, More Energy, MORE ENERGY! Physics needed more energy in its attempt to understand the universe's tiniest parts. But where would it come from? The BIGGEST event in the universe's history, the big bang itself.

"The universe is the poor man's accelerator; experiments don't need to be funded, and all we have to do is collect the experimental data and interpret them properly." This quote from Yakov Zeldovich was almost a mantra for astrophysicists, many of whom were formerly high-energy eral relativity equations to the universe, assuming slightly different conditions than other solutions, and arrived at the notion that the universe might be in a state of expansion. He referred to this idea as a "cosmic egg." Einstein was critical, and pointed to other similar solutions, but he thought they were all unrealistic. Within two years, the situation changed completely. The man responsible: Edwin Hubble.

Edwin Powell Hubble (1889–1953) Mini-Biography[1]

Virginia Lee James married John Powell Hubble in 1884 in Marshfield, Missouri. Edwin Powell Hubble was the third of their eight children. He seemed to have inherited his mother's good looks and his father's athletic build. As a young child, Hubble had no exceptional qualities, but once he learned to read, he devoured the classics and got high grades in school. The only exceptions were deportment and spelling. Hubble's father traveled a lot, but ruled the roost with an iron fist when he was home. After his law practice failed, John Hubble went into the insurance business. He became an executive, responsible for agents and adjusters in several states. The family moved several times, landing in Wheaton, Illinois, in time for Edwin to attend high school. Academically, Hubble was near the top, but his athletic prowess was even higher. He played many sports and once achieved seven first-place showings in a single track meet. At graduation, the superintendent began announcing class honors. "Edwin Hubble, I have watched you for four years and I have never seen you study for ten minutes." He paused for effect. "Here is a scholarship to the University of Chicago."

At Chicago, Hubble toed a fine line. His father thought astronomy was "outlandish," so Hubble had to study his real love quietly. His other effort was preparation for law school, which was far more pleasing to his family. His athletic activities were also curtailed. Hubble loved football and was approached by Amos Alonzo Stagg to play for the University of Chicago. Hubble's father put his foot down, saying there were too many injuries in football. Hubble did some statistical research and pointed out that baseball injuries were just as likely. His father took that as an argument against baseball. Oddly, his father allowed boxing, so Edwin Hubble became a well-regarded amateur heavyweight boxer.

Toward the end of his undergraduate days, the Rhodes scholarship program was instituted, and Hubble became one of the first Rhodes scholars.

In 1910 Hubble arrived in Queen's College, Oxford, set to study jurisprudence. Over the next three years, he acquired not only legal knowledge but also British mannerisms and dress that would last a lifetime. Toward the end of his time in England, his father died, and Hubble came home to help support the family. Living in Louisville, he became a Spanish and physics teacher and basketball coach at New Albany (Indiana) High School, just across the river. After a year, Hubble received a scholarship to study astronomy at Yerkes Observatory and earned his PhD from the University of Chicago in 1917. He accepted a job at the Carnegie Institution's Mt. Wilson Observatory in California, but first served in World War I, attaining the rank of major. Arriving at Mt. Wilson in 1919 still in uniform, Hubble and fellow Missourian Harlow Shapley didn't hit it off well. Just after the 100-inch Hooker Telescope came online, Shapley left to become head of the Harvard Observatory, giving Hubble free access to the world's largest telescope.

By 1924 Hubble had not only made significant observations of galaxies (he never stopped calling them nebulae), but he had found a partner for life. Grace Burke Leib first met Hubble when she visited Mt. Wilson with friends in 1920. A year later, her geologist husband met with an unfortunate mining accident. She and Hubble got together in 1922 and married in 1924. In recognition of his meager salary and his new wife's background, Hubble supposedly offered to give up astronomy to become an attorney, but Grace would have none of it.

At the telescope, Hubble was superb. Studying his nebulae night after night, Hubble took many photographic plates that allowed him to make distance estimates for galaxies, one of the most difficult problems in astronomy. The velocities of these galaxies were determined on the basis of Doppler shifts of their light. (Recall the Doppler effect involving sound?) Many Doppler measurements had already been made by Vesto M. Slipher, so Hubble trained Milton Humason to make additional Doppler measurements. (Humason was quite a story in himself. Having only a fourth-grade education, Humason started at Mt. Wilson as a mule driver, helping to clear the site. He worked his way up to become a night watchman, assistant astronomer, and eventually coauthor with Hubble on important papers.) By 1929 Hubble and Humason had accumulated enough data to show an interesting relationship between a galaxy's speed and its distance from Earth, $v = H_0D$, where v is the galaxy's speed, H_0 is the Hubble constant, and D is the distance to the galaxy. The farther away the galaxy was, the faster it was moving. This became known as *Hubble's law* and provided experimental support for LeMaître's theory because of its consistency with a picture of overall expansion of the universe.

Once the necessity for a static universe was removed, the need for Einstein's "cosmological constant" vanished. Einstein removed the term, calling it "the biggest blunder in my life." In 1931 Einstein came to Mt. Wilson to see the telescope and thank Hubble personally. During the visit, Hubble stuck to Einstein like glue, as the following picture shows.

The tall, dashing Hubble loved publicity, and the Hubbles entertained many Hollywood stars, including Anita Loos, Harpo Marx, Charlie Chaplin, Paulette Goddard, Lillian Gish, Helen Hayes, Frank Capra, Jane Wyatt, George Jessel, Clifford Odets, Leslie Howard, William Randolph Hearst, and Clare Boothe Luce.

Hubble became involved in the design of the 200-inch telescope on Mt. Palomar and was granted first use in 1949. His health deteriorated and he died in 1953.

Humason, Hubble, Einstein, and others

Let's return to the connection between the smallest and largest parts of the universe.

In his 1948 doctoral thesis, Ralph Alpher suggested that all the hydrogen and helium in the universe had been created by nuclear fusion in the first three hundred seconds after the start of the universe's original expansion. Alpher's thesis adviser, George Gamow, added the name of fellow physicist Hans Bethe to the paper, making the author list Alpher, Bethe, and Gamow. Although Bethe didn't actually work on the project, he later extended the ideas to account for nuclear fusion reactions in stars. The big bang theory got its name from Fred Hoyle at about this time, when he referred to it as such the cosmic microwave background radiation that was the leftover of the big bang explosion. This supported the big bang theory so strongly that the steady state theory was discarded.

The only time high-energy particles had existed naturally was during the first few moments of the big bang, just as the matter and energy in the universe began expanding. Conditions during the first small fractions of a second of the big bang were spectacularly hot and dense. Because of the high energy, all families of fundamental particles were present, so the initial moments of the big bang would have provided a great laboratory for testing the Standard Model. Even though that era is inaccessible, predictions about current conditions may be made and compared with measured reality.

For example, if *four* families of fundamental particles existed, the amount of helium formed during the first few minutes of the big bang expansion would constitute more than 26 percent of the present-day universe. *Three* fundamental particle families would produce only a 25 percent helium abundance. Since 25 percent is the helium abundance that has been detected, the Standard Model's inclusion of exactly three families of quarks and leptons has strong experimental support.

So has astrophysics settled the Standard Model, with the exception of the Higgs field? No, it's way worse. Thanks to astrophysics, we have two total unknowns: *dark matter* and *dark energy*.

DARK MATTER

Remarkably, a huge discrepancy regarding the masses of galaxies was discovered in 1933 but went unnoticed for almost forty years. Even more remarkably, the astronomer who first noticed the problem graduated from ETH in Switzerland, as did Einstein, and spent his professional career at Caltech, Mount Wilson, and Mount Palomar like Hubble.

His name was Fritz Zwicky.[2] Born in Bulgaria in 1898, Zwicky went to Switzerland to live with his grandparents at age six and remained a Swiss citizen all his life. Too young for World War I, Zwicky studied theoretical physics at ETH, where he applied quantum mechanics to crystals for his PhD thesis in 1922. In 1925 Zwicky came to the United States on a Rockefeller fellowship, arriving at Caltech because Pasadena's foothills bore some small resemblance to his beloved Alps. Although his sponsor, Robert A. Millikan, expected Zwicky to focus on quantum mechanics, Zwicky became attracted to astronomy. Early in his career, Zwicky studied the cluster of galaxies known as the Coma Berenices Cluster, M100. Using Doppler techniques, Zwicky found the velocities of eight of the galaxies in the Coma cluster, and calculated the mass needed to keep these galaxies gravitationally bound to the cluster. Next, he compared that mass to the cluster's mass based on the light it gives off. The mass needed to emit the observed light was much less than the mass needed to keep the cluster from flying apart. Zwicky called this missing mass "*dunkle materie—dark matter.*" His calculations implied that there had to be much more dark matter than ordinary matter in the Coma cluster. Zwicky's alarming result went unnoticed by other astrophysicists. Perhaps this was because it was published only in German, in a barely read Swiss journal.

During a long and fruitful career, Zwicky held a wide assortment of ideas of checkered quality, all pursued with relentless conviction. Some thought him brilliant, while others regarded him as belligerent. Almost everyone who met Fritz Zwicky had an opinion about him. Perhaps the way he often greeted visitors to Caltech, "Who the devil are you?" should be applied to dark matter. Whatever the reason, Zwicky's dark matter, as noted above, didn't make a big impact on astronomy for almost forty years.

In 1970 Vera Rubin and W. K. Ford studied the rotation of M31 (the Andromeda galaxy) and then more than sixty other spiral galaxies. It turned out that these galaxies were all rotating faster than their visible mass could support. This implied the existence of unseen mass. As more experimental evidence came in, the (visible) matter. Although its details are still unknown, we know it's matter because it exerts gravitational force, and we call it dark because we can't see it.

DARK ENERGY[3]

Although the universe is expanding, the presence of mass should be slowing down the expansion. Two separate groups set out to determine the deceleration of the universe by finding its expansion rate at an earlier time and comparing it to the present rate of expansion. What they found wasn't deceleration at all. Their results showed just the opposite: *acceleration*. The groups were so surprised that, in order to make sure there were no errors, they analyzed the results several times before releasing them. Let's see how the groups accomplished their measurements.

A supernova is an event that marks the end of a large star's lifetime. Gravity compresses nuclei in the star's core so rapidly that an explosion occurs, blowing out the star's atmosphere and some of its core in a spectacular detonation. One particular kind of supernova involves a star of fixed mass and hence always follows the same pattern. A supernova of *Type Ia* occurs when the collapsed core of a former star (a *white dwarf*) receives mass from a companion star (most stars exist as multiples). As soon as it gets enough mass to push it past the white dwarf mass limit (1.4 times the mass of the Sun), the white dwarf reignites and explodes completely. They are so bright that Class Ia supernovae are easily spotted in distant galaxies. Because their mass is fixed, Type Ia supernovae always explode with the same absolute luminosity, so their distance may be determined by measuring their apparent luminosity: *the dimmer the supernova, the more distant it is.*

In 1998 the Supernova Cosmology Project and the High-z Supernova Team analyzed several Type Ia supernovae near peak brightness and determined their distances. Using the Doppler shift technique, they determined the redshifts of the galaxies in which the supernovae were located and compared them with the Hubble relationship. These measurements showed that these distant supernovae were substantially *dimmer* than the Hubble relationship predicted. Since light from these events has taken four to eight billion years to reach us, what we're seeing is that the universe is expanding more rapidly now than in the past. In other words, *the rate of expansion of the universe is accelerating.* The following year, a more distant supernova was found. It turned out to be the most distant one observed so far. Its light originated eleven billion years ago. This supernova is *brighter* than expected. So eleven billion years ago, the early universe's expansion must have been *decelerating* because of gravity. But then, four to eight billion years ago, the universe started to *accelerate* and galaxies began to spread apart at an ever-faster rate.

The strong implication is that whatever is causing the acceleration of the universe now was less important or even absent during the universe's early stages. It showed up about halfway through the universe's expansion and now has become dominant. It's almost like a driver who slows down while approaching a red traffic light, then stomps on the accelerator as soon as the light turns green. Whatever is causing the universe's acceleration must not be mass, because that would decelerate the expansion. Further, it doesn't emit any light or we would have seen it. Cosmologist Michael Turner has given a name to this phantom: *dark energy.*

Oddly enough, although dark energy's identity is still unknown, we have a good idea of how much dark energy exists. Since mass and energy are interconvertible according to $E = mc^2$, there is a certain average mass/energy density in the universe. This density is directly related to the average curvature of space-time. Measuring small ripples (*anisotropies,* to be technical) in the cosmic microwave background radiation allows the curvature of space-time to be determined. After a series of sophisticated measurements, space-time turns out to be *flat.* Adding up the mass/energy density contributions of ordinary matter and dark matter leaves us far short of the amount needed to make it flat, so the missing mass/energy must consist of dark energy.

The following table gives the percentage of the universe's mass/energy for the various categories:

Category	Percent of the Universe's Mass/Energy	Comments
dark energy	73	Causes faster and faster expansion of the universe. Although unseen and of an unknown nature, its powerful effects have been noted.
dark matter	23	Also not yet observed, dark matter accounts for rapid rotation of galaxies and clusters of galaxies.
ordinary	4	The observed bright stars, galaxies, and clusters of galaxies.
neutrinos	< 1	An upper limit for their total mass has been set, but the actual value is not yet determined.

Some may think this situation so bizarre that it might best be described by a quote from the early twentieth-century biochemist and science popularizer J. B. S. Haldane: "Now my own suspicion is that the universe is not only queerer than we suppose, but queerer than we can suppose."

I think this is good news. There is plenty of room for more physics and thus more joy.

AFTERWORD

"I am neither especially clever nor especially gifted. I am only very, very curious."

—Albert Einstein

For almost forty years, I have dedicated myself to helping college students understand physics. A lot of problem solving takes place in college courses, but my students would tell you that is not necessarily the most joyful part of physics. For this book—for adult readers who have wanted to rediscover physics and for curious college students—I have endeavored to extract the most enjoyable parts of this exciting science. (Some of my students would refer to this as *Physics' Greatest Hits*.)

My goal has been to evoke the joy of it.

- **Satisfying Curiosity:** Knowing how various parts of the universe work satisfies a deep human trait: the desire to understand the world around us.
- **Hands-On Experiences:** Many of us take great pleasure in working directly with physical reality. This tinkering may elicit the delight of constructive play that we experienced as children. Often, that Aha! moment occurs in a laboratory setting and provides a lasting insight.
- **Learning about the Fascinating Lives of Physicists:** The people who played key roles in physics' development lived very interesting lives, and some of the details of their lives provide inspiration for all of us.
- **Science's Methodology:** The dynamic interplay between ideas and reality seems so simple, yet it has produced powerful progress. The scientific method has provided us with a compass by which to navigate the uncertainties and mysteries of the universe.

So what was my main purpose throughout these pages? Here's a and dark energy, yet they make up the vast majority of the universe. Suppose an understanding of dark matter or dark energy brings up a possible new technology that includes travel through space to distant galaxies or planets. Suppose our understanding of the most fundamental particles of the universe changes, uncovering a vast new energy source. Suppose there are interactions among dark matter and dark energy that produce such complicated structures that they resemble our most precious possession, life. Are these far-fetched speculations, unlikely to see the light of day? Perhaps. In 1910, when Rutherford discovered the nucleus, how would he have reacted if someone had speculated about the prospect of enormously powerful nuclear bombs?

Physics plays a key role in the future of our civilization. We cannot afford a large disconnect between physics and the rest of the culture. In spite of the complex mathematics and expensive experiments that physics often requires—which we've avoided here as much as possible—physicists have an obligation to help people understand how the universe works. And people, as thinking members of this universe, have the responsibility to work toward an understanding of physics. If you're a qualified optimist, as I am, you'll see the point of this book. How better to approach understanding than through joy?

STUDY GUIDE

Questions and problems help a reader to learn and reinforce the ideas contained in lessons, lectures, and readings.

Many of the questions offered below approach a topic from a slightly different standpoint than the book's discussion, so you will have to read, digest, and summarize in order to answer the questions effectively. Other questions call for applications of physics principles, examples of physics from your own experience, or your reaction to physicists' biographies.

Problems allow you to test your understanding of the concepts in the classical, direct way: If you can solve the problems, you "got the idea." This set of problems is designed to be illustrative rather than exhaustive, focusing on the range of physics ideas, from the smallest to the largest items in the universe.

A tried-and-true problem-solving technique, metric prefixes, conversions, handy constants, and answers to numerical problems are listed at the end of this supplement for your convenience.

Enjoy!

INTRODUCTION—THE SCIENTIFIC METHOD

INT-1. If a curious visitor from another planet asked you to describe the goal of physics, how would you respond?

INT-2. Use an example from your personal experience to demonstrate the workings of the scientific method on a step-by-step basis.

INT-3. Two steps in the scientific method make extensive use of symbols, while two other steps are grounded in reality. Identify each pair of steps and discuss briefly.

INT-4. The interaction between the general and the specific is critical to the workings of the scientific method. Discuss this interplay briefly and [illegible] method in your discussion.

INT-5. Physics and religion both seek truth, but their methods are quite different. Contrast these methods, especially focusing on the role of experimental evidence.

INT-6. Physics graduate students usually declare their major interest in terms of being either a theoretician or an experimentalist. Which would be your choice, and why?

INT-7. Unit conversion errors have ruined many projects. Give an example of unit confusion based on your own experience or research.

INT-8. Of the many levels of mathematics (arithmetic, algebra, geometry, trigonometry, calculus), which level are you comfortable with? What prevents you from understanding the next higher level?

INT-9. Physics is allied with many other fields, such as biophysics, astrophysics, chemical physics, and geophysics. Do a little research on one of these fields and write a short paragraph explaining the role of physics within it.

CHAPTER 1: HOW THINGS MOVE—POSITION, VELOCITY, AND ACCELERATION

Q1-1. Displacement is simply the distance from one location to another. What is the largest displacement you have experienced in a single day? Give your answer in miles and in meters (1 mile = 1609 meters).

Q1-2. "Tailgating" describes following another vehicle too closely. Various driver training courses recommend maintaining a distance interval that corresponds to a time of 3–10 seconds between vehicles. At a speed of 60 MPH (about 28 m/s), how far does your car travel in 3 seconds? In 10 seconds? Since a full-size car has a length of about 5 m, how many car lengths do these distances represent? Is this the way you drive? Explain briefly.

Q1-3. What is the fastest speed you have ever experienced in an automobile? Give your answer in MPH and meters/second (1 MPH = 0.447 m/s). Briefly explain the circumstances.

Q1-4. Acceleration can be either positive or negative. If you were riding a bicycle, explain what you would have to do to achieve both positive and negative acceleration.

Q1-5. How would you explain acceleration, including units, to a friend or a relative?

Q1-6. Analysis of two- and three-dimensional motion requires the use of velocity rather than speed, which was sufficient for one-dimensional motion. State and discuss briefly the difference between velocity and speed.

Q1-7. Air resistance is ignored in the simplified analysis of motion. Discuss what would probably be the effect of air resistance on a soccer ball's trajectory.

Problems (More challenging problems are marked with a ◇.)

P1-1. A proton is ejected from the sun at a speed of 2×10^6 m/s. How long does it take for this proton to reach Earth, a distance of 1.5×10^{11} m away? (Answer in hours.)

P1-2. The western half of California lies on the Pacific Plate, the eastern half on the North American Plate, and the plate boundary is called the San Andreas Fault. The Pacific Plate moves northwest relative to the North American Plate at an average rate of approximately 1 cm/yr (about 0.4 in/yr). Knowing the distance between Los Angeles and San Francisco is 560 km (about 350 miles), how long will it take for the two cities to be at the same latitude? (Answer in years.)

P1-3. A legendary giant supposedly covers 30 km (18.6 miles) in a single stride. If he marches along the equator at 60 steps/minute, over oceans and land alike, find (A) his speed, in km/hr (and MPH), and (B) the time it would take to march completely around the globe, which is 6.37×10^6m in radius (3960 miles). (Answer in minutes.)

P1-5. A turtle (neither mutated nor ninja) starts crawling from rest and accelerates at 6×10^{-4} m/s^2. After 30 seconds at this strenuous rate, find (A) the turtle's velocity and (B) the distance covered.

P1-6. In order to catch a speeder, a police officer must achieve a speed of 40 m/s (almost 90 MPH) within a distance of 1 km (0.62 miles). Presuming the officer starts from rest, how much acceleration is required? Discuss the reasonability of this acceleration in comparison to the acceleration due to gravity.

P1-7. Flix the cat ambles along at a leisurely cat pace of 1 m/s when he suddenly spies an interesting mouse. Flix then accelerates at 2 m/s^2 until he reaches his top speed of 4 m/s. Find (A) how long this acceleration took and (B) how far Flix moved while he was accelerating.

P1-8. An auto traveling at 24 m/s screeches to a stop in 50 m. Find (A) the time required and (B) the auto's average acceleration during this maneuver.

P1-9. An airplane touches down at the extreme end of a 2000 m runway and decelerates uniformly at a rate of 2 m/s^2. The plane stops at the other end of the runway. Find (A) the plane's velocity as it began to land and (B) the amount of time the landing roll required.

P1-10. A spaceship zips past Mars's orbit at a speed of 3×10^3 m/s and accelerates in a straight line until it reaches the orbital path of Jupiter,

5.56×10^{11} m away, when its speed is 4×10^4 m/s. Find (A) the space ship's acceleration and (B) the time required to travel from Mars's orbit to Jupiter's. (Answer in days.)

P1-11◇. P travels due east along a straight highway at a constant speed of 30 m/s. At 9:00 a.m., P passes Exit 17. At precisely the same moment, Q passes Exit 16, traveling due west at a constant 26 m/s. Slightly later, P and Q pass the same point. Knowing the exits are exactly 7 km (4.35 miles) apart, how many minutes past 9:00 a.m. does this meeting occur?

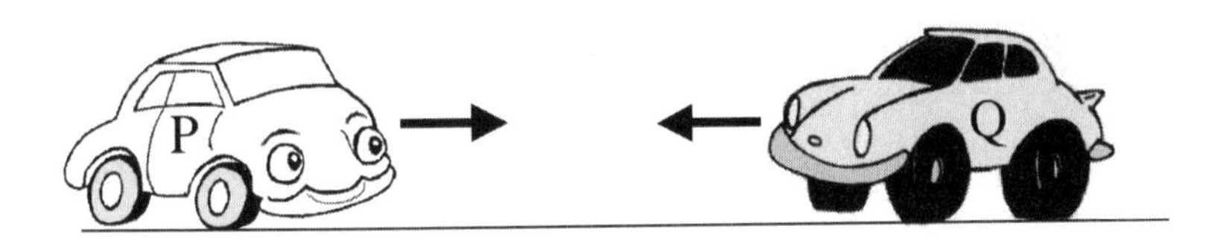

P1-12. Seeing a traffic jam ahead, P slams on the brakes and decelerates from 30 m/s at a constant rate, reaching a complete stop in 8 seconds. How far did the car travel during the braking maneuver?

P1-13◇. Meanwhile, Q continues at 26 m/s west and, at the next overpass, zips past a stationary police officer, who immediately begins accelerating at 3 m/s^2. Eventually, the police officer catches up to Q, who blissfully maintains the previous speed. When the police officer and Q are side by side, the officer motions to Q to pull over. (A) How much time elapsed since the police officer started up, (B) how far did Q travel before being caught, and (C) how fast was the police officer going when Q was finally nabbed?

P1-14. In order to avoid a vicious dog, a particularly clever animal uses a catapult (where did you think the name came from?). If the cat is launched at 8 m/s and an angle of 60°, will he clear the dog, who can leap 2 m straight up? (Hint: maximum height occurs at the time when $v_y = 0$.)

P1-15. A diver makes a running horizontal jump off a cliff located 4 m above the water level below. How fast must the diver be traveling at launch in order to clear a 6 m section of rocks at the water's edge?

P1-16. In World Cup action, a soccer player kicks the ball 9 m/s at an angle of 33° above the horizontal. Find (A) the total time the ball is in the air and (B) the ball's maximum range along the horizontal plane.

P1-17◇. In the Boston Red Sox stadium, Fenway Park, the Green Monster is a wall in left field, roughly 100 m from the batter and 10 m above the batter's hitting point. If the ball leaves the bat at an angle of 40°, what initial speed must it have to clear the Green Monster?

P1-18. A relative of the cricket known as a katydid is capable of launching itself at 3 m/s and at an angle of 37°. How far along the ground would such a leap propel a katydid?

P1-19◇. A stone is dropped into a well and falls freely. Then, 1.6 seconds later, an impatient person throws a stone into the same well at 25 m/s straight downward. The two stones land at exactly the same moment. How deep is the well?

P1-20. Standing in Juliet's garden 6 m below and 8 m along the ground from her balcony, Romeo pitches a pebble to get her attention. If his toss is just right, it will hit the window with zero vertical velocity, minimizing the chance of window breakage. How fast is the pebble traveling just before it hits the window?

P1-21. An airplane traveling horizontally at 160 km/hr and an altitude of 800 m drops medical supplies to an island. (A) How many seconds prior to the plane being directly over the island should the pilot release the supplies, and (B) what is the horizontal distance traveled by the supplies?

CHAPTER 2: WHY THINGS MOVE—FORCES AND THEIR EFFECTS

Q2-1. How did Aristotle regard the role of force in relationship to motion of bodies? Give an example to illustrate.

Q2-2. In terms of the scientific method, what major change came about as a result of the scientific revolution? Explain briefly.

Q2-3. Galileo Galilei had a significant role in the scientific revolution. Explain briefly and give an example of Galileo's actions that are relevant to the scientific method.

Q2-4. Many people mistakenly think that Galileo invented the telescope. Explain the real relationship between Galileo and the telescope.

Q2-5. Explain the meaning of the term *inertia*, and give an example from your own experience.

Q2-6. List and explain briefly three aspects of Newton's life that surprised or impressed you.

Q2-7. If a friend or a relative asked you to explain Newton's first law, how would you respond?

Q2-8. Newton's second law assigned a far different role to force than the [illegible] Contrast the two approaches.

Q2-10. Do Newton's laws apply everywhere in the universe, and for all time, past, present, and future? Explain briefly.

Problems

P2-1. If a skydiver's weight is perfectly balanced by air drag, what happens to the diver's velocity as she falls?

P2-2. A 200 N force acts on a 50 kg ice-skater. Ignoring any friction, find the skater's acceleration.

P2-3. An automobile is accelerated at 2 m/s^2 by a 4000 N net force. Find the auto's mass.

P2-4. If a 6×10^{11} kg asteroid accelerates at 0.5 m/s^2, how much force must be applied?

P2-5. If you push on a wall with a force of 166 N, how hard does the wall push on you?

CHAPTER 3: SOURCES OF FORCES— THE BIGGIES IN NATURE

Q3-1. How did Edmond Halley's 1686 announcement of a new Earth-moon distance influence Newton's theory of gravity?

Q3-2. The terms *mass* and *weight* are often used synonymously but are really quite different. List and explain briefly three significant differences between mass and weight.

Q3-3. Our extensive experience with contact between two bodies all takes place at a macroscopic level, but at the level of molecules, something quite different occurs. Explain briefly.

Q3-4. A snowboarder slides down a steep slope at constant velocity. Draw a free body diagram for this snowboarder, including all forces that act on him.

Q3-5. From your own experience, give an example illustrating the difference between static and kinetic friction.

Q3-6. If an automobile moving along a road at constant velocity experiences a force of 400 N in the forward direction, how much frictional force opposes the car's motion? Explain briefly.

Q3-7. The tension force in a rope is actually electrical in nature. Explain how this works.

Q3-8. Why do strong and weak nuclear forces seldom appear in free body diagrams that analyze the motion of normal-sized bodies?

Q3-9. Rapid maneuvers reported for alleged UFOs would create difficulties for any fragile creatures aboard. Explain why the physics of violent maneuvers would pose such a substantial problem for humans.

Q3-10. If there were such things as astrological forces, and if they were related to gravity (since no other known force works at long distances except electricity), explain why large machinery near a newborn baby might exert a much more significant influence than most planets or stars.

Problems

P3-1. The sun, with a mass of 1.99×10^{30} kg, is located approximately 2.35×10^{20} m from the Milky Way galaxy's center. If 10 percent of the galaxy's 1.15×10^{42} kg mass lies between the sun's orbit and the galaxy center, find the gravitational attractive force between the sun and the entire galaxy. (The mass beyond the sun's orbit exerts zero net gravitational force.)

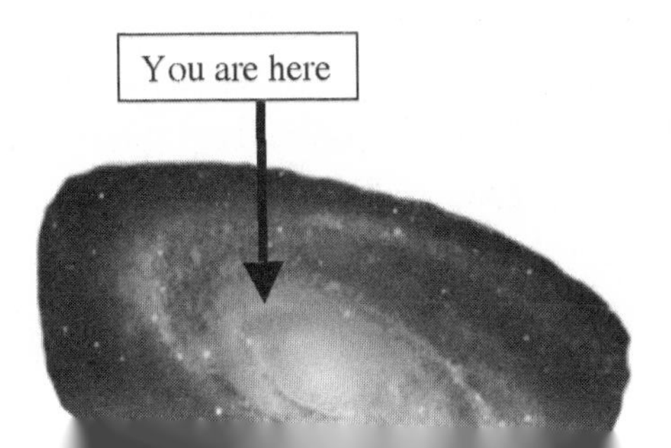

P3-2. A proton (mass = 1.67×10^{-27} kg) in the nucleus of a hydrogen atom exerts a gravitational force on the atom's electron (mass = 9.11×10^{-31} kg) located approximately 5×10^{-11} m away. Find the magnitude of this force.

P3-3. How much gravitational attraction is there between two 60 kg people, seated 1 m apart? Compare this force with the weight of a dust particle, 10^{-5} N, and discuss why ordinary-sized objects aren't strongly affected by gravitational attraction to other ordinary-sized bodies.

P3-4. The mass of an average adult is roughly 75 kg. What is this average adult's corresponding weight?

P3-5. The gravitational pull of the planet Jupiter is so strong that at Jupiter's cloud tops, the acceleration due to gravity is 2.36 times as great as that of Earth. If your mass is 60 kg, what would be your weight on Jupiter? What effects would this have if you tried to move about normally?

P3-6. A 90 kg sofa sits stubbornly on the wrong side of your living room. If the static friction coefficient between the sofa and your carpeting is 0.2, find the amount of force it will take to budge the sofa to start it on its way across the room.

P3-7. A 160 g hockey puck slides horizontally along the ice, and the coefficient of kinetic friction between puck and ice is 0.15. (A) Find the friction force involved. (B) If the puck starts sliding at 3 m/s, how far will it travel before it comes to a stop?

P3-8. (With apologies to Dr. Seuss) A silly old cat sits in a hat. The hat then sits on a green doormat. Cat, hat, mat—draw free body diagrams for all and indicate which forces are equal.

P3-9. An airplane flies along at a constant speed and altitude. Using L for lift, D for drag, T for thrust, and w for weight, draw a free body diagram for the airplane and indicate which forces are equal.

P3-10. The hockey puck from problem 3-7 (same mass and friction) has slowed down to 2 m/s when a 30 N force is applied by a hockey stick. If the force lasts for 0.3 seconds, find (A) the puck's acceleration and (B) the puck's velocity when the force ceases.

P3-11◇. A 3000 kg truck careens down a 6° grade at 20 m/s when its brakes suddenly give out. If you assume a kinetic friction coefficient of 0.05 (rolling resistance + air friction), find the truck's (A) acceleration and (B) speed after it has moved 1 km down the hill. (Answer in m/s and MPH.)

P3-12. A dogsled driver hitches a new dog to his sled to test the dog's pulling power. The harnesses are all set so the dog will exert a purely horizontal force, but the dog refuses to move. Instead, he barks the following message to the driver: "I know physics. The sled will pull back just as hard as I pull it, so the net force will be zero. No force, no acceleration. So why even try?" Using free body diagrams and Newton's second and third laws, explain why the dog's logic is faulty. (He's pulling the driver's leg?)

P3-13◇. An initially resting 300 g car is dragged along a horizontal, frictionless table by a massless cord. The cord is attached to a vertically hanging 60 g mass by a frictionless, massless pulley. If the car starts from rest, how long would it take for it to move 20 cm?

P3-14. Starting from rest, your swell new car, the Zipper Sipper, travels 402 m in 16 seconds. (A) What is its acceleration? (B) If your car's mass is 900 kg, how much force was required to cause this acceleration?

of wire just under Tanya's foot. (B) If Tanya's mass is 60 kg, find the tension in the wire.

P3-16. A 30 kg crate of persimmons slides down a rough 26° incline and accelerates at 3 m/s^2. Find (A) the friction force and (B) the kinetic friction coefficient.

P3-17. Some massive stars end their stellar career by becoming a ball of densely packed neutrons, becoming what is called a *neutron star.* Using $G = 6.67 \times 10^{-11}$ N m^2/kg^2, find the acceleration due to gravity on the surface of this bizarre former star, given that its mass is 4×10^{30} kg and its radius is only 12 km. (Hint: set $F_g = ma_g$.)

P3-18. A 29 kg sled starts from rest at the top of a frictionless hill inclined at an angle of 20° to the horizontal. Find (A) the net force on the sled, (B) the sled's acceleration, and (C) how long it will take the sled to slide 33 m down the hill.

CHAPTER 4: ROUND AND ROUND IT GOES, AND WHERE IT STOPS . . . CIRCULAR MOTION

Q4-1. Traveling in a circular path at constant speed constitutes accelerated motion. Explain briefly how a body may have constant speed and be accelerated at the same time.

Q4-2. As a car travels around a roundabout, what force supplies the centripetal force necessary to maintain circular motion? Explain briefly, including a discussion of possible conditions that might make the force too small to maintain motion in a circle.

Q4-3. The term *centrifugal force* is sometimes used as if it were the same thing as *centripetal force*. Explain the difference between the terms.

Q4-4. How would you explain a geosynchronous satellite to a friend or a relative?

Q4-5. Based on your personal experience, analyze the physics of a rotating carnival ride.

Q4-6. Do some research to find various applications of a centrifuge. Briefly explain your findings.

Q4-7. If a person is rotating a yo-yo in a clockwise horizontal circle just above her head and the string breaks just as the yo-yo is headed west, what will be the resulting motion of the yo-yo? Possible choices: upward, downward, north, south, east, or west. Explain briefly.

Problems

P4-1. Zartan of the jungle crosses a river by grabbing an 8-meter-long vine—which is conveniently hanging from a strong branch—and swinging over the water. Knowing Zartan has a mass of 85 kg (too many coconuts?) and achieves a speed of 4 m/s at the lowest point on the arc of the swing, find (A) his acceleration, and (B) the tension in the vine.

P4-2. A 1200 kg automobile rounds a horizontal curve that is a section of a circle with a radius of 70 m. If the automobile is traveling at 18 m/s, find (A) the force necessary to make this turn and (B) the static friction coefficient needed to keep the automobile from slipping.

P4-3. An oversimplified model of the hydrogen atom has the electron (mass = 9.11×10^{-31} kg) orbiting the nucleus's proton at a distance of

P4-4◇. An automobile rounds a curve banked at 6° to the horizontal. If the friction coefficient between the car's tires and the track is 0.6, how fast can the auto negotiate this 300 m radius portion of a circle without slipping?

P4-5. If the sun ($m = 1.99 \times 10^{30}$ kg) is located approximately 2.35×10^{20} m from the Milky Way galaxy's center, use the gravitational force of attraction between the sun and the inner galaxy (from problem 3-1) to determine the amount of time (years) required for the sun to make one complete orbit around the galaxy's center.

P4-6. A carnival loop-the-loop ride has a radius of 4 m. What is the **minimum** speed of the riders at the top of the loop? (Answer in m/s and MPH.)

P4-7◇. An auto test track is banked so that cars can move around the track at 37 m/s without slipping up or down the track. Knowing that the radius of curvature of the track is 900 m, find the bank angle.

CHAPTER 5: THE ULTIMATE FOUR-LETTER WORD—WORK

Q5-1. How would you explain to a friend or a relative the way physics uses the term *work* differently from ordinary language?

Q5-2. Explain the conditions under which zero work is being done, in a physics sense, even though forces are being exerted or a body is moving.

Q5-3. Explain briefly how negative work could be done.

Q5-4. Briefly describe significant differences between kinetic and potential energy, and give an example of each form of energy being generated by work.

Q5-5. State the principle of conservation of energy as you would explain it to a friend or a relative.

Q5-6. Explain briefly the significant differences between power and energy, including units.

Problems

P5-1. Zartan of the jungle (him again, from problem 4-1) crosses another river by grabbing a 9-meter-long vine and swinging over the water. If the vine starts out at angle of 17° to the vertical, use the principle of conservation of energy to find Zartan's speed at the bottom of the swing.

P5-2. How much work would it take to push a 1200 kg car 200 m (measured along the hill) up a 6° hill, assuming (A) no friction and (B) $\mu = 0.2$?

P5-3. An electron ($m = 9.11 \times 10^{-31}$kg) moves at 5000 km/s toward the front of a tube-type television. When it hits the screen, it is stopped in 12 μs. Find (A) the energy involved and (B) the power.

P5-4◇. A golfer hits a golf ball with a 6-iron while standing on a 35 m high cliff. The golf ball is launched with a speed of 30 m/s and at an angle of 60°. Using energy methods, find (A) the maximum height to which the ball will rise (measured from the ground) and (B) the speed with which the ball hits the ground.

P5-5. A (not-so-super) ball dropped from 50 cm rebounds to a height of 33 cm. Find (A) the fraction of the original energy lost and (B) the speed with which the ball left the ground. Where did the "lost" energy go?

P5-6. A large truck (m = 3300 kg) accelerates from 15 m/s to 27 m/s in a time of 15 seconds. Neglecting any friction or air resistance, find the average mechanical power that the truck's engine must supply. (Answer in watts, and recall that 1 hp = 746 W.)

P5-7. A meteoroid is often a dirty snowball moving at high speeds relative to Earth. If meteoroids approach Earth closely enough to hit its atmosphere, friction heats them and they radiate, causing a spectacular light display. If a 15 g (about marble-sized) meteoroid approaches at 50 km/s, (A) find the meteoroid's kinetic energy. (B) How fast would a 1400 kg auto have to move to achieve the same kinetic energy? Answer in MPH, and comment on this speed.

P5-8. How much work would it take for an ant to push a 0.05 g crumb 50 cm to the edge of a picnic table, knowing that the kinetic friction coefficient between the crumb and table is 0.4?

P5-9. An athlete bench-presses 140 kg a vertical distance of 30 cm. Find the weight's change in gravitational potential energy.

P5-10. An asteroid that slammed into Earth about 65 million years ago (low on the dinosaurs' hit parade) was estimated to be about 8 km in diameter and about 10^{15} kg in mass. If a similar asteroid headed toward Earth now, how much energy would it take to give this intruder a speed of 100 m/s, deflecting its course to miss Earth? (B) Compare this answer to the energy used by the United States in one year, about 10^{20} J. Discuss.

P5-11◊. A roller coaster starts out a distance h above the ground. Ignoring friction and using the principle of conservation of energy, find h, knowing that the roller coaster doesn't leave the track at the top of the 12 m diameter loop. (Hint: if N = 0, centripetal force can only be supplied by weight.)

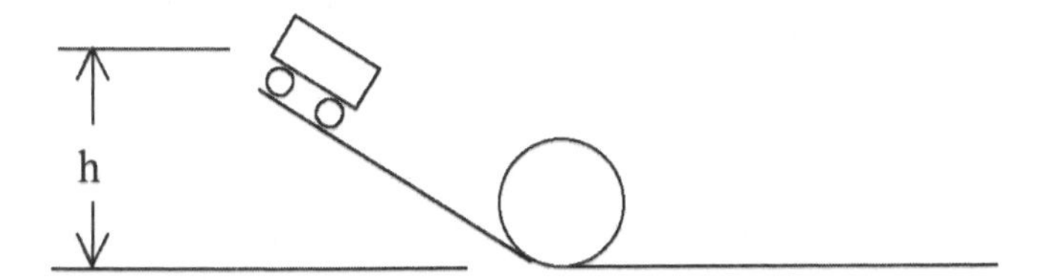

P5-12. A skier starts from rest 7 m above the horizontal and slides down a frictionless slope. The skier then encounters a frictional flat surface ($\mu = 0.3$); how far does the skier travel along the flat surface before she stops?

CHAPTER 6: OOPS! COLLISIONS—IMPULSE AND LINEAR MOMENTUM

Q6-1. How would you explain the physics concept of momentum to a friend or a relative?

Q6-2. Use the concept of momentum conservation to explain how expulsion of gases propels a rocket forward.

Q6-3. In a collision, which of the mechanical forces is involved? Explain briefly.

Q6-4. In a perfectly elastic collision between billiard balls, croquet balls, [illegible] relationship between the velocity of the moving ball before the collision and that of the formerly stationary ball after the collision.

Q6-5. In a railroad switchyard, railroad cars are coupled together by collisions between moving cars and stationary cars. This is an example of perfectly inelastic collisions. Compare the velocity of two joined cars with the velocity of the single car that was originally moving. What would the velocity be if there were two stationary cars? Three stationary cars?

Q6-6. From personal experience, rate an automotive collision in terms of being closer to perfectly elastic or perfectly inelastic. Explain briefly.

Q6-7. How would you explain the physics concept of impulse to a friend or a relative?

Q6-8. Viewing the impulse as a cause, what is the effect on a body's momentum?

Q6-9. In sports that involve striking a ball, explain briefly how follow-through leads to longer contact time and produces more change in velocity.

Problems

P6-1. A baseball pitcher hurls a 400 g baseball toward a batter with a speed of 30 m/s. The batter hits the pitch directly back toward the pitcher, and it travels at 38 m/s. If the collision with the bat required 80 milliseconds, find the amount of force the batter exerted. (Watch signs.)

P6-2. An enormous urban vehicle (EUV), all 2500 kg of it, zips along the road. The driver fails to notice a stopped 1100 kg auto directly ahead. After the inevitable collision, the vehicles stick together and slide to a stop 50 m down the road. Evidence technicians measure the friction coefficient to be 0.7. How fast was the EUV going before the collision (m/s and MPH)? Do you think legal action is likely? Explain briefly.

P6-3. To accomplish a tricky maneuver in space, a 2000 kg instrument package must change its speed from 44 km/s to 33 km/s using a rocket that has a thrust of 4×10^5 N. How long must the engine burn?

P6-4. An NHL hockey player (m = 90 kg) stands motionless on the ice. The 10 kg Stanley Cup is thrown to the player at a speed of 4 m/s. After he catches the cup, how fast does the cup/player combination travel, assuming there is no friction between the player's skates and the ice?

P6-5. An electron (m = 9.11×10^{-31}kg) travels at 4×10^5 m/s. Find the electron's momentum.

P6-6. A huge asteroid (m = 2×10^{19} kg) travels at 8000 m/s. Find this asteroid's momentum.

P6-7. A 46 g golf ball sits on a tee, awaiting its fate—a launch at 40 m/s. If the golfer exerts a 50 N force when the club head strikes the ball, how long must the contact last in order to achieve the desired momentum change?

P6-8. A bug is hovering in the air (v = 0), then suddenly it is banged into by a heavy automobile moving at 20 m/s. If the collision is perfectly elastic (the bug hopes this is the case), the auto continues without any change in velocity after the collision. What is the bug's speed after the collision? (Think carefully.)

P6-9. One of the functions of an automotive air bag is that it lengthens the collision time. Without an air bag, suppose the 100 Ns impulse of a collision is taken up by a fairly rigid dashboard, requiring a time of 15 ms. (A) Find the amount of force exerted on the passenger. (B) If the air bag deploys, the collision time is increased to 85 ms. What is the force in this

P6-10. In a railroad switchyard, a 4000 kg car is sent down the track at 5 m/s. This car meets and couples with a second 4000 kg car, which was initially stationary. (A) What is the speed of the two linked cars after the collision? (B) Suppose two identical cars at rest were coupled to the original car in motion. What is the speed of the three-car combination?

P6-11. If a 4 g bullet emerges from a rifle at 800 m/s, use momentum conservation to find the recoil velocity of the 5 kg rifle.

CHAPTER 7: SPINNING WHEELS, GOT TO GO ROUND

Q7-1. Explain briefly how to find the center of mass of a body and why it is such a significant spot.

Q7-2. In automotive crash tests, how is the center of mass marked and how it is used?

Q7-3. For a rotating body, how is the rotation angle defined? Include its unit.

Q7-4. The radian is actually a nonunit. Explain briefly.

Q7-5. Many calculators have three angle modes: DRG. Explain each mode briefly.

Q7-6. A rotating body's angular velocity has official units of rad/s, but more a commonly used unit is revolutions per minute (RPM). Knowing that 1 RPM = 0.104 rad/s, find the angular velocity in rad/s of an automobile engine idling at 900 RPM.

Q7-7. Briefly explain the concept of angular acceleration, including units.

Problems

P7-1. A hard disk in your computer spins at 5400 RPM. If this disk turns through 12 revolutions before coming to a stop, find (A) the disk's angular acceleration (rad/s^2) and (B) the time it takes to stop.

P7-2. A coin falls onto an old-fashioned record player that is rotating at 45 RPM. The coin lands 6 cm from the center. Find (A) the coin's angular speed in rad/s and (B) its tangential speed in m/s.

P7-3. If the record player in problem 7-2 above starts at rest and reaches its operational speed of 45 RPM in 15 seconds, find (A) the turntable's constant angular acceleration and (B) the number of **revolutions** (NOT radians) the turntable makes during the 15-second warm-up.

P7-4. A potter's wheel accelerates from 2 rad/s to 10 rad/s in 9 seconds. Find (A) the acceleration of this wheel and (B) the angle through which the wheel turned, in units of revolutions.

P7-5. Knowing Earth rotates once every 24 hours, find (A) the rotational speed of Earth in rad/s and RPM and (B) the tangential speed of Earth at the equator, where the radius is 6.37×10^6 m.

P7-6. Initially, a bicycle wheel rotates at 3 RPM, but within 6 seconds, the rotation rate increases to 7 RPM. Find the bicycle wheel's angular acceleration in rad/s^2.

CHAPTER 8: FORCE WITH A TWIST—TORQUE

Q8-1. Briefly explain torque as you would to a friend or a relative, and include at least two examples from ordinary life.

Q8-2. Give a brief explanation of the similarities and differences between

Q8-3. Briefly explain the concept of angular momentum, including similarities to linear momentum.

Q8-4. Give an example illustrating the principle of conservation of angular momentum.

Q8-5. Explain briefly the concept of rotational kinetic energy and how it modifies the key physics principle of energy conservation.

Problems

P8-1. A CD player rotates at a variable speed so that a laser can scan pits and lands on the disc's bottom surface at a constant tangential speed of 1.2 m/s. If the disc has a moment of inertia of 1.2×10^{-4} kg m^2, and the music is first detected when the laser is located 15 mm from the disc's center, find the work done by the motor during this start-up, and assume the disc started from rest.

P8-2. A four-bladed propeller on a cargo aircraft has a moment of inertia of 40 kg m^2. If the propeller goes from rest to 400 RPM in 14 seconds, find (A) the torque required and (B) the number of revolutions turned as it achieves operating speed.

P8-3. A baseball batter angularly accelerates a bat from rest to 20 rad/s in 40 ms. If the bat's moment of inertia is 0.6 kg m^2, find (A) the torque applied by the batter and (B) the angle through which the bat moved, in units of revolutions.

P8-4. An ice-skater spins with her arms and one leg outstretched and achieves a rotational velocity of 2 rad/s. When she pulls her arms in, her moment of inertia decreases to 65 percent of its original value. What is her new rotational rate?

P8-5. The cloud of gas and dust that eventually became the solar system, called the *protosolar nebula*, started out with a small spin rate. But when it shrank (due to gravity, possibly triggered by a nearby supernova), the moment of inertia decreased to only 1 percent of its original value. What happened to the spin rate?

P8-6. When you open a door by pushing on the handle with a force of 20 N, how much torque do you generate, given that the hinges are located 80 cm from the handle?

P8-7. A bicycle and rider with a combined mass of 75 kg zoom along at 8 m/s. If you assume each wheel has a radius of 40 cm and a moment of inertia of 0.3 kg m^2, find the ratio of the rotational kinetic energy to the translational kinetic energy.

P8-8. A plumber uses a torque wrench to loosen a corroded nut. (A) If he exerts a force of 400 N, and the wrench is 15 cm long, how much torque does he generate? (B) What might a plumber have available to help generate more torque easily?

P8-9. If an engine bolt has a maximum fastening torque of 20 Nm, what is the maximum force that may be exerted, using a 15 cm wrench?

P8-10◇. A ball with a diameter of 1 cm rolls down a slope and goes through a loop with a radius of 20 cm, as shown below. Presuming the ball rolls without slipping, and that it has zero normal force at the top of the loop, find the release height necessary to achieve this maneuver. (Use sphere moment of inertia = 2/5 mr^2.)

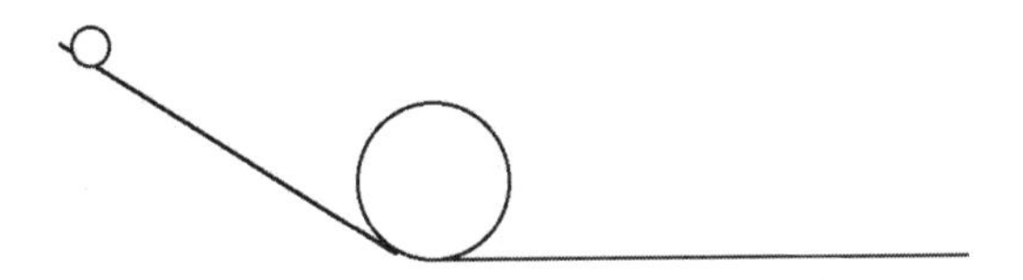

P8-11. A merry-go-round rotates at 3 RPM. If a 25 kg child jumps on the merry-go-round, increasing its moment of inertia by 20 percent, what is the new rotation rate?

P8-12. A 50 g yo-yo is released and travels vertically downward, rolling without slipping. Knowing that the yo-yo's moment of inertia is 9×10^{-5} kg m^2, and that the radius of its center spindle is 0.7 cm, find (A) the torque involved and (B) the angular acceleration of the yo-yo.

CHAPTER 9: THE STRANGE CASE OF THE BODY THAT DOESN'T MOVE—STATIC EQUILIBRIUM

Q9-1. Explain briefly the concept of static equilibrium and include two examples.

Q9-2. State the first condition for static equilibrium and give an example of a body that fulfills this condition but is still not in complete equilibrium.

Q9-3. State and explain briefly the second condition for static equilibrium.

Q9-4. Briefly explain the difference between stable, unstable, and neutral equilibrium, giving examples to illustrate each.

Q9-5. Apply the conditions of static equilibrium to rounded and pointed arches.

Q9-6. Choose an object in static equilibrium for which you can obtain data (likely from the Internet), and draw a dimensioned free body diagram to determine the magnitude of any supporting forces involved.

Problems

P9-1. A 20 kg sign is pulled by a horizontal force such that the single rope (originally vertical) holding the sign now makes an angle of 21° with the vertical and the sign is motionless. Find (A) the magnitude of the tension in the rope and (B) the magnitude of the horizontal force.

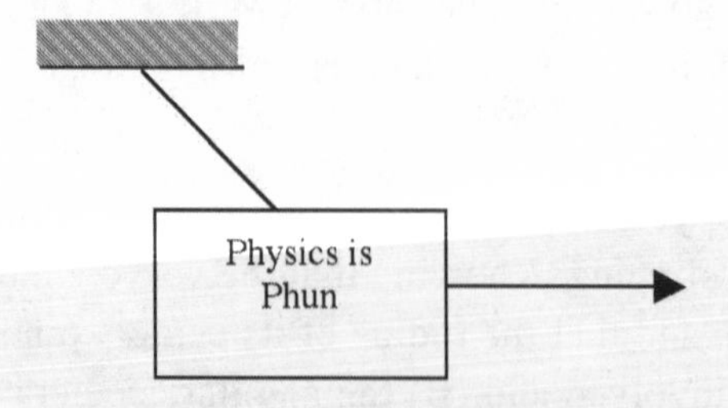

P9-2. A 100 kg athlete doing push-ups places his hands 0.6 m from his center of mass, while his feet are located 1.7 cm from his hands. Find the forces exerted on the athlete's hands and feet by the floor when the athlete is at rest.

P9-3. A 70 kg acrobat stands on a table that has a mass of 40 kg. The top of the table is 4 m wide, and the legs are each 1 m from the edge. How close to the edge can the acrobat stand without causing the table to tip? (Hint: tipping occurs when the far leg support force = 0.)

P9-4◇. Two 75 kg painters stand on a 20 kg, 4-meter-long scaffold, supported at the ends by two ropes, as shown below. If the rightmost painter stands 1 m from the right end, how far from the right end can the other painter stand so that the rope, capable of supporting 1100 N of force, does not break?

P9-5◇. A 2-meter-long ladder leans against a building and forms an angle of 37° with the horizontal. If the ladder and the building have no friction, and the friction coefficient with the ground is 0.4, how far up the ladder could a 90 kg person climb without causing the ladder to slip? (Ignore the ladder's weight.)

CHAPTER 10: GETTING BENT OUT OF SHAPE—ELASTICITY

Q10-1. In an ideal spring, how does the distance the spring stretches relate to the applied force?

Q10-2. Explain briefly the concepts of stress and strain, including units.

Q10-3. Explain briefly the pattern followed by a body undergoing strain,
plastic deformation, and failure.

Q10-4. List and explain briefly six different stresses, and give an example of each.

Q10-5. Explain briefly the concept of fatigue stress.

Q10-6. Before stress analysis, explain how building safety was determined.

Problems

P10-1. An automobile spring compresses 4 cm when a 900 N person stands on the car's bumper. Find the spring constant, k, for the auto's spring.

P10-2. Inside a computer mouse, a delicate spring (k = 8.5 N/m) deflects a distance of 3 mm. How much force is required to cause this deflection?

P10-3. (A) How much stress is exerted on a road by an 18-wheel truck with a mass of 15000 kg, that has a footprint area for each tire of 0.04 m^2? (B) A 1200 kg auto is supported by 4 tires, each with areas 0.015 m^2. How much stress does this cause? (C) Compare the two stresses and discuss. (D) What would happen if the trucker raised 4 of the wheels so they wouldn't touch the ground? This would reduce tire wear, but what happens to the stress on the road?

P10-4. If a 50 kg woman stands only on a single heel of her high-heeled shoes, 1 cm by 1 cm, how much stress acts on the floor directly beneath the heel?

P10-5. Zartan's vine, initially 6 m long, has a Young's modulus of 1×10^8 N/m^2. If Zartan's weight (too many bananas?) causes a stress of 2.5×10^6 N/m^2, what would be the vine's deflection? (Refer to problems 4-1 and 5-1 for more information if necessary.)

P10-6. A 50 kg chandelier is hung from the ceiling by an aluminum wire, initially 2 m long. (A) Knowing the ultimate tensile stress of aluminum is 200×10^6 N/m^2, find the minimum diameter of wire needed. (B) When the full weight of the chandelier pulls the wire, by how much does it stretch? Use 70×10^9 N/m^2 for Young's modulus for aluminum. Does this stretch seem reasonable? Discuss briefly.

P10-7. One of the supports for a bridge bears a force of 3×10^5 N, distributed over a cross-sectional area of 1.3 m^2. (A) Find the stress on the supporting soil. (B) Knowing that Young's modulus for soil is approximately 2×10^6 N/m^2, how deep into the soil does the support sink, when there is 5 m of soil above bedrock?

CHAPTER 11: BACK AND FORTH, BACK AND FORTH . . . SIMPLE HARMONIC MOTION AND WAVES

Q11-1. Based on your own experience, give an example of a familiar object undergoing simple harmonic motion. Explain briefly.

Q11-2. List the two fundamentally different types of waves and briefly explain the differences.

Q11-3. Briefly explain the wave properties of amplitude, crest, node, trough, and wavelength.

Q11-4. Briefly explain a wave's period and frequency.

Q11-5. An earthquake produces four seismic waves. List them and explain briefly each type of wave.

Q11-6. List and explain seven different properties of waves and give an example of each.

Q11-7. List and explain briefly three different kinds of damping and give examples of each.

Q11-8. If a body is forced to oscillate, briefly discuss the way the body's response is based on the relationship between the frequency of the forcing function and the natural frequency of the body.

Problems

P11-1. If your grandfather was a Martian and you wanted to build a clock for him that would tick once every three seconds (clock period = 6 seconds), how long would the pendulum have to be? The acceleration due to gravity on Mars is 38 percent of Earth's gravity.

P11-2. If Zartan swinging on his 6-meter-long vine may be considered to be a simple pendulum (don't let him hear that), find the pendulum's period. (Refer to problems 4-1, 5-1, and/or 10-5 for more information, if necessary.)

P11-3. An auto spring ($k = 2.5 \times 10^4$ N/m) supports a quarter of the car's weight. If you start the 1200 kg car oscillating, find its period.

P11-4. A mild spring for which k = 50 N/m supports a mass of 2 kg. Find the period when this system is set into oscillation.

P11-5. One of the strings on a violin has a tension of 70 N and a linear mass density of 6 g/m. Find the speed of transverse waves on this string.

P11-6. A passing speedboat generates waves that cause a fisherman's bobber to oscillate up and down. If the waves (wavelength = 5 m) travel at 3 m/s, find the bobber's (A) frequency and (B) period.

P11-7. A 3 kg mass fastened to the end of a 50 cm massless string constitutes a pendulum. If the string makes an angle of 13° with the vertical when the pendulum bob is released, find (A) the pendulum's oscillation frequency and (B) the pendulum's oscillation frequency for a 4 kg bob.

P11-8. A piano string is so badly out of tune that the tension must be increased by 10 percent; find the percentage increase of the frequency.

CHAPTER 12: HEY, LISTEN, WHAT'S THAT SOUND?

Q12-1. Explain briefly how the structure of matter determines the different speeds of sound in solids, liquids, and gases.

Q12-2. What is the normal sound-frequency range that human beings can hear? Explain how this range is altered by the aging process.

Q12-3. What is meant by infrasound and ultrasound? Give examples of each.

Q12-4. Explain briefly the decibel scale, including the reasons for its adoption.

Q12-5. Using any relevant diagrams, explain briefly how standing waves are set up in vibrating strings.

Q12-6. In terms of frequencies of musical notes generated, explain briefly how vibrating air columns are similar to, but different from, vibrating strings.

Q12-7. When two different instruments play the same note, both sounds contain the same set of frequencies, but the human ear can distinguish one instrument from another. Explain this briefly.

Q12-8. The Doppler effect depends on relative motion between the source and the receiver of the sound. Explain this briefly and give an example from your experience.

Q12-9. Explain the Mach number briefly, including the way it relates to supersonic airplanes.

Q12-10. Reverberation time is a good measure of the acoustic properties of a concert hall. Explain briefly and include the factors that influence this quantity.

Problems

P12-1. If an unfingered violin string normally plays A (440 Hz), what frequency would it play if you placed your finger 1/5 of the way down the string? (Hint: 4/5 if the string is available for standing waves.)

P12-2. If your neighbor's "quiet" party generates sound of intensity 3×10^{-5} W/m^2 at your house, (A) find the sound intensity level in dB. (B) If the neighbor's discussion about your complaint causes an increase of 3 dB, find the new sound intensity.

P12-3. A 90-centimeter-long organ tube, open at both ends, resonates at two successive harmonics of 760 Hz and 950 Hz. Find (A) the fundamental frequency and (B) the speed of sound in the tube.

P12-4. A referee's whistle sounds at 2000 Hz. Since you know the speed of sound is 340 m/s, find the apparent frequency if you're moving at 40 m/s (A) toward the whistle and (B) away from the whistle.

P12-5. In a large soccer stadium, suppose you are sitting 200 m away from the ball as it is kicked. (A) Find the time that elapses before the sound of the ball being kicked arrives at your ear. (B) Since light travels at 300 Mm/s, you see the kick much earlier than you hear it. Find the time it takes for light to travel to you, compare the two times, and comment on your result.

P12-6. The driver of the car next to you in traffic is playing extremely loud music, and you make hand gestures to get his attention. If the new sound intensity is 30 percent less than before, find the change in intensity level, in dB.

P12-7. A railroad crossing warning signal sounds at 120 Hz. (A) If you are approaching the crossing at 3 percent of the speed of sound, find the frequency you would hear. (B) If you were going away from the crossing at 3 percent of the speed of sound, find the frequency you would hear.

P12-8. An automobile's exhaust makes a sound with an intensity level of 95 dB. (A) Find this sound's intensity in W/m^2, using $I_0 = 10^{-12}$ W/m^2. (B) After installing a new muffler, the intensity (W/m^2) is reduced to only 10 percent of its original value. What is the new intensity level in dB?

P12-9. In a particular organ pipe open at both ends, successive harmonics are heard at 360 Hz and 450 Hz. (A) What is this organ pipe's fundamental frequency, and (B) since the speed of sound is 340 m/s, how long is the pipe?

P12-10. In an organ pipe open at one end and closed at the other, the fundamental frequency emitted is 240 Hz. Find (A) the next overtone frequency and (B) the length of the pipe.

CHAPTER 13: GOOEY AND GASSY— FLUIDS AT THEIR FINEST

Q13-1. Explain briefly why pressure varies with depth in a fluid. Use any free body diagrams necessary.

Q13-2. Briefly explain density, including why the density of a particular material is constant, regardless of the amount of material involved.

Q13-3. Explain briefly the concept of buoyant force and give an example from your own experience.

Q13-4. Briefly explain the concept of surface tension.

Q13-5. The equation of continuity relates fluid mass flow rate at two different points in a flowing fluid. Explain how the cross-sectional area and fluid velocity are involved.

Q13-6. List and explain briefly three aspects of Daniel Bernoulli's life that surprised or impressed you.

Q13-7. List and explain briefly each of the terms in the Bernoulli equation.

Q13-8. Explain how top spin alters a tennis ball's or a golf ball's trajectory.

Q13-9. Explain briefly how high winds make a chimney "draw" better.

Q13-10. The aerodynamic drag force depends on what quantities? Briefly explain.

Q13-11. Explain briefly the concept of terminal velocity, familiar to skydivers.

Problems

P13-1◇. For a new television series *Stupidity Factor*, contestants are dropped into the ocean (density = 1030 kg/m^3) along with a Styrofoam block (density = 300 kg/m^3) that is 2 m by 3 m by 20 cm thick.

If too many contestants climb aboard the block and it sinks below the water surface, they are declared "stupid" and are abandoned at sea. Find the maximum number of 70 kg contestants that the block can hold. (No fractional contestants, please. They hate it when that happens.)

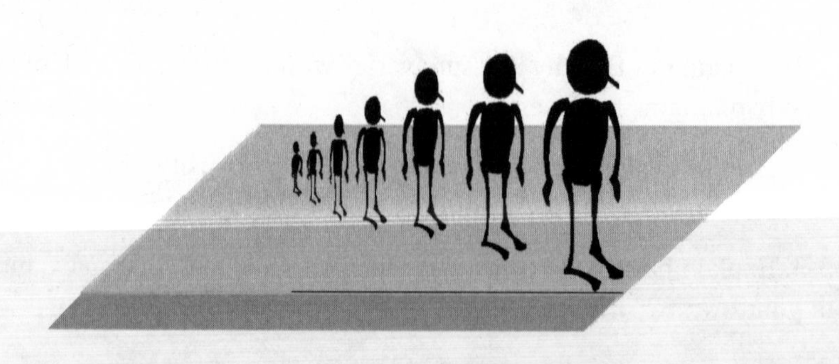

P13-2. Knowing that the standard air pressure at sea level is 101 kPa, find the static pressure during a Category 3 hurricane, which features wind speeds of 200 km/hr. (Use air density = 1.3 kg/m^3.)

P13-3. Dikes or levees act as barriers to contain large amounts of water. An old fairy tale concerns a boy who saved Holland from flooding using his finger to plug a hole in a leaking dike. Consider that this hole was 4 m below sea level (density = 1030 kg/m^3), and the boy's finger had a cross-sectional area of 3 cm^2, then use the atmospheric pressure of 101 kPa to find the amount of force the boy had to exert, and comment on the reasonableness of your answer.

P13-4. Lift is produced when air flows faster across the top of a wing than the bottom. If a particular aircraft has air flowing at 140 m/s over the top and 130 m/s over the bottom, use the wing area of 80 m^2 and air density of 1.3 kg/m^3 to find the amount of lift force generated.

P13-5. Water in a water tower stands a full 50 m above underground pipes that carry the water to nearby homes. If you consider that the water in the tower is at atmospheric pressure, use Bernoulli's relation to find (A) the speed of water when there is a water main leak, and (B) the height above the pipes to which the water would rise if it shot vertically.

P13-6. Consider a situation in which 30 m/s winds blow over the top of a flat-roofed house that is 3 m by 20 m. (A) If the air pressure inside is normal atmospheric pressure, find the net force trying to push the roof up. Assume the roof's weight is 1500 N. (B) If the roof is held on by nails, each of which would require 200 N of force to pull, how many nails would be required to keep the roof in place? (Use air density = 1.3 kg/m^3.)

P13-7. In an automotive service center, the lift works by hydraulic pressure transmitted by a piston. If the piston's cross-sectional area is 0.04 m^2, how much pressure would be required to lift a 1800 kg car?

P13-8. Water flows through a normal garden hose at 2 m/s. If a nozzle at the end of the hose reduces the cross-sectional area to 5 percent of its original value, how fast does the water emerge from the hose nozzle?

P13-9. Knowing Earth's radius to be 6.38 Mm and its mass to be 5.98×10^{24} kg, (A) find the overall density of Earth. (B) Since the density of surface rocks tends to be between 2000 kg/m^3 and 3000 kg/m^3, what does that imply about the density of material beneath the surface? (Use sphere volume = $4/3\ \pi r^3$.)

CHAPTER 14: EVEN PERFECTION HAS ITS FLAWS—THE IDEAL GAS

Q14-1. Explain briefly the two fundamentally different ways to analyze gases. Historically, which analytical technique was developed first?

Q14-2. In the 1600s and 1700s, devices were invented that measured two large-scale properties of gases. Name the two quantities measured and give a brief explanation about how the measurement devices worked.

Q14-3. Experiments were performed on gases for which each of three variables were held constant, in turn. Name each of the variables, the experimenters involved, and the resulting properties of gases revealed.

Q14-4. State and briefly explain Avogadro's hypothesis.

Q14-5. Show how the perfect gas law may be obtained by applying Newton's second law to a particle that collides perfectly elastically with the walls of a box.

Q14-6. Using the perfect gas law in the form $pV = nRT$, explain briefly each term involved, including its units.

Q14-7. How must the perfect gas analysis be modified when attempting to account for the properties of real gases?

Q14-8. Briefly explain how a plasma differs from a gas.

Problems

P14-1. If you have a tank that contains 30 kg of oxygen (O_2) at STP (0°C, 101 kPa, R = 8.314 J/molK), (A) find the tank's volume. (B) If 10 kg of oxygen is added to this tank without changing the temperature, find the new pressure. (Use O_2 molecular mass = 32.)

P14-2. A hospital patient breathes oxygen from a cylinder that contains 0.5 m^3 of O_2 at 60 atm of pressure. (A) What volume would the O_2 occupy at normal atmospheric pressure, at the same temperature? (B) If the patient breathes 4 L/min at atmospheric pressure, how many hours will the oxygen in the tank last?

P14-3. A tank of helium used to fill balloons starts out at a gauge pressure of 7 atm. After filling a few balloons (at constant temperature), the gauge pressure falls to 4 atm. Find the fraction of the original gas remaining.

P14-4. When a furnace ignites and raises the temperature in a house from 15°C to 21°C (59°F to 70°F), what fraction of the original air molecules are pushed out? (The pressure remains atmospheric since houses aren't airtight.)

P14-5. Nine moles of a perfect gas at 20°C and 1 atm are suddenly transferred to a new container with double the original volume. Presuming the temperature remains constant, find the new pressure.

P14-6. If an enclosed volume of a perfect gas is heated, raising its temperature from 10°C to 20°C (50°F to 68°F), by what factor does the pressure change?

P14-7. If a perfect gas's pressure is increased by 20 percent and the volume decreases by 10 percent, what is the new temperature, given that the old temperature was 0°C (32°F)?

P14-8. If a perfect gas at 20°C (68°F) expands its volume by a factor of 1.1, but the pressure remains the same, find the new temperature in degrees Celsius.

CHAPTER 15: SOME LIKE IT HOT—THERMODYNAMICS

Q15-1. List and explain briefly early ideas about the nature of heat.

Q15-2. Explain briefly the two theories of heat from the 1700s: caloric and phlogiston.

Q15-3. List and explain briefly three things that surprised or impressed you in the biography of Benjamin Thompson, Count Rumford.

Q15-4. Explain briefly how the caloric theory was put to rest by Joule in 1849.

Q15-5. What happens to the heat energy that is added to a solid when it remains solid?

Q15-6. What happens to the heat energy that is added to a solid when it turns into a liquid?

Q15-7. Explain briefly the action of a bimetallic strip.

Q15-8. Explain briefly how calorimetry experiments are designed to work.

Q15-9. State and explain briefly the first law of thermodynamics.

Q15-10. On a pV diagram, draw processes that show constant pressure, constant volume, constant temperature, and zero heat energy input.

Q15-11. Draw a diagram that illustrates the overall energy flow for a generalized heat engine.

Q15-12. Explain briefly the coefficient of performance of a refrigerator.

Q15-13. Briefly explain the second law of thermodynamics.

Q15-14. Explain briefly the third law of thermodynamics.

Q15-15. What is conduction? Give an example from your experience.

Q15-16. the concept of convection and provide an example from your experience.

Q15-17. Briefly explain the concept of radiation and give an example from your experience.

Problems

P15-1◇. Several billionaires commission the building of a gold band that completely circles Earth at the equator. However, due to global warming, it is predicted that the band will heat up by 7.2°C (13°F) within the next ten years, while Earth will retain its original diameter. Find the band's altitude, presuming the expansion is uniform. (Use $\alpha_{gold} = 14 \times 10^{-6}$°$C^{-1}$.)

P15-2. If a pedestrian bridge is made of aluminum ($\alpha = 22.5 \times 10^{-6}$ 1/°C), how much variation is there in the length of the 60 m span when the temperature varies from –44°C to +39°C? Comment on the reasonableness of your answer.

P15-3. A roadway is made up of 20-meter-long concrete slabs, for which $\alpha = 12 \times 10^{-6}$ 1/°C. If the slabs just touch at the maximum temperature expected, +60°C, how wide will the gap be when the temperature falls to – 45°C?

P15-4. A 30 kg block of ice originally at –9°C is heated and eventually turns into steam at 113°C. Using specific heats of ice = 2.1, liquid water = 4.186, and steam = 2.01, all in units of kJ/kg K; and heats of transformation of ice to water = 333.7, and water to steam = 2256, in units of kJ/kg; find the amount of heat energy required, in units of J.

P15-5◇. A 25 g lead bullet is shot at a temperature of 20°C and hits a heavy metal plate, where it melts completely. Knowing the specific heat of lead to be 0.13 kJ/kg K, its melting point to be 327°C, and its heat of transformation to be 22.9 kJ/kg, find the weapon's muzzle velocity. Presume that 80 percent of the bullet's kinetic energy stays in the bullet.

P15-6. In fits of frustration, people sometimes bang on the side of their computer when it malfunctions. If a 4 kg hammer hits the monitor five times at 10 m/s, and the 20 kg computer's specific heat is 1500 J/kg K, find the amount of temperature increase the computer will experience. (Feel better now?)

P15-7. A whale's 6-centimeter-thick blubber has a conductivity of 0.012 W/mK. If this whale's resting metabolism puts out 550 W, its internal temperature is 38°C, and surface area is 45 m^2, what external temperature can it stand without any change in internal metabolic rate?

P15-8. An ideal monatomic gas is subjected to a constant volume process from point A (5 atm, 2 L) to point B (1.5 atm, 2 L), then a constant pressure process from point B to point C (1.5 atm, 6 L). (A) Show these processes on a pV diagram and label the diagram. (B) Find the work done, in Joules, during each process (1 L atm = 101.3 J).

P15-9. A good-sized country uses 6×10^{14} Joules of energy each day to accomplish (useful?) work. The country's entire energy generating system is 21 percent efficient. Find (A) the amount of heat energy extracted from high-temperature reservoirs and (B) the amount of heat energy rejected to the environment each day.

P15-10. An enterprising inventor proposes a heat engine to drive a ship based on using the difference in temperature between surface water (18°C on the average) and water 15 m below the surface (7°C). He claims an efficiency of 37 percent for this engine. Find the maximum efficiency possible, and discuss whether you would invest money in this scheme.

P15-11. If 30 MJ of heat energy is supplied to water at 100°C, how much liquid water can be converted to vapor? (The latent heat of water is 2260 kJ/kg.)

P15-12. The exterior walls of a house are made of a 10-centimeter-thickness of layers of wood, insulation, and glass, with a combined thermal conductivity of 0.06 J/sm°C. If you know that the total surface area of all four side walls is 100 m^2, and the exterior temperature is 2°C, how many 60 W light bulbs would it take to maintain an interior temperature of 20°C, if you presume all the bulbs' energy goes into heat? (Ignore ceiling and floor.)

P15-13. If a 30 kg marble sphere rolls off a 20-mile-high cliff, and 40 percent of the energy of the collision with the ground is kept within the sphere, how much will its temperature rise, when the specific heat of marble is 860 J/kg°C?

P15-14. Two L of an ideal gas at an initial pressure of 3 atm (absolute) is allowed to (A) expand at constant temperature until the pressure is 1.2 atm. The gas is then (B) compressed at constant pressure to its initial volume, then (C) is returned to its original pressure by heating at constant volume. On a pV diagram, sketch these three processes. Make sure the axes are labeled and the processes are appropriately lettered.

P15-15. A particular power plant operates at 68 percent of its maximum (Carnot) efficiency between temperatures of 500°C and 20°C. If the plant's useful power output is 6 MW, find the rate of exhaust heat energy the plant discharges.

P15-16. If a refrigerator's coefficient of performance (COP) is listed as 3.0, and it extracts 25 J of heat energy from the cold reservoir (the inside of the fridge), find (A) the amount of work required and (B) the amount
(the room air)

P15-17. What is the maximum COP of a refrigerator operating between the temperatures of 0°C inside and 22°C outside?

P15-18. If the sun is considered an ideal blackbody radiator (e = 1), find the power per unit area radiated, given that the sun's surface temperature is 5800 K, and using $\sigma = 5.67 \times 10^{-8}$ W/m^{2}K^4.

CHAPTER 16: CHARGE IT—ELECTRICITY AT REST

Q16-1. How were Thales of Miletus and the material amber involved in the discovery of electricity?

Q16-2. Who was Queen Elizabeth I's personal physician, and how was he involved in the study of electricity?

Q16-3. The Leyden jar was very important in the development of electricity. Explain briefly.

Q16-4. List and explain briefly three things from Benjamin Franklin's biography that surprised or impressed you.

Q16-5. List and explain briefly three things about the life of Charles Coulomb that surprised or impressed you.

Q16-6. List and explain briefly similarities and differences between the electric force between two charges and the gravitational force between two masses.

Q16-7. Explain the concept of an electric field as if you were addressing a friend or a relative.

Q16-8. What are dipoles, and what is the difference between a natural dipole and an induced dipole? Give an example of an induced dipole in action.

Q16-9. What needs to be done to charges in order to construct a battery? Explain briefly.

Q16-10. Briefly explain electric potential, including its unit.

Q16-11. What is a capacitor and how does it work?

Q16-12. What functions do dielectrics serve in capacitors?

Q16-13. Briefly explain the term *breakdown voltage* and apply this idea to air.

Problems

P16-1. Using Coulomb's law, find the electrical force between two electrons (charge 1.6×10^{-19} C) located a distance 10^{-10} m apart. (Use $k = 9 \times 10^9$ Nm^2/C^2.) Compare this force with the weight of an electron using $w = mg$, given that the electron mass is 9.11×10^{-31}kg and $g = 9.8$ m/s^2.

P16-2. A magician's trick involves charging a 50 kg box with 2×10^{-3} C of charge, then charging a platform with the same charge. When the box is released, it "floats" how far above the platform?

P16-3. The Bohr model for the hydrogen atom says that the electron orbits the single-proton nucleus because of electrostatic attraction. If a particular excited state has the electron orbiting at a distance of 3.2×10^{-10} m, then find the electron's speed, presuming a circular orbit.

P16-4. Earth and the sun exert a gravitational attractive force on each other of 3.55×10^{22} N. If an evil genius decided to free Earth from the sun by charging both to generate an electrical force equal to the gravitational force, how much charge would be needed? (Use Earth-sun distance = 1.5×10^{11} m.)

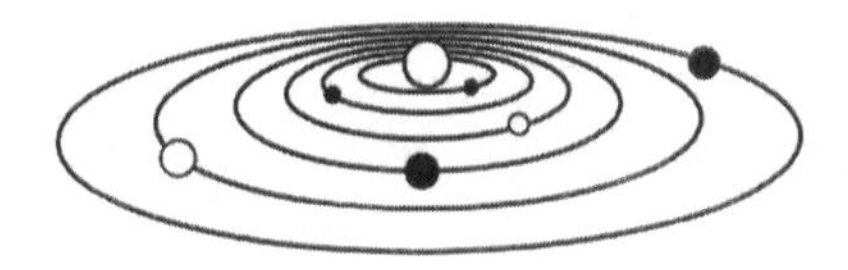

P16-5. The two protons in a helium nucleus are separated by a distance of approximately 1.1×10^{-15} m. How much electrical repulsive force is there between these protons?

P16-6. Two 5 g spheres are suspended from a point above by a nonconducting thread 40 cm long. If the spheres are separated by 7 cm because of electrostatic repulsion, find the charge on each, knowing they are

P16-7. If 9 V is maintained across a 200 μF capacitor, how much energy is stored?

P16-8. If a lightning strike can be approximated by considering a cloud and the ground as a parallel plate capacitor, (A) find the capacitance of such a system if the cloud's dimensions are 8 km by 9 km, at a height of 600 m. (Use $\varepsilon_0 = 8.85 \times 10^{-12}$ F/m.) (B) If 30 C of charge sit on the cloud, find the energy stored in this system.

P16-9. A hazard warning flasher consists of a light bulb and a 500 μF capacitor. When the bulb lights up, 40 Joules of energy are converted to light and heat. (A) What voltage must be applied across the capacitor? (B) What is the capacitor's initial charge?

CHAPTER 17: WHO LET THE CHARGES OUT?—ELECTRIC CURRENT

Q17-1. Explain the similarities and differences between a skier going down and then back up a hill, and an electric charge traveling through an electric circuit.

Q17-2. Explain briefly the concept of resistance, including its unit.

Q17-3. Briefly explain what little is known about the physics of lightning, including which parts of clouds are positive, which are negative, and how many lightning strikes occur on average.

Q17-4. There are several different kinds of lightning. List them and explain each briefly.

Q17-5. The effects of electrical current on humans depend on the amount of current involved. List the various effects and the corresponding currents.

Q17-6. For safety's sake, list and explain briefly four ways to limit electrical current passing through the body.

Problems

P17-1. Batteries power much of our modern mobile lifestyle. AA batteries are often rated in units of milliampere hours (mAh). This unit is actually a measure of the charge that the battery contains, and may be converted directly to Coulombs by the formula 1 mAh = 3.6 C. How many Coulombs of charge are contained in a AA battery rated at 1500 mAh?

P17-2. If 35 Coulombs of charge flow through a wire in two minutes, how many amps of current are flowing?

P17-3. If 2 mA of current flow in your MP3 player, how long will it take for 1 C of charge to flow?

P17-4. In a 12 V automotive electrical system, how much current flows through a 50 Ω resistor?

P17-5. If 5 amps of current flow through a 20 Ω resistor, how much voltage drives the circuit?

P17-6. In the near future, automotive electronics systems will be driven by 42 Volt batteries. If a resistor draws 3 A in this system, how much resistance is involved?

P17-7. In a 9 V circuit, how much power is dissipated by a resistor, if 0.2 amps of current flow?

P17-8. Laptop computers generate a lot of power. This power must be dissipated by heat sinks on the circuit boards that transfer heat energy to the case, and eventually, to your lap. If a particular computer runs on 14 V, and 2 A of current flow, how much power must be dissipated?

P17-9. Old-style dimmers consisted of a variable resistor, so you could dial in more resistance to lower the current and hence dim the light. If the increased total resistance caused the current to be reduced by 50 percent, how much power would be dissipated, as a fraction of the original power to the bulb alone?

CHAPTER 18: SILENT STRUGGLES IN THE WIRES—RESISTANCE

Q18-1. Explain briefly what is happening in solids at the submicroscopic level, (A) normally and (B) when a strong electric field is present.

Q18-2. Briefly explain superconductivity as you would to a friend or a relative.

Q18-3. Explain similarities and differences between semiconductors and conductors.

Q18-4. Use the difference in resistivity of conductors and insulators to explain why electrical currents travel long distances through wires rather than short distances through insulation.

Q18-5. State and explain briefly Kirchhoff's rules that are used for analyzing multiloop electrical circuits.

Q18-6. In an electrical circuit containing a battery, a resistor, and a capacitor, once the circuit is energized, what effect does the resistor have on the changing voltage across the capacitor?

Problems

P18-1. If a 20 Ω resistor and a 300 Ω resistor were connected in series, what single resistor could be used as a replacement for them and would have the same effect in the electrical circuit?

P18-2. If a 20 Ω resistor and a 300 Ω resistor were connected in parallel, what single resistor could be used as a replacement for them and would have the same effect in the electrical circuit?

P18-3. If a 20 μF capacitor and a 300 μF capacitor were connected in series, what single capacitor could be used as a replacement for them and would have the same effect in the electrical circuit?

P18-4. If a 20 μF capacitor and a 300 μF capacitor were connected in parallel, what single capacitor could be used as a replacement for them and would have the same effect in the electrical circuit?

P18-5. If you had three resistors, all of the same resistance, R, (A) what would the equivalent resistance be if they were connected in series? (B) Determine the equivalent resistance of four equal resistors in series. (C) Do you see the pattern developing? How about N resistors in series?

P18-6. If you had three resistors, all of the same resistance, R, (A) what would the equivalent resistance be if they were connected in parallel? (B) Determine the equivalent resistance of four equal resistors in parallel. (C) Do you see the pattern developing? How about N resistors in parallel?

P18-7. Resistors are manufactured in standard sizes, so if you needed a 75 Ω resistor, how would you achieve this value using 50 Ω resistors?

P18-8. If you had three capacitors, all of the same capacitance, C, (A) what would the equivalent capacitance be if they were connected in series? (B) Determine the equivalent capacitance of four equal capacitors in series. (C) Do you see the pattern developing? How about N capacitors in series?

P18-9. If you had three capacitors, all of the same capacitance, C, (A) what would the equivalent capacitance be if they were connected in parallel? (B) Determine the equivalent capacitance of four equal capacitors in parallel. (C) Do you see the pattern developing? How about N capacitors in parallel?

P18-10. Capacitors are manufactured in standard sizes, so if you needed a 150 μF capacitor, how would you achieve this value using 100 μF capacitors?

P18-11. A crow sits on a copper ($\rho = 1.67 \times 10^{-8}$ Ωm) wire with a diameter of 2 cm, which carries a current of 70 amps. If the bird's feet are 11 cm apart, (A) find the potential difference between the bird's feet. (B) If the bird ate away the wire's insulation, how much current would flow through the bird, whose resistance is 0.4 M Ω?

P18-12◇. In the circuit shown, R_1 = 800 Ω, R_2 = 500 Ω, R_3 = 200 Ω, and V = 9 V. Find the equivalent resistance of the whole circuit and the current through each resistor.

P18-13◇. In the capacitor network shown, C_1 = 300 nF, C_2 = 400 nF, C_3 = 500 nF, and V = 12 V. Find the equivalent capacitance, the charge on each capacitor, and the voltage across each capacitor.

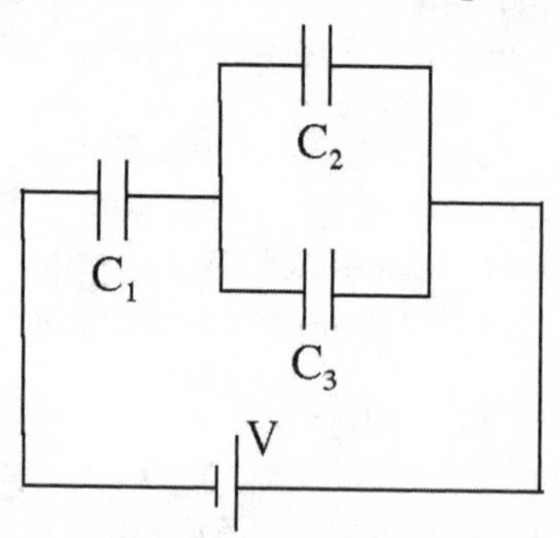

P18-14. Find the resistance of a 3-meter-long copper wire with cross-sectional area 2×10^{-5} m^2. Use 1.67×10^{-8} Ωm for the resistivity of copper.

P18-15. If a human being has a resistance of 1.5 MW, what length copper wire of a cross-sectional area of 2.5×10^{-5} m^2 would be required in order to achieve the same value of resistance? (Use 1.67×10^{-8} Ωm for the resistivity of copper.) Compare your answer to a common length.

P18-16. In a circuit containing a 200 Ω resistor, a 400 μF capacitor, and a 6 V battery in series, find the voltage across the capacitor after (A) 50 ms and (B) 120 ms.

P18-17. In a circuit containing a 600 Ω resistor, a 300 μF capacitor, and a 9 V battery in series, find the time necessary for the voltage across the capacitor to become 99 percent of its maximum value.

CHAPTER 19: APPROACH/AVOIDANCE—MAGNETISM

Q19-1. List and explain briefly five magnetic devices found in the popular culture.

Q19-2. Are stationary charges affected by a magnetic field? What determines the direction of the magnetic force on a moving charge? Explain briefly.

Q19-3. List and explain briefly three things about the life of Nikola Tesla that surprised or impressed you.

Q19-4. Explain briefly how auroras work, including both the role of Earth's magnetic field, and the auroras of other planets.

Q19-5. Briefly explain the original discovery of magnetism.

Q19-6. Explain briefly the role of poles in magnetism.

Q19-7. What is the cause of all magnetic fields? Explain briefly.

Q19-8. Explain how moving charges lead to magnetic effects for iron atoms. Include in your explanation the idea of domains and the Curie temperature.

Q19-9. Explain briefly the dynamo theory of planetary magnetism.

Problems

P19-1. A 42 mT magnetic field points due west. If a proton (mass = 1.67×10^{-27} kg) of kinetic energy 9×10^{-12} J enters this field in an upward direction, find the magnetic force acting on the proton, in magnitude and direction.

P19-2. An electron moves (at 7 Mm/s) directly away from a wire that carries a current of 40 A. When the electron is 5 cm away from the wire, find the **magnitude and direction** of the magnetic force on the electron due to the magnetic field of the wire. (Use $\mu_o = 4 \pi \times 10^{-7}$ Tm/A.)

P19-3. What magnetic force would be experienced by an electron ($q = 1.6 \times 10^{-19}$ C) moving at 3×10^6 m/s perpendicular to Earth's magnetic field, 550 μT?

P19-4. Find the magnitude of the magnetic field a distance of 10 centimers from a wire carrying a current of 3 amps.

P19-5. Find the magnitude of the magnetic field at the center of a 10 centimenter diameter loop carrying a current of 3 amps.

P19-6. Find the magnitude of the magnetic field in the center of a 500 turn, 15-centimeter-long solenoid carrying a current of 3 amps.

P19-7. A constant potential difference accelerates several singly charged isotopes of palladium, which are then shot into a magnetic field in a mass spectrograph. The isotope of mass 106 moves in a circular arc of diameter 44 cm. Another track shows up at 42.74 cm. Find the atomic mass of the second isotope.

CHAPTER 20: WE ARE FAMILY—ELECTRICITY FROM MAGNETISM

Q20-1. Describe briefly how the interactions among Humphry Davy, William Wollaston, and Michael Faraday produced the law of induction.

Q20-2. What actions did Faraday (and Joseph Henry) perform to induce electric current to flow?

Q20-3. State Lenz's law and explain briefly.

Q20-4. Explain briefly how mutual inductance works and why it might pose a problem for circuit designers.

Q20-5. Briefly explain self-inductance and give an example.

Q20-6. Briefly explain the operation of an electric generator.

Q20-7. Explain briefly the differences between AC and DC, as you would to a friend or a relative.

Problems

P20-1. According to Faraday's law, if a 200 turn, 10^{-3} m^2 cross-sectional area coil is immersed in a magnetic field that increases by 6 T/s, how much voltage is induced?

P20-2. In order to generate an induced voltage, a coil with a cross-sectional area of 2×10^{-4} m^2 is placed in a 0.4 T magnetic field, then rotated 90° within 3 ms so that zero magnetic field lines go through the coil. This is called a *flip coil*. How many turns must a flip coil have in order to induce 120 V?

P20-3. You are given two circuits located near each other, with a mutual inductance coefficient of 300 mH. If one circuit's current varies at a rate of 2 A/s, how much voltage will be induced in the other circuit?

P20-4. In two adjacent circuits, an induced voltage of any more than 11 mV will cause unacceptable interference with one circuit's function. If the current varies by as much as 3 A/s, what is the maximum mutual inductance coefficient that is acceptable?

P20-5. A particular circuit's self-inductance is 14 mH. How much back voltage is induced if the current varies by 0.08 A/s?

P20-6. In a circuit containing a 6 V battery, a switch, a 500 mH inductor, and a 20 W resistor, how much current will flow 30 μs after the switch is thrown?

P20-7. In an LR series circuit, where L = 600 mH and R = 70 W, how long after the switch is thrown will the current reach 99 percent of its maximum value?

CHAPTER 21: BIG-TIME ELECTRICITY—AC

Q21-1. Explain briefly what factors determine power losses when electric current flows through wires.

Q21-2. Explain briefly the workings of electrical transformers, both step-up and step-down.

Q21-3. In a step-up transformer, compare the secondary voltage and current to the corresponding primary values. Explain briefly how transformers minimize losses in long-distance electrical circuits.

Q21-4. List and explain briefly three things about the life of Thomas Alva Edison that surprised or impressed you.

Q21-5. List and explain briefly three things about the life of George Westinghouse that surprised or impressed you.

Q21-6. Explain briefly how the bizarre electrical execution of William Kemmler related to the competition between Edison and Westinghouse.

Q21-7. Explain briefly why the electrical system chosen for Niagara Falls was AC, not DC.

Q21-8. Give a brief explanation of the operation of the modern power grid, including the number of generators, the total watts produced, the voltage in transmission lines, and the variation in demand.

Q21-9. In practical AC circuits, describe briefly the role of fuses and/or circuit breakers.

Q21-10. Explain briefly the operation of a GFCI.

Q21-11. What is the relationship between kWh and Joules? Explain briefly.

Q21-12. Explain briefly the difference between peak and rms voltage.

Q21-13. In an AC circuit that contains a resistor, a capacitor, and an inductor, impedance replaces the DC concept of resistance. Explain briefly the difference between the two quantities.

Q21-14. An AC circuit with a variable frequency generator, a resistor, and a capacitor is called a high-pass filter. Explain briefly.

Q21-15. An AC circuit with a variable frequency generator, a resistor, and an inductor is called a low-pass filter. Explain briefly.

Q21-16. An AC circuit with a variable frequency generator, a resistor, an inductor, and a capacitor has a resonant frequency at which the maximum current flows. Explain briefly.

Problems

P21-1. Doorbells usually operate on 16–24 V, so the 110 V household voltage must be stepped down to accommodate. Find the ratio of the number of turns on the secondary to the number of turns on the primary to step 110 V down to 16 V.

P21-2. If a transformer has 110 V AC on the primary side, and a secondary/primary turns ratio of 0.05, what is this transformer's secondary voltage?

P21-3. For long-distance transmission of electricity, voltages of 1–26 kV at the generator are stepped up to high levels, such as 230–768 kV. In stepping up from 1 kV to 768 kV, if there are 200 turns of transformer wire on the primary, how many turns will the secondary contain?

P21-4. Ideal transformers transfer 100 percent of the power from the primary to the secondary. Real transformers fall only slightly short of this goal. Assuming an ideal transformer, find the current in the primary for a step-down transformer that changes the 768 kV primary to 69 kV, 1000 A for distribution.

P21-5. The function of those little black transformers that clog your power strips is to step down line voltage to run small electronic gadgets. Suppose a fax machine uses an ideal transformer with a primary voltage of 110 V and 300 primary windings. If the secondary voltage is 12 V, (A) find the number of turns on the secondary. (B) If the fax machine at idle draws enough current to dissipate 6 W, find the current that must flow in the primary.

P21-6. Energy from the power grid is charged on the basis of kWh = 3.6 million Joules. If your computer consumes 300 W of power, and you use it 5 hours/day, 30 days/month, how much does your computer usage cost per month, assuming electrical energy costs 11 ¢/kWh?

P21-7. RMS values represent a type of average for the sine function that describes the voltage and current in AC, and all listed values are RMS. What is the peak voltage corresponding to 110 V RMS?

P21-8. An electric blow-dryer rated at 1500 W operates on the 110 V power grid. If you assume this blow-dryer has no capacitance or inductance, find (A) its resistance and (B) the current drawn.

P21-9. In an RLC circuit, a 110 V 60 Hz generator powers a 40 Ω resistor, a 50 mH inductor, and a 60 μF capacitor. Find (A) the impedance and (B) the current in the circuit.

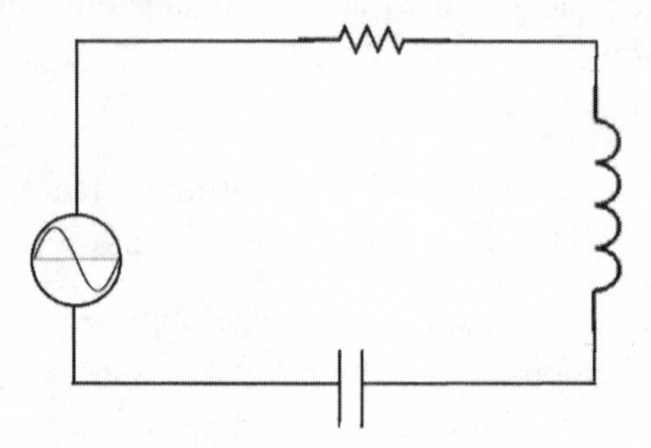

P21-10. Find the resonant frequency of a circuit that has a 400 mH inductor and a 30 nF capacitor.

CHAPTER 22: THE LIGHT DAWNS—ELECTROMAGNETIC WAVES

Q22-1. If two charges are separated by a substantial distance, what must be done to one charge to exert a significant influence on the other?

Q22-2. Both Beeckman and Galileo suggested unsuccessful experimental procedures for determining the speed of light. Explain their ideas and why they were not successful.

Q22-3. Explain briefly the history of experimental measurements of the speed of light.

Q22-4. Briefly explain how Hertz broadened the electromagnetic spectrum.

Q22-5. In order of increasing frequency, list the waves that make up the electromagnetic spectrum and give an example of each type of wave.

Problems

P22-1. (A) A Canadian FM radio station (CLAW) broadcasts lobster information at 95.9 MHz. Find the wavelength of this station's wave. (B) What would be the frequency of a station for which the carrier wave was 400 m long?

P22-2. An AM radio station has a wavelength of 300 m. Find this station's frequency.

P22-3. The microwave portion of the spectrum has wavelengths ranging from 1 mm to 1 m. Find the corresponding frequencies.

CHAPTER 23: MIRROR, MIRROR ON THE WALL—REFLECTION AND REFRACTION OF LIGHT

Q23-1. At a submicroscopic level, explain briefly how an electromagnetic wave attempts to oscillate molecules, atoms, electrons, and nuclei. How do the results of these efforts depend on frequencies?

Q23-2. Distinguish four different cases of light waves forcing matter to oscillate, giving details of the interaction and what is happening at a visible level.

Q23-3. At the submicroscopic level, what is happening when a light wave is reflected?

Q23-4. If light is incident on a reflecting surface at an angle of 35° to the normal, at what angle will the light be reflected?

Q23-5. Explain briefly what differences in molecular properties lead to transmission versus absorption of light.

Q23-6. Even though light travels only at one speed, c, light's average speed within a medium is always less than c. Explain how this is possible.

Q23-7. Explain how Snell's law governs refraction. You may need a diagram or an analogy.

Q23-8. What molecular properties lead to the dispersion of light?

Q23-9. Explain briefly the phenomenon of polarization, including the properties of molecules that are involved.

Q23-10. Draw a ray diagram for a person standing in front of a plane mirror.

Q23-11. Draw a ray diagram for an object placed more than two focal lengths in front of a convex mirror.

Q23-12. Draw a ray diagram for an object placed more than two focal lengths in front of a concave mirror.

Q23-13. Draw a ray diagram for an object placed more than two focal lengths in front of a converging lens.

Q23-14. Draw a ray diagram for an object placed more than two focal lengths in front of a diverging lens.

Q23-15. Explain briefly why the sun appears almost white at noon, but quite orange, or even red at sunrise or sunset. How does this relate to the color of the sky?

Q23-16. Contrast the operation of a camera and the human eye, in terms of the lens and the image location. Include in your explanation how adjustments are made in each, and how the human eye's capabilities change with age.

Q23-17. List the additive primary colors and the subtractive primary colors, and give a brief explanation of why there are two different sets of primaries.

Problems

P23-1. If light enters a plane mirror at an angle of 44° to the mirror's normal, what is the angle of the reflected ray?

P23-2. If the average speed of light in a medium is 2.2×10^8 m/s, what is the medium's index of refraction?

P23-3. A fish (in the water) sees the sun by looking at an angle of 33° to the vertical. Given that the index of refraction of water is 1.33, find the angle the sun makes with the horizon. If the fish is located in the mid-Northern hemisphere and the time is around noon, what is the season, and why do you think so?

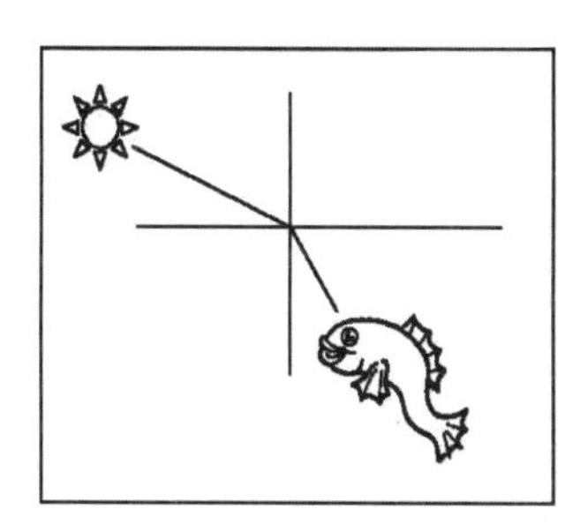

P23-4. An object placed 24 cm from a 15-centimeter-focal-length concave mirror produces an image how far from the mirror? Draw a ray diagram to illustrate.

P23-5. If an object sits 33 cm in front of a 55-centimeter-focal-length convex mirror, locate the image.

P23-6. A 6-centimeter-high object is placed 43 cm from a converging lens of focal-length of 24 cm. (A) Find the image's position and (B) draw a ray diagram approximately to scale.

P23-7. Reading glasses sold at the drugstore are designed to allow the wearer to read something that is placed 25 cm from the glasses, even though, without glasses, he would have to hold the object farther away to see it clearly (called the near point.) If a person's uncorrected near point is 80 cm, find the power of corrective glasses, measured in diopters.

CHAPTER 24: LIGHT'S STRANGEST TRICKS—DIFFRACTION AND INTERFERENCE

Q24-1. Briefly explain the phenomenon of diffraction and give an example.

Q24-2. Describe the physical setup of Young's double-slit experiment.

Q24-3. Explain briefly why the result of Young's double-slit experiment supported the wave theory of light over the particle theory.

Q24-4. Explain briefly how thin film interference works and give an example.

Q24-5. Briefly describe the attitude of major physicists in the late 1800s and give an example.

Q24-6. In the late 1800s, only a few thorns were present in an overall rosy picture of physics. Briefly describe these thorns.

Problems

P24-1. An interference pattern is produced on a screen 3 m away from a pair of narrow slits located 5×10^{-4} m apart. If the distance on the screen between adjacent bright fringes is 0.4 cm, find the wavelength of the light involved.

P24-2. A 633 nm wavelength laser produces interference patterns separated by 3 cm on a screen located 2 m from the diffraction grating. How far apart are the slits in the grating?

CHAPTER 25: EINSTEIN'S PRODIGIOUS EFFORTS—SPECIAL AND GENERAL RELATIVITY

Q25-1. Briefly describe the Michelson-Morley interferometer experiment and explain why its result was so perplexing.

Q25-2. List and explain briefly three things about the life of Albert Einstein that surprised or impressed you.

Q25-3. List and explain the two postulates of Einstein's special theory of relativity.

Q25-4. Applying Einstein's postulates to a simple case, explain the relativity of simultaneity.

Q25-5. Why do measurement disagreements between observers seldom show up in ordinary life?

Q25-6. List and explain briefly two bits of experimental evidence that support the special theory of relativity.

Q25-7. Describe briefly the way Einstein developed the general theory of relativity, including the role of his friend Marcel Grossmann.

Q25-8. Describe briefly the experimental evidence that supports the general theory of relativity.

Q25-9. Shortly after the general theory of relativity was published, astronomer Willem de Sitter had some discussions with Einstein that moved Einstein to modify the theory. What was the basis of the controversy, and what change did Einstein make?

Problems

P25-1. An unstable particle called bullroarium has a lifetime of 6×10^{-10} seconds, measured at rest. If a bullroarium particle zips past a stationary observer at .67c, find the stationary observer's measurements for (A) bullroarium's lifetime and (B) the distance it travels during this time.

P25-2. A broken-down spaceship (Star Wreck?) flies across an international soccer field (regulation length = 100 m) at a speed of 0.55c. (A) How long does the spaceship pilot measure the field to be? (B) From the spaceship pilot's perspective, how long does it take to make this dash? (C) How long does this spaceship's flight take, according to ground observers?

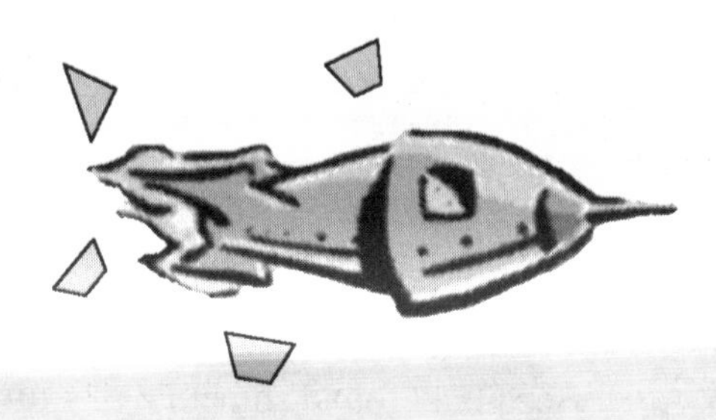

P25-3. If an electron travels at 40 percent of the speed of light, then collides with a phosphor molecule in a television screen, how much mass would be sensed by an observer stationed on the molecule?

P25-4. The *Apollo 10* spacecraft traveled at 11 km/s just prior to its reentry, setting a speed record for humans. Suppose advanced technology raises the speed record by a factor of 100, to 1.1×10^6 m/s. If a time interval of 1 hour elapsed on a ground-based clock, by how many seconds would the ship's clock differ from one hour? (Carry as many sig figs as your calculator will permit.)

P25-5. If a meter stick zips past a ground-based observer at 75 percent of the speed of light, how long would the ground-based observer measure for the stick's length?

P25-6. An extremely underweight person wishes to add 5 percent to his mass, as measured by someone else. How fast would he have to travel to gain 5 percent according to a fixed observer?

P25-7. If a 70 kg human body could be converted entirely into energy, and the energy sold at the rate of 12 ¢/kWh, how much money could be raised? (For comparison, the US national debt is approximately $\$9 \times 10^{12}$.)

P25-8. How much mass would be required to be converted to energy to supply the US energy consumption for one year, 10^{20} J? Compare this to an object of known mass.

CHAPTER 26: MATTER'S INNARDS—ATOMS AND QUANTUM MECHANICS

Q26-1. What did Wilhelm Conrad Röntgen discover in his 1895 experiments? Explain briefly.

Q26-2. In 1896, A. H. Becquerel found something he didn't quite expect. Explain briefly.

Q26-3. J. J. Thomson's 1897 experiments led to the discovery of a new particle and the development of a theory about atoms. Explain briefly.

Q26-4. Max Planck proposed what he thought was a math-based solution to one of physics' thorny problems, but his idea led to a whole new branch of physics. Explain briefly.

Q26-5. There was a clear pattern to Einstein's series of papers from 1905. Identify this pattern and explain briefly.

Q26-6. Briefly explain the experimental efforts that led to the Rutherford model of the atom.

Q26-7. In 1913, Niels Bohr applied Planck's idea to the hydrogen atom with spectacular success. Explain briefly.

Q26-8. Einstein and Bohr had a friendly debate that took place over many years. Explain briefly the substance of the debate and give a quote that illustrates the character of the discussion.

Q26-9. State and explain briefly the Heisenberg Uncertainty Principle. Is this likely to be noticed in ordinary life? Answer yes or no and explain your answer briefly.

Q26-10. Define and explain briefly the de Broglie wavelength.

Q26-11. J. J. Thomson's son, G. P. Thomson, worked on another aspect of the same particle his father discovered, yet there was a substantial irony involved. Identify the irony and explain briefly.

Q26-12. So, is light/matter a particle or a wave? Give physics' current understanding of this knotty problem, using analogies if appropriate.

Q26-13. P. A. M. Dirac and C. D. Anderson both worked on the same particle. Name the particle and explain briefly what each one did.

Q26-14. The cyclotron was the brainchild of Ernest O. Lawrence. Describe the operation of a cyclotron and explain briefly its advantages over cosmic rays.

Problems

P26-1. If the uncertainty in momentum of an electron in the first Bohr orbit of the hydrogen atom is 2×10^{-24} kgm/s, find its uncertainty in position.

P26-2. In trying to pinpoint the position of an electron, the uncertainty in position is reduced to 10^{-12} m. What is the corresponding uncertainty in momentum?

P26-3. A 58 g tennis ball is served at 40 m/s. (A) Find the de Broglie wavelength of the tennis ball. (B) Compare this length to the diameter of a proton (about 10^{-15} m). Would it be possible to observe any wave properties of the tennis ball? Explain briefly.

P26-4. (A) Find the de Broglie wavelength of an electron traveling at 2×10^6 m/s. (B) Does this length come close to any familiar dimensions? Explain briefly.

CHAPTER 27: ATOMS' INNARDS—NUCLEAR PHYSICS

Q27-1. If the atom could be enlarged to the size of a football field, and the electrons were represented by gnats swooping around the edges of the field, how large would the nucleus be, and what would be its mass relative to electrons?

Q27-2. Briefly describe the strong nuclear force, including the particles it affects, its strength, and its dependence on distance. Use any analogy to help this description.

Q27-3. Since there are energy levels in the nucleus, explain how nucleons can move to higher energy levels and what happens when they return to lower energy levels.

Q27-4. Briefly explain isotopes, including the difference between stable and unstable ones.

Q27-5. List and explain briefly radioactive decay products.

Q27-6. List and explain briefly three devices used to detect radioactive decay.

Q27-7. The bombardment of the nucleus by neutrons produced interesting results. Explain briefly, including a comment by German chemist Ida Noddack.

Q27-8. Explain briefly the tangled project started by Lise Meitner, Otto Hahn, and Fritz Strassman, especially including their unexpected results and Meitner's interpretation.

Q27-9. Leo Szilard, Albert Einstein, and Franklin Delano Roosevelt had an interaction that led to the Manhattan Project. Briefly explain the beginnings of this project, showing the role of each.

Q27-10. List and explain briefly three things about the life of Enrico Fermi that surprised or impressed you.

Q27-11. List and explain briefly your view of the successes and failures of the Manhattan Project.

Q27-12. List and explain briefly the differences between fission and fusion.

Problems

P27-1. If the half-life of palladium 100 is 3.63 days, what percentage of the original amount is left after three weeks?

P27-2. An unknown radioactive element is found to decay such that only 2 percent remains after six hours. What is the unknown element's half-life?

P27-3. The fictitious radioactive element felinium has a half-life of 5.9 years. If 12 percent of a sample remains, how much time has elapsed?

CHAPTER 28: DOWN TO THE NITTY-GRITTY—THE STANDARD MODEL OF THE UNIVERSE'S SMALLEST CONSTITUENTS

Q28-1. Explain briefly the operation of that staple of modern high-energy physics, the cloud chamber.

Q28-2. Explain briefly the operation of the synchrotron and list the results obtained by its use.

Q28-3. Quarks were first theorized, then found experimentally. Briefly explain these developments.

Q28-4. Quantum electrodynamics theory was developed by Feynman and Tomonaga in the 1950s. Draw and explain briefly Feynman diagrams, which illustrate how this theory applies to electrons and photons.

Q28-5. What are virtual photons, and how do they "carry" the electromagnetic interaction between electrons? Use Feynman diagrams and the Heisenberg Uncertainty Principle in your explanation.

Q28-6. The strong interaction is thought of as an exchange of virtual gluons. Explain briefly.

Q28-7. Earlier, particles were thought to interact with each other by action-at-a-distance forces. But the Standard Model represents a radical departure from these ideas. Explain briefly.

Q28-8. Apply the Standard Model to the atom in order to obtain physics' modern view of the atom.

Q28-9. Briefly explain the evidence for exactly three families of quarks, leptons, and bosons.

Q28-10. Briefly explain the evidence supporting the Standard Model.

Q28-11. Describe the Higgs field, and explain briefly how the Higgs field might fit into the Standard Model.

Q28-12. Currently, one of the most interesting problems in experimental physics is the search for a Higgs particle. Explain briefly how this search is being conducted and what will happen to the Standard Model, depending on the results.

Q28-13. Briefly describe three possible theories that may ultimately replace the Standard Model.

CHAPTER 29: BUT WAIT, THERE'S EVEN MORE—THE UNIVERSE'S BIGGEST CONSTITUENTS

Q29-1. In 1927, Georges LeMaître proposed a theory of the universe's beginning, which he referred to as a "cosmic egg." Explain this theory briefly.

Q29-2. List and explain briefly three things about the life of Edwin Hubble that surprised or impressed you.

Q29-3. In his 1948 doctoral dissertation, Ralph Alpher analyzed the building of nuclei. Explain how this idea was developed into the Big Bang theory.

Q29-4. Arno Penzias and Robert Wilson conducted an experiment in 1965 that provided experimental evidence that supported the Big Bang theory. Briefly describe the experiment and discuss how the result provided support.

Q29-5. Fritz Zwicky had an unusual personality but conducted valuable experiments regarding dark matter in the Coma Berenices galaxy. Briefly describe his personality, his experiments, and the reason his results were largely ignored for almost forty years.

Q29-6. Thanks to a series of experiments conducted in 1970, the existence of dark matter was put on a solid footing. Describe briefly who performed the experiments and why the results were so well accepted.

Q29-7. Explain briefly what happens to produce a Type Ia supernova.

Q29-8. Experimental studies of Type Ia supernovae by two different groups in 1998 led to a remarkable conclusion about the universe's expansion. Explain the studies' results, the inference about the universe's expansion, and the name given to the cause of the unusual expansion.

Q29-9. Another recent set of experiments led to the conclusion that the geometry of the overall universe is flat. Briefly describe the experiments and their conclusion.

Q29-10. Knowing the universe's geometry, the amounts of normal matter, dark matter, and dark energy can be estimated fairly accurately. List the percentages of each constituent of the universe and give your reaction to this news.

A TRIED-AND-TRUE PROBLEM-SOLVING TECHNIQUE

1. Read the problem statement carefully—sometimes English is harder than physics.
2. Draw a picture (even if it's crude) illustrating the problem, using any physics-specific diagrams that apply (free body, energy conservation, pV, heat engine, ray, etc.).
3. Identify knowns and unknowns from the problem statement.
4. Select possibly relevant physics principles in general equation format.
5. Simplify the equations—still in symbolic form—based on the specific problem, then manipulate them to solve for the unknown quantity—still symbolic (no numbers yet).
6. Substitute given values, including units, into the final equation(s) and calculate.
7. Check the answer to make sure it seems reasonable, based on the problem given.
8. If the answer isn't reasonable, repeat steps 4–7. If it is, breathe a sigh of relief.

To watch some experiments and problem-solving examples firsthand, simply visit my YouTube page: http://www.youtube.com/user/artwiggins.

METRIC PREFIXES

The standard system for measuring physical quantities for scientific purposes is the metric system (officially, SI), which measures length in meters, mass in kilograms, and time in seconds. More detailed information may be found on the World Wide Web, at http://physics.nist.gov/cuu/index.html.

Conversions within the metric system are simplified by the use of powers of ten prefixes that precede the basic unit to make it smaller or larger. For example, a millimeter = mm = 10^{-3} m and a kilometer = km = 10^3 m. The following table lists the twenty most common metric prefixes, their shorthand symbol, and their corresponding power of ten:

Metric Prefix	Shorthand Symbol	Power of Ten
yotta	Y	10^{24}
zetta	Z	10^{21}
exa	E	10^{18}
peta	P	10^{15}
tera	T	10^{12}
giga	G	10^{9}
mega	M	10^{6}
kilo	k	10^{3}
hecto	h	10^{2}
deka	da	10^{1}
deci	d	10^{-1}
centi	c	10^{-2}
milli	m	10^{-3}
micro	μ	10^{-6}
nano	n	10^{-9}
pico	p	10^{-12}
femto	f	10^{-15}
atto	a	10^{-18}
zepto	z	10^{-21}
yocto	y	10^{-24}

CONVERSIONS

British System Quantity	Metric System Equivalent
1 inch	2.54 cm
1 foot	30.5 cm
1 mile	1609 m
1 pound (weight)	0.454 kg (mass)
1 mile per hour (MPH)	0.447 m/s
1 pound (weight)	4.45 newtons (N)
1 horsepower (hp)	746 Watts

More details and conversions may be found at http://en.wikipedia.org/wiki/ Conversion_of_units.

CONSTANTS

Many problems have necessary constants given in the problem statement, but here are a few handy ones:

Quantity	Approximate Value	Quantity	Approximate Value
Acceleration due to gravity, g	$9.80\ m/s^2$	Gravitational force constant, G	$6.67 \times 10^{-11}\ Nm^2/kg^2$
Earth radius (mean)	$6.38 \times 10^6\ m$	Earth mass	$5.98 \times 10^{24}\ kg$
Earth-sun distance (mean)	$1.50 \times 10^{11}\ m$	Earth-moon distance (mean)	$3.84 \times 10^8\ m$
Electron charge	$e = -1.60 \times 10^{-19}\ C$	Electron mass	$9.11 \times 10^{-31}\ kg$
Coulomb's law constant	$k = 9 \times 10^9\ Nm^2/C^2$	Magnetic field constant	$\mu_o = 4\pi \times 10^{-7}\ Tm^2/A^2$
Speed of light, c	$3 \times 10^8\ m/s$	Planck's constant, h	$6.63 \times 10^{-34}\ Js/cycle$

Additional constants may be found at http://physics.nist.gov/cuu/Constants/index.html.

Problem	Answer	Problem	Answer	Problem	Answer
P1-1	20.8 hrs	P1-2	5.6×10^7 yrs	P1-3	(A) 1.08×10^5 km/h (B) 22.2 min
P1-4	65 MPH	P1-5	(A) 0.018 m/s (B) 0.27 m	P1-6	0.8 m/s^2
P1-7	(A) 1.5 sec (B) 3.75 m	P1-8	(A) 4.17 sec (B) –5.76 m/s^2	P1-9	(A) 89.4 m/s (B) 44.7 sec
P1-10	(A) 1.43×10^{-3} m/s^2 (B) 299 days	P1-11	2.08 minutes past 9:00 a.m.	P1-12	120 m
P1-13	(A) 17.3 sec (B) 451 m (C) 52 m/s	P1-14	2.45 m	P1-15	6.64 m/s
P1-16	(A) 1 sec (B) 7.55 m	P1-17	33.6 m/s	P1-18	0.883 m
P1-19	42.5 m	P1-20	7.23 m/s	P1-21	(A) 12.8 sec (B) 568 m
P2-1	v is constant	P2-2	4 m/s^2	P2-3	2000 kg
P2-4	3×10^{11} N	P2-5	166 N	P3-1	2.76×10^{20} N
P3-2	4.06×10^{-47} N	P3-3	2.4×10^{-7} N	P3-4	735 N

P3-5	1390 N	P3-6	176 N	P3-7	(A) 0.235 N (B) 3.06 m
P3-8	**?**	P3-9	T = D, L = w	P3-10	(A) 186 m/s^2 (B) 57.8 m/s
P3-11	(A) 0.538 m/s^2 (B) 38.9 m/s = 85.9 MPH	P3-12	**?**	P3-13	0.495 sec
P3-14	(A) 3.14 m/s^2 (B) 2830 N	P3-15	(A) **?** (B) 7360 N	P3-16	(A) 38.9 N (B) 0.147
P3-17	1.85×10^{12} m/s^2	P3-18	(A) 97.2 N (B) 3.35 m/s^2 (C) 4.44 sec	P4-1	(A) 2 m/s^2 (B) 1000 N
P4-2	(A) 5550 N (B) 0.472	P4-3	(A) 8.82×10^{-8} N	P4-4	47.1m/s = 105 MPH
P4-5	259 million years	P4-6	6.26 m/s = 14 MPH	P4-7	8.82°
P5-1	2.78 m/s	P5-2	(A) 2.46×10^5 J (B) 7.14×10^5 J	P5-3	(A) 1.14×10^{-13} J (B) 9.49×10^{-17} W
P5-4	(A) 69.4 m (B) 39.8 m/s	P5-5	(A) 34 % (B) 2.54 m/s	P5-6	5.54×10^4 W = 74.3 hp
P5-7	(A) 1.88×10^7 J (B) 366 MPH	P5-8	9.8×10^{-5} J	P5-9	412 J

P5-10	(A) 5×10^{18} J (B) 5 %	P5-11	15 m	P5-12	23.3 m
P6-1	340 N	P6-2	37.7 m/s = 84.4 MPH	P6-3	55 sec
P6-4	0.4 m/s	P6-5	3.64×10^{-25} kgm/s	P6-6	1.6×10^{23} kgm/s
P6-7	36.8 ms	P6-8	40 m/s	P6-9	(A) 6670 N (B) 1180 N
P6-10	(A) 2.5 m/s (B) 1.67 m/s	P6-11	0.64 m/s	P7-1	(A) –2120 rad/s^2 (B) 0.267 sec
P7-2	(A) 4.71 rad/s (B) 0.283 m/s	P7-3	(A) 0.314 rad/s^2 (B) 5.63 revs	P7-4	(A) 0.889 rad/s^2 (B) 8.59 revs
P7-5	(A) 7.27×10^{-5} rad/s $= 6.94 \times 10^{-4}$ RPM (B) 463 m/s = 1040 MPH	P7-6	0.0698 rad/s^2	P8-1	0.384 J
P8-2	(A) 120 Nm (B) 46.7 revs	P8-3	(A) 300 Nm (B) 0.0637 revs	P8-4	3.08 rad/s
P8-5	100 times	P8-6	16 Nm	P8-7	0.05
P8-8	60 Nm	P8-9	133 N	P8-10	54 cm

P8-11	2.5 RPM	P8-12	(A) 3.43×10^{-3} Nm (B) 38.1 rad/s^2	P9-1	(A) 210 N (B) 75.2 N
P9-2	634 N, 346 N	P9-3	1.57 m	P9-4	1.55 m
P9-5	0.568 m	P10-1	2.25×10^4 N/m	P10-2	2.55×10^{-2} N
P10-3	(A) 2.04×10^5 N/m^2 (B) 1.96×10^5 N/m^2 (C) ? (D) 2.63×10^5 N/m^2	P10-4	4.9×10^6 N/m^2	P10-5	15 cm
P10-6	(A) 1.77 mm (B) 5.71 mm	P10-7	(A) 2.31×10^5 N/m^2 (B) 0.577 m	P11-1	3.4 m
P11-2	4.92 sec	P11-3	0.688 sec	P11-4	1.26 sec
P11-5	108 m/s	P11-6	(A) 0.6 Hz (B) 1.67 sec	P11-7	(A) 0.705 Hz (B) same as (A)
P11-8	4.88 %	P12-1	550 Hz	P12-2	(A) 74.8 dB (B) 6.03×10^{-5} W/m^2
P12-3	(A) 190 Hz (B) 342 m/s	P12-4	(A) 2235 Hz (B) 1765 Hz	P12-5	(A) 0.588 sec (B) 6.67×10^{-7} sec

P12-6	1.55 dB less	P12-7	(A) 124 Hz (B) 116 Hz	P12-8	(A) 3.16×10^{-3} W/m^2 (B) 85 dB
P12-9	(A) 90 Hz (B) 1.89 m	P12-10	(A) 720 Hz (B) 0.354 m	P13-1	12
P13-2	99 kPa	P13-3	42.4 N	P13-4	1.39×10^5 N
P13-5	(A) 31.3 m/s (B) 50 m	P13-6	(A) 3.36×10^4 N (B) 168 nails	P13-7	4.41×10^5 Pa
P13-8	40 m/s	P13-9	5.5×10^3 kg/m^3	P14-1	(A) 21 m^3 (B) 135 kPa
P14-2	(A) 30 m^3 (B) 125 hrs	P14-3	0.625	P14-4	0.0204
P14-5	0.5 atm	P14-6	1.035	P14-7	364 K
P14-8	49 °C	P15-1	645 m	P15-2	11.2 cm
P15-3	25.2 mm	P15-4	9.16×10^7 J	P15-5	396 m/s
P15-6	0.0333 °C	P15-7	–23.1 °C	P15-8	608 J
P15-9	(A) 2.86×10^{15} J (B) 2.26×10^{15} J	P15-10	0.0378	P15-11	13.3 kg
P15-12	18 bulbs	P15-13	0.0913 °C	P15-14	?

P15-15	8.21 MW	P15-16	33.3 J	P15-17	12.4
P15-18	6.42×10^{7} W/m^{2}	P16-1	(A) 2.3×10^{-8} N 8.93×10^{-30} N (B) ?	P16-2	8.57 m
P16-3	8.89×10^{5} m/s	P16-4	2.98×10^{17} C	P16-5	190 N
P16-6	4.84×10^{-8} C	P16-7	8.1×10^{-3} J	P16-8	(A) 1.06×10^{-6} F (B) 4.24×10^{6} J
P16-9	(A) 400 V (B) 0.2 C	P17-1	5400 C	P17-2	0.292 A
P17-3	500 sec	P17-4	0.24 A	P17-5	100 V
P17-6	14 Ω	P17-7	1.8 W	P17-8	28 W
P17-9	50 %	P18-1	320 Ω	P18-2	18.8 Ω
P18-3	18.8 μF	P18-4	320 μF	P18-5	(A) 3R (B) 4R (C) NR
P18-6	(A) R/3 (B) R/4 (C) R/N	P18-7	1 in series with 2 in parallel	P18-8	(A) C/3 (B) C/4 (C) C/N
P18-9	(A) 3C (B) 4C (C) NC	P18-10	1 in series with 2 in parallel	P18-11	(A) 4.09×10^{-4} V (B) 1.02×10^{-9} A

P18-12	$R_{eq} = 943\ \Omega$ $i_1 = 9.54$ mA $i_2 = 2.73$ mA $i_3 = 6.81$ mA	P18-13	$C_{eq} = 225$ nF $Q_1 = 2700$ nC $Q_2 = 1200$ nC $Q_3 = 1500$ nC $V_1 = 9$ V $V_2 = V_3 = 3$ V	P18-14	$2.51 \times 10^{-3}\ \Omega$
P18-15	2.25×10^9 m (56 times around the world)	P18-16	(A) 2.79 V (B) 4.66 V	P18-17	0.829 sec
P19-1	6.99×10^{-13} N	P19-2	1.79×10^{-16} N opposite current	P19-3	2.64×10^{-16} N
P19-4	6×10^{-6} T	P19-5	6×10^{-6} T	P19-6	1.26×10^{-2} T
P19-7	100 amu	P20-1	1.2 V	P20-2	4,500
P20-3	0.6 V	P20-4	3.67×10^{-3} H	P20-5	1.12 mV
P20-6	3.6×10^{-4} A	P20-7	0.0395 sec	P21-1	0.145
P21-2	5.5 V	P21-3	1.54×10^5 turns	P21-4	89.8 A
P21-5	(A) 32.7 turns (B) 0.0545 A	P21-6	$4.95	P21-7	156 V
P21-8	(A) 8.07 Ω (B) 13.6 A	P21-9	(A) 47.4 Ω (B) 2.32 A	P21-10	1450 Hz

P22-1	(A) 3.13 m (B) 7.5×10^5 Hz	P22-2	1 Mhz	P22-3	3×10^{11} Hz 3×10^8 Hz
P23-1	44°	P23-2	1.36	P23-3	43.6°
P23-4	40 cm	P23-5	–20.6 cm	P23-6	54.3 cm
P23-7	2.75 D	P24-1	6.67×10^{-7} m	P24-2	4.22×10^{-5} m
P25-1	(A) 8.08×10^{-10} sec (B) 0.162 m	P25-2	(A) 83.5 m (B) 5.06×10^{-7} sec (C) 6.06×10^{-7} sec	P25-3	9.94×10^{-31} kg
P25-4	0.0242 sec	P25-5	66.1 cm	P25-6	30.5 % of c
P25-7	$210 billion	P25-8	1110 kg (same mass as a small car)	P26-1	2.64×10^{-11} m
P26-2	5.28×10^{-23} kgm/s	P26-3	2.86×10^{-34} m	P26-4	3.64×10^{-10} m
P27-1	1.81 %	P27-2	1.06 hrs	P27-3	18.1 yrs

ADDITIONAL RESOURCES

Arthur W. Wiggins and C. M. Wynn, *The Five Biggest Unsolved Problems in Science* (Hoboken, NJ: John Wiley & Sons, 2003).

Charles M. Wynn and A. W. Wiggins, *The Five Biggest Ideas in Science* (New York: John Wiley & Sons, 1997).

Douglas C. Giancoli, *Physics,* 6th ed. (Upper Saddle River, NJ: Pearson Education Inc, 2005).

Science fiction authors: Poul Anderson, Isaac Asimov, Steven Baxter, Greg Benford, Ray Bradbury, David Brin, John Brunner, Arthur C. Clarke, Hal Clement, John Cramer, Greg Egan, Robert L. Forward, Robert Heinlein, James Hogan, Damon Knight, Geoffrey Landis, Ursula LeGuin, Paul Nahin, Larry Niven, Pellligrino and Zebrowski, Frederik Pohl, Charles Sheffield, A. E. van Vogt, and Vernor Vinge

NOTES AND CREDITS

INTRODUCTION

1. Experimental apparatus

http://www.sciplus.com/
http://www.ustoy.com/novelty/default.htm
http://www.orientaltrading.com/

CHAPTER 1

1. Quotes (many other chapters also)

http://en.thinkexist.com/quotes/top/

Chapter 2

Newton, pp. 37, 41

1. http://www.gap-system.org/~history/Biographies/Newton.html
   http://en.wikipedia.org/wiki/Isaac_Newton
2. Shoulders of giants quote: Letter from Isaac Newton to Robert Hooke, February 5 1675/76.

Chapter 7

1. Crash test dummies

http://www.nhtsa.gov/
http://auto.howstuffworks.com/crash-test.htm

2. Jack Nicklaus quote

http://www.golftoday.co.uk/noticeboard/quotes/nicklaus.html

Chapter 8

1. Giant impact hypothesis: formation of the Moon

http://en.wikipedia.org/wiki/Giant_impact_hypothesis

Chapter 9

1. Architects with physics connections: Antoni Gaudi

http://www.greatbuildings.com/buildings/Casa_Batllo.html
http://en.wikipedia.org/wiki/Antoni_Gaudí

Frank Gehry

http://en.wikipedia.org/wiki/Frank_Gehry
http://www.arcspace.com/gehry_new/

Zaha Hadid

http://en.wikipedia.org/wiki/Zaha_Hadid
http://www.guggenheim.org /hadid/index.html

Frank Lloyd Wright

http://www.delmars.com/wright/flwright.htm
http://en.wikipedia.org/wiki/Frank_Lloyd_Wright

CHAPTER 11

1. Tacoma Narrows Bridge

http://physics.kenyon.edu/coolphys/FranklinMiller/protected/tacoma.html

CHAPTER 13

1. Daniel Bernoulli

Ioan James, *Remarkable Physicists from Galileo to Yukawa* (Cambridge: Cambridge University Press, 2004), pp. 31–35.

http://www-history.mcs.st-and.ac.uk/Biographies/Bernoulli_Daniel.html

http://library.thinkquest.org/22584/temh3007.htm
http://www.bookrags.com/Daniel_Bernoulli

2. Curveball videos

http://www.metacafe.com/watch/20570/now_this_is_a_curve_ball/
http://video.google.com/videoplay?docid=-2291168642414807392

Experiment 16

Soda bottle crush

Cathy Cobb and Monty L. Fetterolf, *The Joy of Chemistry* (Amherst, NY: Prometheus Books, 2005), p. 151.

CHAPTER 14

1. Title

This aphorism is often said by my friend and colleague, acclaimed artist Kegham Tazian.

2. Moles

Cathy Cobb and Monty L. Fetterolf, *The Joy of Chemistry* (Amherst, NY: Prometheus Books, 2005), pp. 148–49.

CHAPTER 15

1. Benjamin Thompson, Count Rumford

Ioan James, *Remarkable Physicists from Galileo to Yukawa* (Cambridge: Cambridge University Press, 2004), pp. 36–46.

http://en.wikipedia.org/wiki/Benjamin_Thomson,_Count_Rumford

http://www.dartmouth.edu/~library/Library_Bulletin/Apr1995/King_Rumford.html

http://www.bookrags.com/Benjamin_Thompson

http://www.famousamericans.net/benjaminthompsonrumford/

2. Specific heat and phase change

Cathy Cobb and Monty L. Fetterolf, *The Joy of Chemistry* (Amherst, NY: Prometheus Books, 2005), p. 204.

3. Entropy

Cathy Cobb and Monty L. Fetteroff, *The Joy of Chemistry* (Amherst NY: Prometheus Books, 2005), pp.196–97.

4. Paraphrased laws of thermodynamics

http://en.wikiquote.org/wiki/Transwiki:Quotes_&_humor_(thermodynamics)

CHAPTER 16

1. Benjamin Franklin

http://en.wikipedia.org/wiki/Benjamin_Franklin

http://www.answers.com/topic/benjamin-franklin

Philip Dray, *Stealing God's Thunder* (New York: Random House, 2005), pp. 18–30.

2. Charles A. Coulomb

http://www-groups.dcs.st-andrews.ac.uk/~history/Biographies/Coulomb.html

3. Electric field diagrams

www.cco.caltech.edu/~phys1/ java/phys1/EField/EField.htm

www.glenbrook.k12.il.us/GBSSCI/PHYS/Class/estatics/u8l4c.html

4. Electrical bonds between molecules

Cathy Cobb and Monty L. Fetterolf, *The Joy of Chemistry* (Amherst, NY: Prometheus Books, 2005), pp. 132–39.

CHAPTER 17

1. Georg Simon Ohm

http://www-gap.dcs.st-and.ac.uk/~history/Biographies/Ohm.html

Ioan James, *Remarkable Physicists from Galileo to Yukawa* (Cambridge: Cambridge University Press, 2004), pp. 107–12.

2. Lightning pictures

https://thunderstorm.vaisala.com/tux/jsp/explorer/explorer.jsp

http://www.lightningphotography.com/photos.html

http://thunder.msfc.nasa.gov/

Chapter 19

1. Tesla

http://en.wikipedia.org/wiki/Nikola_Tesla

Cecyle S. Neidle, *Great Immigrants* (New York: Twayne, 1973), pp. 133–62.

Jill Jonnes, *Empires of Light* (New York: Random House, 2003), pp. 87–115.

2. Auroras

http://www.spaceweather.com/

http://saturn.jpl.nasa.gov/news/features/saturn-story/space-based-obs.cfm

http://hubblesite.org/newscenter/archive/releases/1996/32/

Experiment 23

Shake flashlight

http://www.msscweb.org/Public/Forever%20Flashlight%20R2%20sans%20path.pdf

Chapter 21

1. Thomas Edison

http://en.wikipedia.org/wiki/Thomas_Edison

Jill Jonnes, *Empires of Light* (New York: Random House, 2003), pp. 51–85.

2. George Westinghouse

Jill Jonnes, *Empires of Light* (New York: Random House, 2003), pp. 117–39 and 165–213.

http://en.wikipedia.org/wiki/George_Westinghouse

Experiment 24

Semiconductors

Cathy Cobb and Monty L. Fetterolf, *The Joy of Chemistry* (Amherst, NY: Prometheus Books, 2005), pp. 115–18.

Chapter 22

1. Experimentino 14

http://www.physics.umd.edu/ripe/icpe/newsletters/n34/marshmal.htm

http://superpositioned.com/articles/2006/03/09/measure-the-speed-of-light-with-chips

Chapter 23

1. Interaction of light with matter

Richard P. Feynman, *QED: The Strange Theory of Light and Matter* (Princeton, NJ: Princeton University Press, 1988). (For further reading.)

2. Rainbow

http://en.wikipedia.org/wiki/Rainbow

3. Mirages and Looming

http://mintaka.sdsu.edu/GF/mirages/mirintro.html

http://www .atoptics.co.uk/

4. Primary colors

http://en.wikipedia.org/wiki/Primary_color

Chapter 24

1. Animal plumage colors

http://hyperphysics.phy-astr.gsu.edu/hbase/vision/peacock.html

http://webexhibits.org/causesofcolor/15C.html

2. Thin films

http://hyperphysics.phy-astr.gsu.edu/hbase/phyopt/thinfilm.html

3. Soap bubbles (Experimentino 19)

http://en.wikipedia.org/wiki/Soap_bubble

http://www.exploratorium.edu/ronh/bubbles/bubbles.html

CHAPTER 25

1. Albert Einstein

http://en.wikipedia.org/wiki/Albert_Einstein

http://www-history.mcs.st-andrews.ac.uk/Biographies/Einstein.html

http://www.time.com/time/time100/poc/magazine/albert_einstein5a.html

http://www.einstein-website.de/z_biography/print/p_biography.html

http://www.aip.org/history/einstein/essay-einsteins-third-paradise.htm

http://www.einstein.caltech.edu/vol10_intro.htm

Ronald W. Clark, *Einstein: His Life and Times* (New York: Random House Value Publishing, 1995), pp. 32–70.

A. Einstein and M. Grossmann, *Entwurf einer verallgemeinerten Relativitätstheorie und einer Theorie der Gravitation* (Zeitschrift für Mathematik und Physik, 1914), 62:225.

Walter Isaacson, Einstein: His Life and Universe (New York: Simon & Schuster, 2007).

CHAPTER 26

1. Einstein-Bohr exchange

http://en.wikipedia.org/wiki/Solvay_Conference

CHAPTER 27

1. Enrico Fermi

Laura Fermi, *Atoms in the Family: My Life with Enrico Fermi* (Chicago: University of Chicago Press, 1995). (For further reading.)

http://www.time.com/time/time100/scientist/profile/fermi.html

http://www.fnal.gov/pub/about/whatis/enricofermi.html

http://en.wikipedia.org/wiki/Enrico_Fermi

http://www.aip.org/pt/vol-55/iss-6/p38.html

2. Manhattan Project

Richard Rhodes, *The Making of the Atomic Bomb* (New York: Simon & Schuster, 1986). (For further reading.)

http://www.childrenofthemanhattanproject.org/HISTORY/ERC-1.htm

General Advisory Committee Minority Annex, http://www.pbs.org/wgbh/amex/bomb/filmmore/reference/primary/extractsofgeneral.html

CHAPTER 28

1. Richard P. Feynman

Richard P. Feynman, *Surely You're Joking, Mr. Feynman!* (New York: W. W. Norton, 1985).

Feynman, *What Do You Care What Other People Think?* (New York: W. W. Norton, 2001).

CHAPTER 29

1. Edwin Hubble

http://antwrp.gsfc.nasa.gov/diamond_jubilee/1996/sandage_hubble.html

http://en.wikipedia.org/wiki/Edwin_Hubble

http://www.time.com/time/time100/scientist/profile/hubble.html

http://query.nytimes.com/gst/fullpage.html?res=990CE3DF1439F930A3575AC0A963958260&sec=&pagewanted=print

2. Fritz Zwicky

http://scienzapertutti.lnf.infn.it/biografie/zwicky-bio.html
http://www.dynamical-systems.org/zwicky/Zwicky-e.html

Fritz Zwicky, *Die Rotverscheibung von Extragalaktischen Nebeln, Helvetica Physica Acta* 6, no. 110 (1933).

3. Dark energy

http://www.sciencenews.org/articles/20010407/bob14.asp
http://www.fnal.gov/pub/ferminews/ferminews04-01-01/p5.html

ILLUSTRATION CREDITS

Cartoons by Sidney Harris: cover and pages 10, 37, 68, 86, 150, 172, 192, 207, 229, 253, 256, 269, 299, 313, 320, 329, 330, 333, 343, 323.

Graphic drawings by Eugene Mann: pages 20, 27, 35, 45, 55, (*top*), 67, 75, 83, 95, 107, 115, 123, 131, 145, 161, 171, 191, 221, 241, 251, 267, 277, 293, 323, 364, 375, 381, 384, 395.

Graphic drawings by Melanie Rohrbeck: pages 358, 360.

Photographs by Barbara Mann Wiggins: pages 22, 24, 43, 64, 130, 144, 290, 311.

Photograph on page 302: Courtesy of the Albert Einstein Archives, the Jewish National and University Library, the Hebrew University of Jerusalem, Israel.

Photograph on page 321: Courtesy of the Lawrence Berkeley National Laboratory.

Photograph on page 345: Reproduced by permission of the Huntington Library, San Marino, California.

Photograph on page 347: Courtesy of the Carnegie Institution of Washington.

Photograph on page 349: By Floyd Clark/courtesy Caltech Archives.

INDEX

Page numbers in *italic* indicate experiments and experimentations